DECADAL SURVEY OF CIVIL AERONAUTICS

Foundation for the Future

Steering Committee for the Decadal Survey of Civil Aeronautics

Aeronautics and Space Engineering Board

Division on Engineering and Physical Sciences

NATIONAL RESEARCH COUNCIL
OF THE NATIONAL ACADEMIES

THE NATIONAL ACADEMIES PRESS
Washington, D.C.
www.nap.edu

THE NATIONAL ACADEMIES PRESS 500 Fifth Street, N.W. Washington, DC 20001

NOTICE: The project that is the subject of this report was approved by the Governing Board of the National Research Council, whose members are drawn from the councils of the National Academy of Sciences, the National Academy of Engineering, and the Institute of Medicine. The members of the committee responsible for the report were chosen for their special competences and with regard for appropriate balance.

This study was supported by Contract No. NNH05CC15C between the National Academy of Sciences and the National Aeronautics and Space Administration. Any opinions, findings, conclusions, or recommendations expressed in this publication are those of the author(s) and do not necessarily reflect the views of the organizations or agencies that provided support for the project.

International Standard Book Number 0-309-10158-1

Available in limited supply from the Aeronautics and Space Engineering Board, 500 Fifth Street, N.W., Washington, DC 20001, (202) 334-2858.

Additional copies of this report are available from the National Academies Press, 500 Fifth Street, N.W., Lockbox 285, Washington, DC 20055; (800) 624-6242 or (202) 334-3313 (in the Washington metropolitan area); Internet, www.nap.edu.

Printed in the United States of America

THE NATIONAL ACADEMIES

Advisers to the Nation on Science, Engineering, and Medicine

The **National Academy of Sciences** is a private, nonprofit, self-perpetuating society of distinguished scholars engaged in scientific and engineering research, dedicated to the furtherance of science and technology and to their use for the general welfare. Upon the authority of the charter granted to it by the Congress in 1863, the Academy has a mandate that requires it to advise the federal government on scientific and technical matters. Dr. Ralph J. Cicerone is president of the National Academy of Sciences.

The **National Academy of Engineering** was established in 1964, under the charter of the National Academy of Sciences, as a parallel organization of outstanding engineers. It is autonomous in its administration and in the selection of its members, sharing with the National Academy of Sciences the responsibility for advising the federal government. The National Academy of Engineering also sponsors engineering programs aimed at meeting national needs, encourages education and research, and recognizes the superior achievements of engineers. Dr. Wm. A. Wulf is president of the National Academy of Engineering.

The **Institute of Medicine** was established in 1970 by the National Academy of Sciences to secure the services of eminent members of appropriate professions in the examination of policy matters pertaining to the health of the public. The Institute acts under the responsibility given to the National Academy of Sciences by its congressional charter to be an adviser to the federal government and, upon its own initiative, to identify issues of medical care, research, and education. Dr. Harvey V. Fineberg is president of the Institute of Medicine.

The **National Research Council** was organized by the National Academy of Sciences in 1916 to associate the broad community of science and technology with the Academy's purposes of furthering knowledge and advising the federal government. Functioning in accordance with general policies determined by the Academy, the Council has become the principal operating agency of both the National Academy of Sciences and the National Academy of Engineering in providing services to the government, the public, and the scientific and engineering communities. The Council is administered jointly by both Academies and the Institute of Medicine. Dr. Ralph J. Cicerone and Dr. Wm. A. Wulf are chair and vice chair, respectively, of the National Research Council.

www.national-academies.org

DECADAL SURVEY OF CIVIL AERONAUTICS

Steering Committee

PAUL G. KAMINSKI (NAE), *Chair*, Technovation, Inc., Fairfax Station, Virginia
WILLIAM W. HOOVER, *Co-chair*, U.S. Air Force (retired), Williamsburg, Virginia
INDERJIT CHOPRA, University of Maryland, College Park
EUGENE E. COVERT (NAE), Massachusetts Institute of Technology, Cambridge
ALAN C. ECKBRETH, Connecticut Academy of Science and Engineering, Hartford
THOMAS M. HARTMANN, Lockheed Martin Aeronautics Company, Palmdale, California
ILAN KROO (NAE), Stanford University, Stanford, California
NANCY G. LEVESON (NAE), Massachusetts Institute of Technology, Cambridge
IVETT A. LEYVA, Microcosm, Inc., El Segundo, California
AMY PRITCHETT, Georgia Institute of Technology, Atlanta
EDMOND L. SOLIDAY, United Airlines (retired), Valparaiso, Indiana
JOHN VALASEK, Texas A&M University, College Station
DAVID VAN WIE, Johns Hopkins University, Laurel, Maryland
ROBERT WHITEHEAD, Aerospace Consultant, Henrico, North Carolina
DIANNE S. WILEY, The Boeing Company, Huntington Beach, California

Panel A: Aerodynamics and Aeroacoustics

DAVID VAN WIE, *Chair*, Johns Hopkins University, Laurel, Maryland
PAUL BEVILAQUA (NAE), Lockheed Martin Aeronautics Company, Palmdale, California
CHARLES BOCCADORO, Northrop Grumman Corporation, El Segundo, California
THOMAS CORKE, University of Notre Dame, Notre Dame, Indiana
ILAN KROO (NAE), Stanford University, Stanford, California
ROBERT LIEBECK (NAE), The Boeing Company, Huntington Beach, California
DAN MARREN, Arnold Engineering Development Center—White Oak, Silver Spring, Maryland
STEPHEN RUFFIN, Georgia Institute of Technology, Atlanta
FREDRIC H. SCHMITZ, University of Maryland, College Park
JOHN SULLIVAN, Purdue University, West Lafayette, Indiana
KAREN WILLCOX, Massachusetts Institute of Technology, Cambridge

Panel B: Propulsion and Power

ALAN C. ECKBRETH, *Chair*, Connecticut Academy of Science and Engineering, Hartford
ROBERT BAKOS, ATK GASL, Ronkonkoma, New York*
MEYER J. BENZAKEIN (NAE), Ohio State University, Columbus
JAMES L. BETTNER, Rolls-Royce Corporation (retired), Pittsboro, Indiana
DAVID E. CROW (NAE), University of Connecticut, Glastonbury
MEHRDAD EHSANI, Texas A&M University, College Station
JEFFREY W. HAMSTRA, Lockheed Martin Aeronautics Company, Fort Worth, Texas
IVETT A. LEYVA, Microcosm, Inc., El Segundo, California
TIMOTHY LIEUWEN, Georgia Institute of Technology, Atlanta
LOURDES QUINTANA MAURICE, Federal Aviation Administration, Washington, D.C.
JAMES C. McDANIEL, JR., University of Virginia, Charlottesville
TRESA M. POLLOCK (NAE), University of Michigan, Ann Arbor
WILLIAM TUMAS, Los Alamos National Laboratory, Los Alamos, New Mexico

*Resigned October 12, 2005.

Panel C: Materials and Structures

DIANNE S. WILEY, *Chair*, The Boeing Company, Huntington Beach, California
SATYA N. ATLURI (NAE), University of California at Irvine*
GREGORY CARMAN, University of California at Los Angeles
INDERJIT CHOPRA, University of Maryland, College Park
JANET DAVIS, Rockwell Scientific, Thousand Oaks, California
RAVI B. DEO, Northrop Grumman Corporation, El Segundo, California
PRABHAT HAJELA, Rensselear Polytechnic Institute, Troy, New York
MARK K. HINDERS, College of William & Mary, Williamsburg, Virginia
ROBERT SCHAFRIK, GE Aircraft Engines, Cincinnati, Ohio
NANCY R. SOTTOS, University of Illinois, Urbana
GREGORY WASHINGTON, Ohio State University, Columbus
TERRENCE A. WEISSHAAR, Defense Advanced Research Projects Agency, Arlington, Virginia

Panel D: Dynamics, Navigation, and Control, and Avionics

NANCY G. LEVESON (NAE), *Chair*, Massachusetts Institute of Technology, Cambridge
RICHARD ABBOTT, Lockheed Martin Aeronautics Company, Palmdale, California
CLARK R. BADIE, Honeywell International, Phoenix, Arizona
JEFFERY ERICKSON, Boeing Phantom Works, Huntington Beach, California
EPHRAHIM GARCIA, Cornell University, Ithaca, New York
CHARLES L. GUTHRIE, Northrop Grumman Integrated Systems, Long Beach, California
ELLIS F. HITT, Strategic Systems Solutions, Inc., Westerville, Ohio
JAMES C. NEIDHOEFER, Aerotonomy, Inc., Fort Oglethorpe, Georgia
DARRYLL J. PINES, Defense Advanced Research Projects Agency, Arlington, Virginia
JAMES RANKIN, Avionics Engineering Center, Ohio University, Athens
JASON L. SPEYER (NAE), University of California at Los Angeles
JOHN VALASEK, Texas A&M University, College Station

Panel E: Intelligent and Autonomous Systems, Operations and Decision Making, Human Integrated Systems, and Networking and Communications

EDMOND L. SOLIDAY, *Chair*, United Airlines (retired), Valparaiso, Indiana
ELLA ATKINS, University of Maryland, College Park
TAMER BASAR (NAE), University of Illinois, Urbana
THOMAS Q. CARNEY, Purdue University, West Lafayette, Indiana
JOHN-PAUL CLARKE, Georgia Institute of Technology, Atlanta
MICHAEL DeWALT, Aviation Systems Certification Services, Eastbound, Washington
FRANK L. FRISBIE, Apptis Inc., Washington, D.C.
ANDREW LACHER, MITRE Corporation, McLean, Virginia
RAYMOND R. LaFREY, Massachusetts Institute of Technology Lincoln Laboratory (retired), New Braunfels, Texas
CARL McCULLOUGH, U.S. Air Force Headquarters, Washington, D.C.
AMY PRITCHETT, Georgia Institute of Technology, Atlanta
DONALD W. RICHARDSON, West Palm Beach, Florida
NADINE SARTER, University of Michigan, Ann Arbor

*Resigned January 19, 2006.

Staff

ELIZABETH ALBRIGO, Senior Program Assistant
ALAN ANGLEMAN, Study Director (from December 2005)
ANNA FARRAR, Financial Associate
JAMES GREGORY, NRC Science and Technology Policy Fellow (through December 2005)
KAREN HARWELL, Study Director (through December 2005)
GEORGE LEVIN, Director, Aeronautics and Space Engineering Board
HEIDI MURRAY, Senior Program Assistant
JOHN SISLIN, Program Officer
KERRIE SMITH, Co-Study Director
JOHN WENDT, Senior Program Officer

AERONAUTICS AND SPACE ENGINEERING BOARD

RAYMOND S. COLLADAY, *Chair*, Lockheed Martin Astronautics (retired), Golden, Colorado
WILLIAM L. BALLHAUS, BAE Systems National Security Solutions, Reston, Virginia
CHARLES F. BOLDEN, JR., Jack and Panther, LLC, Houston, Texas
EDWARD M. BOLEN, National Aviation Association, Washington, D.C.
ANTHONY J. BRODERICK, Aviation Safety Consultant, Catlett, Virginia
JOHN-PAUL CLARKE, Georgia Institute of Technology, Atlanta
PHILIP M. CONDIT (NAE), The Boeing Company, Redmond, Washington
ROBERT L. CRIPPEN, Thiokol Propulsion (retired), Palm Beach Gardens, Florida
RICHARD M. GOODY (NAS), Harvard University (emeritus), Falmouth, Massachusetts
PRESTON HENNE (NAE), Gulfstream Aerospace Corporation, Savannah, Georgia
WILLIAM W. HOOVER, U.S. Air Force (retired), Williamsburg, Virginia
SYDNEY MICHAEL HUDSON, Rolls-Royce North America (retired), Indianapolis, Indiana
JOHN M. KLINEBERG, Space Systems/Loral (retired), Redwood City, California
ILAN KROO (NAE), Stanford University, Stanford, California
MOLLY K. MACAULEY, Resources for the Future, Washington, D.C.
FORREST S. McCARTNEY, Lockheed Martin Astronautics (retired), Indian Harbour Beach, Florida
ELON MUSK, Space Exploration Development Corporation–SpaceX, El Segundo, California
AMY PRITCHETT, Georgia Institute of Technology, Atlanta
ROBERT R. RANKINE, JR., Hughes Space and Communications Company (retired), Clearwater, Florida
DEBRA L. RUB-ZENKO, The Boeing Company, Anaheim, California
CYNTHIA SAMUELSON, Logistics Management Institute, McLean, Virginia
PETER STAUDHAMMER (NAE), University of Southern California, La Quinta
HANSEL E. TOOKES II, Raytheon International, Inc. (retired), Palm Beach Gardens, Florida
RAY VALEIKA, Delta Airlines, Inc. (retired), Powder Springs, Georgia
ROBERT S. WALKER, Wexler & Walker Public Policy Associates, Washington, D.C.
ROBERT E. WHITEHEAD, National Institute of Aerospace, Henrico, North Carolina
THOMAS L. WILLIAMS, Northrop Grumman, El Segundo, California

Staff

GEORGE LEVIN, Director

Preface

The air transportation system is important to the economic vitality, public well-being, and national security of the United States. The aerospace industry has historically made a large contribution to the positive balance of trade for the U.S. economy. In 2005, it had a $37 billion positive balance of trade, of which $29 billion was for civil aeronautics.[1] In addition, the United States has had a long history as the unchallenged world leader in civil and military aeronautics, though this position is now in jeopardy in areas such as research capability, technological expertise, and the performance of civil aircraft and air traffic management systems.

With leadership comes opportunity, particularly with regard to setting international standards for aircraft certification and operations. A position of continued leadership would allow the United States to ensure that viable global standards continue to be established for the application of emerging technologies and operational concepts. Without such standards the global aviation market and the global transportation system will be fractured into separate fiefdoms ruled by national and regional aviation authorities acting independently. This would impede the ability of passengers and cargo to move seamlessly—and safely—from country to country. The United States needs "world-class science and engineering—not simply as an end in itself, but as the principal means of creating new jobs for its citizenry as a whole as it seeks to prosper in the global marketplace of the 21st century."[2] Strong action is needed to ensure that U.S. leadership continues to assure the future of the domestic and global air transportation systems.[3]

The National Aeronautics and Space Administration (NASA) is explicitly chartered to preserve the role of the United States as a leader in aeronautics technology. To pursue that goal, NASA contracted with the National Research Council's Aeronautics and Space Engineering Board (ASEB) to complete a decadal survey of civil aeronautics, to prioritize research projects to be undertaken in the next 10 years. For the last 50 years, the National Research Council has conducted decadal surveys in astronomy. The idea of conducting a decadal survey of aeronautics originated in discussions among the ASEB, the Office of Management and Budget, and congressional committees with an interest in civil aviation. Although this study takes special note of NASA's priorities for civil aeronautics research, it also identifies national priorities for non-NASA researchers. Additionally, the study points out synergies between civil aeronautics research and research objectives associated with national defense, homeland security, and the space program.

In FY 2004, NASA's budget for aeronautics was just over $1 billion. NASA's aeronautics budget for FY 2006 was $884 million, and it will be reduced to $724 million in FY 2007 if Congress accepts the President's budget. If that happens, in just 3 years NASA's budget for aeronautics will have sustained a reduction of 32 percent, even as NASA's total budget increases by 9 percent. This budgetary trend will make it increasingly difficult for NASA to build a solid foundation for the future. However, regardless of the overall funding level, NASA's aeronautics program should focus on the key strategic objectives,

---

[1]D. Napier. 2005. *2005 Year-End Review and 2006 Forecast—An Analysis.* Arlington, Va.: Aerospace Industries Association (AIA). Available online at <www.aia-aerospace.org/stats/yr_ender/yrendr2005_text.pdf>.

[2]National Research Council. 2005. *Rising Above the Gathering Storm: Energizing and Employing America for a Brighter Economic Future,* p. 30. Washington, D.C.: The National Academies Press. Available online at <http://fermat.nap.edu/catalog/11463.html>.

[3]National Research Council. 2003. *Securing the Future of U.S. Air Transportation: A System in Peril,* p. 11. Washington, D.C.: The National Academies Press. Available online at <http://fermat.nap.edu/catalog/10815.html>.

themes, and high-priority research and technology challenges described herein. The present survey was completed in parallel with ongoing efforts to create a national policy on aviation and separate efforts by NASA Headquarters to assess the aeronautics program. The authors of this report are confident that all three efforts will work toward the common goal of assuring that long-term national investments in aeronautics research and technology substantially improve the air transportation system and achieve other appropriate national objectives.

Paul Kaminski, *Chair*
Decadal Survey of Civil Aeronautics Steering Committee

Acknowledgment of Reviewers

This report has been reviewed in draft form by individuals chosen for their diverse perspectives and technical expertise, in accordance with procedures approved by the Report Review Committee of the National Research Council (NRC). The purpose of this independent review is to provide candid and critical comments that will assist the institution in making its published report as sound as possible and to ensure that the report meets institutional standards for objectivity, evidence, and responsiveness to the study charge. The review comments and draft manuscript remain confidential to protect the integrity of the deliberative process. We wish to thank the following individuals for their review of this report:

Tony Broderick, Aviation Safety Consultant
Dianne Chong, The Boeing Company
Raymond Colladay, Lockheed Martin Astronautics (retired)
John Douglass, Aerospace Industries Association of America
Alan Epstein, NAE, Massachusetts Institute of Technology
Kenneth Rosen, NAE, General Aero-Science Consultants, LLC
David Schmidt, University of Colorado
Gunter Stein, NAE, Honeywell (retired)
Tom Williams, Northrop Grumman

Although the reviewers listed above have provided many constructive comments and suggestions, they were not asked to endorse the conclusions or recommendations, nor did they see the final draft of the report before its release. The review of this report was overseen by Sheila E. Widnall, NAE, Massachusetts Institute of Technology. Appointed by the NRC, she was responsible for making certain that an independent examination of this report was carried out in accordance with institutional procedures and that all review comments were carefully considered. Responsibility for the final content of this report rests entirely with the authoring committee and the institution.

Contents

APPENDIXES

Tables, Figures, and Boxes

TABLES

FIGURES

BOXES

Executive Summary

The U.S. air transportation system is a key contributor to the economic vitality, public well-being, and national security of the United States. The next decade of U.S. civil aeronautics research and technology (R&T) development should provide a foundation for achieving four high-priority Strategic Objectives:

- Increase capacity.
- Improve safety and reliability.
- Increase efficiency and performance.
- Reduce energy consumption and environmental impact.

Civil aeronautics R&T should also consider two lower-priority Strategic Objectives:

- Take advantage of synergies with national and homeland security.
- Support the space program.

The purpose of the Decadal Survey of Civil Aeronautics is to develop a foundation for the future—a decadal strategy for the federal government's involvement in civil aeronautics, with a particular emphasis on the National Aeronautics and Space Administration's (NASA's) research portfolio. A quality function deployment (QFD) process was used to identify and rank 89 R&T Challenges in relation to their potential to achieve the six Strategic Objectives listed above.[1] That process produced a list of 51 high-priority R&T Challenges that must be overcome to further the state of the art (see Table ES-1). These high-priority Challenges are equally divided among five R&T Areas:

- Area A: Aerodynamics and aeroacoustics.
- Area B: Propulsion and power.
- Area C: Materials and structures.
- Area D: Dynamics, navigation, and control, and avionics.
- Area E: Intelligent and autonomous systems, operations and decision making, human integrated systems, and networking and communications.

Advances in these Areas would have a significant, long-term impact on civil aeronautics. Accordingly, federal funds, facilities, and staff should be made available to advance the high-priority R&T Challenges in each Area.

Five Common Themes summarize threads of commonality among the 51 high-priority R&T Challenges:

- Physics-based analysis tools to enable analytical capabilities that go far beyond existing modeling and simulation capabilities and reduce the use of empirical approaches.
- Multidisciplinary design tools to integrate high-fidelity analyses with efficient design methods and to accommodate uncertainty, multiple objectives, and large-scale systems.
- Advanced configurations to go beyond the ability of conventional technologies and aircraft to achieve the Strategic Objectives.
- Intelligent and adaptive systems to significantly improve the performance and robustness of aircraft and the air transportation system as a whole.
- Complex interactive systems to better understand the nature of and options for improving the performance of the air transportation system, which is itself a complex interactive system.

These Themes are not an end in themselves; they are a means to an end. Each Theme describes enabling approaches that will contribute to overcoming multiple Challenges in the five R&T Areas. Exploiting the synergies identified in each

[1] QFD is a group decision-making methodology often used in product design.

TABLE ES-1 Fifty-one Highest Priority Research and Technology Challenges for NASA Aeronautics, Prioritized by R&T Area

A Aerodynamics and Aeroacoustics	B Propulsion and Power	C Materials and Structures	D Dynamics, Navigation, and Control, and Avionics	E Intelligent and Autonomous Systems, Operations and Decision Making, Human Integrated Systems, Networking and Communications
A1 Integrated system performance through novel propulsion–airframe integration	B1a Quiet propulsion systems	C1 Integrated vehicle health management	D1 Advanced guidance systems	E1 Methodologies, tools, and simulation and modeling capabilities to design and evaluate complex interactive systems
A2 Aerodynamic performance improvement through transition, boundary layer, and separation control	B1b Ultraclean gas turbine combustors to reduce gaseous and particulate emissions in all flight segments	C2 Adaptive materials and morphing structures	D2 Distributed decision making, decision making under uncertainty, and flight-path planning and prediction	E2 New concepts and methods of separating, spacing, and sequencing aircraft
A3 Novel aerodynamic configurations that enable high performance and/or flexible multi-mission aircraft	B3 Intelligent engines and mechanical power systems capable of self-diagnosis and reconfiguration between shop visits	C3 Multidisciplinary analysis, design, and optimization	D3 Aerodynamics and vehicle dynamics via closed-loop flow control	E3 Appropriate roles of humans and automated systems for separation assurance, including the feasibility and merits of highly automated separation assurance systems
A4a Aerodynamic designs and flow control schemes to reduce aircraft and rotor noise	B4 Improved propulsion system fuel economy	C4 Next-generation polymers and composites	D4 Intelligent and adaptive flight control techniques	E4 Affordable new sensors, system technologies, and procedures to improve the prediction and measurement of wake turbulence
A4b Accuracy of prediction of aerodynamic performance of complex 3-D configurations, including improved boundary layer transition and turbulence models and associated design tools	B5 Propulsion systems for short takeoff and vertical lift	C5 Noise prediction and suppression	D5 Fault-tolerant and integrated vehicle health management systems	E5 Interfaces that ensure effective information sharing and coordination among ground-based and airborne human and machine agents
A6 Aerodynamics robust to atmospheric disturbances and adverse weather conditions, including icing	B6a Variable-cycle engines to expand the operating envelope	C6a Innovative high-temperature metals and environmental coatings	D6 Improved onboard weather systems and tools	E6 Vulnerability analysis as an integral element in the architecture design and simulations of the air transportation system
A7a Aerodynamic configurations to leverage advantages of formation flying	B6b Integrated power and thermal management systems	C6b Innovative load suppression, and vibration and aeromechanical stability control	D7 Advanced communication, navigation, and surveillance technology	E7 Adaptive ATM[a] techniques to minimize the impact of weather by taking better advantage of improved probabilistic forecasts
A7b Accuracy of wake vortex prediction, and vortex detection and mitigation techniques	B8 Propulsion systems for supersonic flight	C8 Structural innovations for high-speed rotorcraft	D8 Human–machine integration	E8a Transparent and collaborative decision support systems
A9 Aerodynamic performance for V/STOL and ESTOL, including adequate control power	B9 High-reliability, high-performance, and high-power-density aircraft electric power systems	C9 High-temperature ceramics and coatings	D9 Synthetic and enhanced vision systems	E8b Using operational and maintenance data to assess leading indicators of safety
A10 Techniques for reducing/mitigating sonic boom through novel aircraft shaping	B10 Combined-cycle hypersonic propulsion systems with mode transition	C10 Multifunctional materials	D10 Safe operation of unmanned air vehicles in the national airspace	E8c Interfaces and procedures that support human operators in effective task and attention management
A11 Robust and efficient multidisciplinary design tools				

[a]ATM, air traffic management; V/STOL, vertical and/or short takeoff and landing; ESTOL, extremely short takeoff and landing.

Common Theme will enable NASA's aeronautics program to make the most efficient use of available resources.

Even if individual R&T Challenges are successfully overcome, two key barriers must also be addressed before the Strategic Objectives can be accomplished:

- *Certification.* As systems become more complex, methods to ensure that new technologies can be readily applied to certified systems become more difficult to validate. NASA, in cooperation with the Fedeal Aviation Administration (FAA), should anticipate the need to certify new technology before its introduction, and it should conduct research on methods to improve both confidence in and the timeliness of certification.
- *Management of change, internal and external.* Changing a complex interactive system such as the air transportation system is becoming more difficult as interactions among the various elements become more complex and the number of internal and external constraints grows. To effectively exploit R&T to achieve the Strategic Objectives, new tools and techniques are required to anticipate and introduce change.

This report also encourages NASA to do the following:

- Create a more balanced split in the allocation of aeronautics R&T funding between in-house research (performed by NASA engineers and technical specialists) and external research (by industry and/or universities). As of January 2006, NASA seemed intent on allocating 93 percent of NASA's aeronautics research funding for in-house use.
- Closely coordinate and cooperate with other public and private organizations to take advantage of advances in cross-cutting technology funded by federal agencies and private industry.
- Develop each new technology to a level of readiness that is appropriate for that technology, given that industry's interest in continuing the development of new technologies varies depending on urgency and expected payoff.
- Invest in research associated with improved ground and flight test facilities and diagnostics, in coordination with the Department of Defense and industry.

The eight recommendations formulated by the steering committee and set forth in Box ES-1 summarize action necessary to properly prioritize civil aeronautics R&T and achieve the relevant Strategic Objectives. This report should provide a useful foundation for the ongoing effort in the executive branch to develop an aeronautics policy. In addition, even though the scope of this study purposely did not include specific budget recommendations, it should support efforts by Congress to authorize and appropriate the NASA aeronautics budget.

BOX ES-1

Recommendations to Achieve Strategic Objectives for Civil Aeronautics Research and Technology

1. NASA should use the 51 Challenges listed in Table ES-1 as the foundation for the future of NASA's civil aeronautics research program during the next decade.
2. The U.S. government should place a high priority on establishing a *stable* aeronautics R&T plan, with the expectation that the plan will receive sustained funding for a decade or more, as necessary, for activities that are demonstrating satisfactory progress.
3. NASA should use five Common Themes to make the most efficient use of civil aeronautics R&T resources:

   - Physics-based analysis tools
   - Multidisciplinary design tools
   - Advanced configurations
   - Intelligent and adaptive systems
   - Complex interactive systems

4. NASA should support fundamental research to create the foundations for practical certification standards for new technologies.
5. The U.S. government should align organizational responsibilities as well as develop and implement techniques to improve change management for federal agencies and to assure a safe and cost-effective transition to the air transportation system of the future.
6. NASA should ensure that its civil aeronautics R&T plan features the substantive involvement of universities and industry, including a more balanced allocation of funding between in-house and external organizations than currently exists.
7. NASA should consult with non-NASA researchers to identify the most effective facilities and tools applicable to key aeronautics R&T projects and should facilitate collaborative research to ensure that each project has access to the most appropriate research capabilities, including test facilities; computational models and facilities; and intellectual capital, available from NASA, the Federal Aviation Administration, the Department of Defense, and other interested research organizations in government, industry, and academia.
8. The U.S. government should conduct a high-level review of organizational options for ensuring U.S. leadership in civil aeronautics.

1

Introduction

IMPORTANCE OF U.S. CIVIL AVIATION

Aviation plays an important role in supporting the preeminent economic, political, and military positions of the United States. U.S. air carriers move more passengers and cargo than those of any other country. U.S. industry is also a leader in manufacturing aircraft and air traffic management (ATM) systems. Globally, the United States has more general aviation and business aircraft than the rest of the world combined (GAMA, 2000, and Lubitz, 1997). In addition, far more commercial air transportation operations occur within the United States than within any other country. The size and efficiency of the U.S. air transportation system help the United States compete in the global economy by providing a transportation infrastructure that often responds quickly to changes in market demand and the various needs of the public, industry, and government at all levels (national, state, and local). An efficient air transportation system enables the rest of the economy to benefit from the efficiencies of just-in-time manufacturing. Seamless links between U.S. and global air transportation systems enable U.S. manufacturers to operate efficiently even with global supply chains, and it allows foreign manufacturers to include U.S. suppliers in their supply chains. Air cargo also helps e-commerce live up to its potential by delivering goods quickly. However, U.S. manufacturers' share of the global market for civil aeronautics is shrinking in the face of foreign competition. Aviation is a technology-intensive field, and maintaining global leadership will be impossible without continued investments in research and technology (R&T) by government and industry.

The air transportation system includes passenger and cargo airlines; general aviation, including business aviation; and the national airspace system, including airports, ATM facilities, and operational elements of the Federal Aviation Administration (FAA). U.S. civil aviation includes all of the above, plus manufacturers and research organizations in government, industry, and academia. Civil aviation benefits the United States in terms of the economy, public well-being, and national security, including homeland security. An affordable air transportation system makes the short travel times of aviation readily available to business and leisure travelers, improving the quality of life for all who choose to travel by air or who benefit from quick delivery of air freight. However, for the purpose of this report, the primary mission of the air transportation system, which is to provide efficient air transportation, is considered to be distinct from the national security and homeland security missions of the Department of Defense (DoD) and the Department of Homeland Security (DHS), respectively.

Growth in air travel comes at a cost in terms of noise for residents of communities around airports and in terms of aircraft emissions locally, regionally, and globally. Aeronautics R&T has reduced the noise and emissions produced by individual aircraft and has significantly reduced the total environmental impacts compared to what they would have been without new aircraft that are quieter, more efficient, and create fewer emissions than earlier generations. Advanced technologies have also substantially improved safety, so that even with substantial increases in air travel, accidents involving large civil transports tend to be increasingly infrequent. Even so, additional research is needed to discern, monitor, and eliminate or reduce the underlying causes and other factors that contribute to aircraft accidents. In addition, research can continue to reduce the environmental impact of individual aircraft, it can offset the environmental impact of increases in domestic and global air travel, and it may even reduce the local, regional, or global impact of air transportation, despite continued growth in air travel. Although the performance of large civil transports is of primary interest to the overall operation of the air transportation system, research can also address issues with other classes of aircraft. For example, the accident rate of general aviation aircraft is much higher than the accident rate of large civil transports, and the high noise levels of rotorcraft inhibit their ability to increase the capacity of the air transportation system.

In decades past, advances in military aviation were the source of many advances in civil aviation, most notably the swept-wing jet transport. More recently, military aviation R&T development funds have been reduced, and the rate at which new military aircraft are developed has greatly declined. In some cases, advances in civil aviation are being transferred to military applications, and dominance of the skies will be greatly affected by the results of civil aeronautics research. A more capable air transportation system could also enhance homeland security. For example, a next-generation air transportation system that uses a network-based approach to communications and the exchange of information would allow surveillance data collected from various air traffic sensors to provide the same comprehensive operational picture to all systems users and monitors, including the DHS and the North American Aerospace Defense Command. The air transportation system of the future should also accommodate routine operations of unmanned air vehicles (UAVs), which are taking an ever larger role in military aviation and will likewise be important to homeland security.

U.S. civil aviation is too important to allow the future to be defined solely by short-term market forces, which are unlikely to produce an efficient system that responds appropriately to user needs. Individual elements of the U.S. air transportation system are owned and operated by competitive companies, government agencies, and private citizens, each with their own motivations, resources, and limitations. Today and in the future, the U.S. air transportation system will not be able to meet the expectations of government, industry, and the public unless ATM equipment and procedures—which generally are owned, controlled, and operated by the federal government—are designed, implemented, and operated as efficiently as possible. In addition, market forces do not provide individual companies with a positive return on investments for research in many areas that are important to public well-being, such as safety, noise, emissions, speed, and basic research. Companies cannot make a business case for supporting an appropriate level of research in these areas, especially when the risk is high and/or a long research program is required to develop commercial applications. NASA, the FAA, and other government agencies must support key noncompetitive and precompetitive research to ensure that the U.S. air transportation system continues to benefit the United States. This is consistent with traditional practices of the FAA and NASA and the legislative charters for these agencies.

PERSPECTIVES

The U.S. air transportation system can be viewed from four perspectives:

- *Operational.* How does the system function in terms of different phases of operation (takeoff, en route, approach, and landing) and different geographical areas of operation?
- *Aircraft and ground systems.* What are the effects on the overall system of changes in the design and performance of individual aircraft and ground facilities, as well as the systems and subsystems that are incorporated within and among various aircraft and facilities?
- *Organizational.* How do manufacturers, airlines, pilots, controllers, customers, regulators, and other stakeholders (some with common interests and some with conflicting interests) function together to operate the air transportation system of today and to develop the air transportation system of the future? Also, how well does the current and future air transportation system meet the needs of stakeholders, individually and collectively?
- *International.* How does the U.S. air transportation system interact with a global economy, international aviation authorities, and international corporations that are interactive, interdependent, and integrated?

Efforts to improve the existing air transportation system—and to develop the next-generation air transportation system—should take a holistic approach that integrates all of the above perspectives and recognizes that the U.S. air transportation system is a complex interactive system that is more than the sum of its parts.[1]

ORIGIN OF THE STUDY

For the last 50 years, the National Research Council (NRC) has conducted decadal surveys in astronomy, prioritizing research projects to be undertaken in the next 10 years.[2] When the latest astronomy survey was released in 2001 (NRC, 2001), all of the large and many of the moderate-sized programs recommended in the preceding report (NRC, 1991) had been enacted. More recently, NASA commissioned additional decadal surveys in the fields of solar and space physics (NRC, 2003a), planetary science (NRC, 2003c), and Earth science (NRC, 2005). The recently

[1]As used in this report, complex interactive systems (or a system of systems) refer to adaptive systems consisting of a large, widespread collection or network of independent systems functioning together to achieve a common purpose. Complex interactive systems are distinguished from large, monolithic systems by the independent functioning of their components, which provides freedom for existing components to evolve and new components to emerge independent of a central configuration control authority. Complex interactive systems also tend to be distributed over a large geographic extent and require effective communications and coordination protocols for the various components to interact efficiently (Maier, 2006).

[2]The research strategies outlined in these reports are decadal surveys in the sense that they are based on thoughtful examinations of research requirements over the subsequent 10 years.

launched and highly publicized mission to Pluto was consistent with the recommendations contained in the 2003 planetary science decadal survey.

The idea of conducting a decadal survey of aeronautics originated in discussions among the NRC's Aeronautics and Space Engineering Board, the Office of Management and Budget, and congressional committees with an interest in civil aviation. Recognizing the potential value of such a study, NASA subsequently contracted with the Aeronautics and Space Engineering Board to carry out the study. Although the study focuses on civil aviation, it recognizes and calls out specific synergies that exist with national defense, homeland security, and the space program.

PURPOSE OF THE SURVEY

As detailed in Appendix G, the purpose of the Decadal Survey of Civil Aeronautics was to develop a decadal strategy for federal aeronautics research. The NRC was charged by NASA with providing guidelines for investment in aeronautics R&T, with a particular emphasis on NASA's research portfolio in each of five R&T Areas:

- Area A: Aerodynamics and aeroacoustics.
- Area B: Propulsion and power.
- Area C: Materials and structures.
- Area D: Dynamics, navigation, and control, and avionics.
- Area E: Intelligent and autonomous systems, operations and decision making, human integrated systems, and networking and communications.

The NRC appointed five panels, each with the expertise necessary to examine one of these Areas, along with a steering committee to oversee the work of the panels and prepare this report based on inputs from the panels as well as information gathered directly by the steering committee. The membership of the steering committee included the five panel chairs and one other member of each panel (see Appendix H).

This report describes research necessary to further the state of the art in the five R&T Areas (see Chapter 3). Advances in these Areas would have a significant long-term impact on national aeronautics, and research in these Areas is consistent with NASA's legislative charter, as described in the National Aeronautics and Space Act of 1958, as amended. Accordingly, federal funds, facilities, and staff should be made available to advance each Area.

This report also includes guidance on how federal resources allocated for aeronautics research should be distributed between in-house and external organizations, how aeronautics research can take advantage of advances in cross-cutting technology funded by federal agencies and private industry, and how far along the development and technology readiness path federal agencies should advance key aeronautics technologies, and it provides a set of overall findings and recommendations to provide a cumulative, integrated view of civil aeronautics research challenges and priorities (see Chapter 5). Lessons learned from other federal agencies appear in Appendix F. In accordance with the statement of task, this report does not include specific budget recommendations.

STRATEGIC OBJECTIVES FOR U.S. CIVIL AERONAUTICS RESEARCH

The existence of an explicit national aeronautics policy on R&D would have greatly facilitated the formulation of an aeronautics research strategy, because it would have defined the strategic objectives that should be used to shape future aeronautics research. In the absence of a stated national aeronautics policy, the steering committee identified and defined six Strategic Objectives that should motivate and guide the next decade of civil aeronautics research in the United States:[3]

- Capacity.
- Safety and reliability.
- Efficiency and performance.
- Energy and the environment.
- Synergies with national and homeland security.
- Support to space.

Capacity is the maximum amount of people and goods that can be moved through the air transportation system per unit time regardless of environmental conditions. The air transportation system of the future will need to double capacity over the next 10 to 35 years (NRC, 2003b).[4]

Safety and reliability refer to the ability of the air transportation system to meet expectations with regard to reductions in fatalities, injuries, loss of goods, and equipment damage or malfunction. The risk of accidents must be continually reduced so that the number of accidents will remain steady or decrease even as the number of flight operations increases substantially.

Efficiency and performance refer to achieving maximum utilization of the air transportation system so that available resources (aircraft, facilities, fuel, etc.) can provide as much service as possible (moving aircraft, passengers, and cargo). This requires an air transportation system with enhanced capabilities that improve mobility, access, and flexibility and reduce travel time and costs. The goal is to increase substantially air transportation system capacity per unit resource.

[3]Strategic Objectives and other key terminology used in this report are described in Box 1-1 and illustrated in Figure 1-1.

[4]This range of outcomes is equivalent to annual growth rates of 2.0 to 7.2 percent. An annual growth rate of 7.2 percent would double demand in 10 years, triple demand in 15 years, quadruple demand in 20 years, and increase demand sixfold in 25 years.

BOX 1-1
Terminology

This report uses the following terminology to create the framework for a decadal plan for civil aeronautics:

- *R&T Area.* Five Areas were identified that encompassed the R&T of greatest relevance to the air transportation system (see Chapter 2).
- *R&T Challenge.* For each Area, a set of key Challenges was identified and prioritized (see Chapter 3).
- *Strategic Objective.* The Strategic Objectives described in the first section of this chapter were used as the primary criteria for assessing the national importance of each R&T Challenge.
- *Why NASA?* Four criteria (supporting infrastructure, mission alignment, lack of alternative sponsorship, and appropriate level of risk) were used to determine how appropriate it is for NASA to address each R&T Challenge. The scores assigned to these four criteria were averaged to create a single "Why NASA?" score for each Challenge.
- *R&T Thrust.* Thrusts describe threads of commonality among R&T Challenges within each Area (see Chapter 3).
- *Common Theme.* Common Themes are used to group cross-cutting Challenges from more than one R&T Area (see Chapter 4). These Themes do not encompass all the high-priority Challenges, because some high-priority Challenges did not have closely linked challenges in other Areas.
- *Milestone.* Milestones for each Challenge are included in the detailed descriptions that appear in Appendixes A through E. These milestones are intended to indicate levels of achievement that demonstrate important advances in capability rather than detailed programmatic progress.

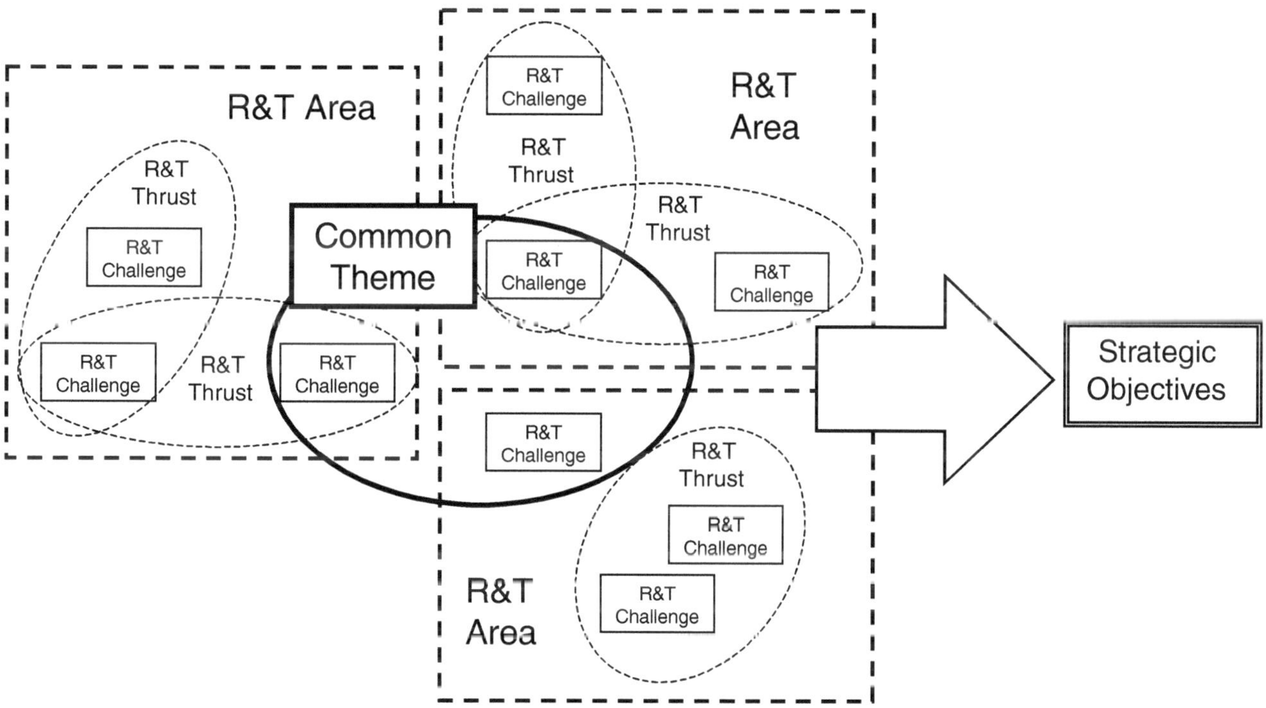

FIGURE 1-1 Terminology breakdown tree.

Energy and the environment refer to minimizing the negative impact of air transportation on Earth, its atmosphere, and its natural resources. This objective also includes the search for alternative fuels should petroleum-derived fuels become a constraint on air transportation. The goal is to reduce noise, emissions, and hazardous waste products (such as coolants and retired aircraft components), as well as fuel use per passenger seat mile and cargo ton mile.

Synergy with national and homeland security refers to the added value of specific aeronautical research when it helps to achieve the first four goals while also helping to achieve the goals of the DoD and the DHS. Because the steering committee had to define priorities for aeronautics R&T at NASA, this report focuses on civil rather than national or homeland security aeronautics research. This objective acknowledges that a great deal of civil aeronautics research also has national and homeland security applications. The goal is to transfer research results to DoD and DHS, as appropriate.

Support to space refers to the added value of specific aeronautical research if it helps to achieve the first four Strategic Objectives while also helping to achieve the goals of NASA's space program, including access to space, space exploration, reentry, and aeronautics as they relate to the performance of vehicles in non-Earth atmospheres. Results of research on relevant topics, such as hypersonics and operations in extreme (or alien) environments, would be transferred to NASA's space program.

The future of the air transportation system should be guided by quantifiable goals (NRC, 2003b). The federal government, however, does not have quantifiable goals related to the Strategic Objectives. Quantifiable goals are included in the strategic research agenda that is guiding aeronautics research in Europe. For example, European research goals for 2020 include the following (ACARE, 2004):

- Reduce fuel consumption and CO_2 emissions by 50 percent.
- Reduce perceived external noise by 50 percent.
- Reduce oxides of nitrogen (NO_x) by 80 percent.

Goals unsupported by funded and approved R&T programs, however, are little more than aspirations, and U.S. efforts to define quantifiable goals for the future should be coordinated with R&T planning efforts to reach the desired end state, consisting of credible goals and a properly directed R&T program.

REFERENCES

Advisory Council for Aeronautics Research in Europe (ACARE). 2004. Strategic Research Agenda 2. Vol 1. Available online at <www.acare4europe.com/html/background.shtml>.

General Aviation Manufacturers Association (GAMA). 2004. General Aviation Statistical Databook. Available online at <www.generalaviation.org/dloads/2004StatisticalDatabook.pdf>.

Lubitz, K. 1997. Study of Aircraft Accidents in Canada from 1987 to 1996. Aurora, Ontario: Ultralight Pilots Association of Canada. Available online at <www.challenger.ca/upac-accident-study/>.

Maier, M. 2006. Architecting Principles for Systems-of-Systems. White Paper. The Information Architects Cooperative (TiAC). Available online at <www.infoed.com/Open/PAPERS/systems.htm>.

National Research Council (NRC). 1991. The Decade of Discovery in Astronomy and Astrophysics. Washington, D.C.: National Academy Press. Available online at <http://fermat.nap.edu/catalog/1634.html>.

NRC. 2001. Astronomy and Astrophysics in the New Millennium. Washington, D.C.: National Academy Press. Available online at <www.nap.edu/catalog/9839.html>.

NRC. 2003a. New Frontiers in the Solar System. Washington, D.C.: The National Academies Press. Available online at <http://books.nap.edu/catalog/10432.html>.

NRC. 2003b. Securing the Future of U.S. Air Transportation: A System in Peril. Washington, D.C.: The National Academies Press. Available online at <http://fermat.nap.edu/catalog/10815.html>.

NRC. 2003c. The Sun to the Earth—and Beyond. Washington, D.C.: The National Academies Press. Available online at <http://books.nap.edu/catalog/10477.html>.

NRC. 2005. Earth Sciences and Applications from Space: Urgent Needs and Opportunities to Serve the Nation (Interim Report). Washington, D.C.: The National Academies Press. Available online at <www.nap.edu/catalog/11281.html>.

2

Process for Integration and Prioritization

STUDY PROCESS

The study began in September 2005 with a joint kick-off meeting between the steering committee and its supporting panels in order to hear directly from NASA and other federal entities the primary purpose of the study. Committee and panel members were also briefed on a recent report of the National Institute of Aerospace (NIA, 2005) and an earlier NRC study, *Aeronautical Technologies for the Twenty-first Century* (NRC, 1992).

At the second steering committee meeting, in November 2005, representatives from industry and academia were consulted in a roundtable discussion. The steering committee then developed a framework for the study, Strategic Objectives, and guidelines for the panels. It developed a quality function deployment (QFD) process, described below, for the panels to use. It identified, defined, and weighted the Strategic Objectives as well as the Why NASA? criteria. Finally, it outlined some basic rules and conventions for completing the prioritization process, all of which are summarized below.

From November 2005 through January 2006, each panel held a series of meetings. The panels identified and consulted a broad range of experts with backgrounds in industry, government, and academia. Many of them were able to attend panel meetings (see Appendix I). Working among themselves, the panel members then developed lists of research topics, called R&T Challenges. In some cases, these lists were very long, exceeding 100 items for a single R&T Area. Because it was not feasible to describe and prioritize so many Challenges in detail, the panels winnowed their lists, first by dropping those that seemed to have very little relevance to the Strategic Objectives. The number of Challenges was further reduced (to a total of 89 among all five Areas) by increasing the breadth of many of them, so that several very specific R&T topics could be collected into a single Challenge. Each panel, working under the oversight of the Steering Committee, then used the QFD methodology to relate these Challenges to the Strategic Objectives, generating a list of the 10 highest-priority Challenges within their Area.[1] All five panels considered issues related to subsonic, supersonic, and hypersonic flight regimes; infrastructure; transformation of the air transportation system; workforce; and education.

At the final meeting of the steering committee, in February 2006, it compiled inputs from the panels, vetted the prioritized list for each R&T Area, resolved conflicts in scoring among panels that had considered similar technologies, identified common themes among R&T Challenges from more than one R&T Area, and reached consensus on the overall content of the report, including summary findings and recommendations.

PRIORITIZATION

The steering committee directed the panels to use a modified QFD approach to rank the R&T Challenges they identified. QFD is a group decision-making methodology often used in product design. It is very useful for evaluating choices given a specific set of values. Cross-cutting research tends to rank highly, because it helps achieve multiple Objectives. The QFD scores described in this report for each R&T Challenge have no absolute, quantitative value. Rather, the QFD process serves as an organizational system that consistently evaluates each R&T Challenge and clearly conveys the rationale for the priority assigned to it. It is a qualitative process that utilizes the judgments and wisdom of informed experts to achieve a collective ranking of disparate objects.

[1]As noted in Chapter 3, the 11 highest-priority Challenges are identified for R&T Area A.

National Priority

The QFD process for this study used a matrix like the one shown in Table 2-1. The primary evaluation criteria are the six Strategic Objectives.[2] The R&T Challenges to be prioritized appear in the left-hand rows.[3] Each panel, as a group, scored each R&T Challenge with respect to the individual Objectives, based on its relevance and impact. Possible scores are limited to 1, 3, or 9. As is often done in QFD exercises, a nonlinear scale is used to magnify the differences in technologies to help delineate the most critical ones. A score of 1 implies that the Challenge has little or no relevance to the Objective. A 3 implies that the Challenge has moderate relevance and impact. A 9 implies that the Challenge has major relevance and impact.

The steering committee assigned each of the six Strategic Objectives a weight of 1, 3, or 5 to convey its relative importance to U.S. civil aeronautics research. The committee believes that the first two Objectives, capacity and safety and reliability, are the most critical because of their broad impact on the air transportation system as a whole, the vital importance of safety, and need to meet growing demand, and assigned them a weight of 5. The next two Objectives, efficiency and performance and energy and the environment, directly affect certain stakeholders and indirectly affect the public as a whole through their secondary effects on capacity and safety and reliability. They are considered to be slightly less important overall and are assigned a weight of 3. Finally, synergy with national and homeland security and support to space are assigned a weight of 1. Neither of these Objectives falls directly under the purview of civil aeronautics. Even so, security and the space program are important to the nation, and all other things being equal, civil aeronautics research that also provides benefits for these two Objectives should be of somewhat higher priority than comparable research that does not provide benefits for them.

The weight for each Strategic Objective (1, 3, or 5) is multiplied by the relevance and impact score (1, 3, or 9), which describes the impact on that Objective of research in a particular R&T Challenge. The sum of those products for each R&T Challenge then becomes the national priority score for that R&T Challenge. That score is a measure of the relative overall value to the nation of conducting research to overcome that particular R&T Challenge.

NASA Priority

Every R&T Challenge that has a high national priority does not necessarily become a high priority for NASA's civil aeronautics research program. To determine the NASA priority scores, each R&T Challenge is given a Why NASA? score, which is multiplied by the national priority score to arrive at a NASA priority. The Why NASA? score for each R&T Challenge is the average of the scores (1, 3, or 9) in the four Why NASA? columns on the right-hand side of the QFD tables. These scores evaluate each R&T Challenge in terms of the following:

- Supporting infrastructure
- Mission alignment
- Lack of alternative sponsors
- Appropriate level of risk

The scores used to assess priorities are based on the current situation, which will change. For example, this study did not attempt to predict how NASA expertise and facilities in various areas might degrade or mature, how NASA's aeronautics mission might be redefined, how the priorities of other research organizations might change, or how advances in the state of the art might change the risk associated with specific R&T Challenges. Changes in any of these factors will change the Why NASA? scores, which will directly change the NASA priority scores.

Supporting infrastructure

Supporting infrastructure refers to whether NASA already possesses the facilities, resources, and expertise to conduct research related to an R&T Challenge. A score of 1 implies that NASA has little or no relevant infrastructure. A score of 3 implies that NASA has infrastructure that is relevant but not unique. That is, industry, academia, or non-NASA federal agencies possess similar infrastructure or could obtain it easily. A score of 9 implies that NASA has infrastructure that is both relevant and unique.

Mission alignment

Mission alignment refers to whether research related to the R&T Challenge falls under NASA's charter, as defined in the National Aeronautics and Space Act of 1958 (As Amended). Relevant portions of the Space Act appear in Box 2-1. A score of 1 implies that the Challenge has little or no relevance to any item in the charter. A score of 3 implies that it has some relevance to and impact on one item in the char-

[2]The QFD matrix used in this study (see Table 2-1 and the QFD matrices in Chapter 3) is a simplified form of the table (sometimes called a house of quality) that is used in a standard QFD assessment. The QFD matrix for this study has also been rotated 90 degrees from the orientation normally used to display a QFD table. The Strategic Objectives in this study take the place of the customer requirements that appear in a standard QFD table, the R&T Challenges take the place of key product and process characteristics, and the Why NASA? composite score takes the place of risk level. The national priority scores are equivalent to the absolute importance rankings in a standard QFD table, and the NASA priority scores are equivalent to risk-weighted importance.

[3]Each Challenge is designated by the letter of the Area to which is belongs and by its NASA priority ranking in that Area. Thus, the R&T Challenge with the highest NASA priority in the aerodynamics and aeroacoustics R&T Area is designated A1. If two Challenges in that Area were to tie for second place, they would be listed alphabetically and designated A2a and A2b, and the next highest priority Challenge would be designated A4.

TABLE 2-1 Sample QFD Prioritization

		Strategic Objective							Why NASA?					
		Capacity	Safety and Reliability	Efficiency and Performance	Energy and the Environment	Synergies with Security	Support to Space	National Priority	Supporting Infrastructure	Mission Alignment	Lack of Alternative Sponsors	Appropriate Level of Risk	Why NASA Composite Score	NASA Priority Score
R&T Challenge	Weight	5		3		1			1/4 each					
X1	R&T Challenge 1	9	9	3	3	1	1	110	3	9	3	9	6.0	660
X2	R&T Challenge 2	3	9	3	9	1	1	98	1	9	3	9	5.5	539
X3	R&T Challenge 3	1	1	1	3	9	9	40	9	9	9	9	9.0	360

BOX 2-1
NASA's Mission as Reflected by Selected Items from the National Aeronautics and Space Act of 1958 (As Amended)

Section 102. Declaration of policy and purpose

(d) The aeronautical and space activities of the United States shall be conducted so as to contribute materially to one or more of the following objectives:

(1) The expansion of human knowledge of the Earth and of phenomena in the atmosphere and space;
(2) The improvement of the usefulness, performance, speed, safety, and efficiency of aeronautical and space vehicles;
(3) The development and operation of vehicles capable of carrying instruments, equipment, supplies, and living organisms through space;
(4) The establishment of long-range studies of the potential benefits to be gained from, the opportunities for, and the problems involved in the utilization of aeronautical and space activities for peaceful and scientific purposes;
(5) The preservation of the role of the United States as a leader in aeronautical and space science and technology and in the application thereof to the conduct of peaceful activities within and outside the atmosphere;
(6) The making available to agencies directly concerned with national defense of discoveries that have military value or significance, and the furnishing by such agencies, to the civilian agency established to direct and control nonmilitary aeronautical and space activities, of information as to discoveries which have value or significance to that agency;
(7) Cooperation by the United States with other nations and groups of nations in work done pursuant to this Act and in the peaceful application of the results thereof;
(8) The most effective utilization of the scientific and engineering resources of the United States, with close cooperation among all interested agencies of the United States in order to avoid unnecessary duplication of effort, facilities, and equipment; and
(9) The preservation of the United States preeminent position in aeronautics and space through research and technology development related to associated manufacturing processes.

ter. A score of 9 implies that the Challenge has great relevance to and impact on at least one item in the charter or some relevance to and impact on multiple items.

Lack of alternative sponsors

Lack of alternative sponsors refers to whether other sponsors are able and willing to perform the necessary research. NASA should not be repeating research that is (or should be) done by others. A score of 1 implies that if NASA did not do the research, some other organization would do it, or does already. A score of 3 implies that if NASA did not do the research, it would be done but not be developed to an adequate level of maturity, or it would lack aeronautical focus. A score of 9 implies that if NASA did not do the research, it would not be done.

Appropriate level of risk

Appropriate level of risk refers to whether the level of risk associated with an R&T Challenge is appropriate for a NASA research project. For example, NASA should not pursue incremental research that is of such low risk that industry could easily complete the research. Nor should NASA pursue research of great theoretical promise if the scientific and technical hurdles are so high that it has very little chance of success. A score of 1 implies that the Challenge is either very low risk (such that industry could pursue it) or extremely high risk (such that there is only a small chance of seeing any benefit without unforeseen revolutionary breakthroughs). A score of 3 implies that the Challenge either has low risk or very high risk. A score of 9 implies that it has moderate to high risk, which is a good fit to NASA's level of risk tolerance. All NASA research should be expected to progress toward established goals, but innovation is not possible without tolerance for failure, and the pursuit of moderate- and high-risk technology is appropriate for the nation's center of excellence for aeronautics.

NEXT STEPS

The top 10 R&T Challenges for each Area, in priority order, are discussed in Chapter 3. All the Challenges are discussed in Appendixes A to E, which also contain specific milestones. The technical discussions and milestones included in this report are intended to be advisory, as it was not feasible to complete a rigorous, comparative assessment of all of the research options that might be associated for each of the 89 Challenges. The committee believes that the best approach for selecting specific research projects to fund would be for NASA to solicit proposals from industry and academia at the level of the individual Challenges.

Comparing Priorities Among Different R&T Areas

The QFD process appears to be a rigorous quantitative process, with strict, laid-out criteria for each score. However, while each panel could consistently distinguish between what deserves a 3 and what deserves a 9, for example, some variations from panel to panel were inevitable. Furthermore, QFD is an iterative process. After initially scoring each R&T Challenge, panel members examined their results, assessed the justifications for each score for internal consistency and accuracy, and then adjusted some scores and justifications, as appropriate.

Once each panel completed the QFD process for its R&T Area, the steering committee reviewed the results and raised issues for the panels to reconsider to assure that the results were generally consistent when two panels had similar R&T Challenges. In the end, the panels and the steering committee concurred that (1) the Strategic Objectives were properly defined and weighted and (2) the Challenges were correctly scored and prioritized. Thus, although the steering committee reserved the right to change QFD scores without the concurrence of the panels, it did not find such action necessary.

The steering committee could have attempted to create a single integrated priority list of the R&T Challenges from all five R&T Areas. However, it was not practical for the committee to make extensive pairwise comparisons to assure that the scores for each R&T Challenge from each panel were consistent with the scores for dissimilar R&T Challenges from other panels. The steering committee also considered the value of having a single list of priorities and satisfied itself that (1) the results from each panel were generally consistent and well justified; (2) the high-priority R&T Challenges in each R&T Area were, indeed, high-priority items that should be included in NASA's aeronautics R&T program; and (3) the ultimate purpose of prioritizing R&T Challenges is presumably to determine which Challenges will be funded, and that determination will depend upon budgetary factors that were beyond the scope of this study (see Appendix G).

Given the above considerations, instead of creating an integrated, prioritized list of R&T Challenges from all five panels, the steering committee decided that the best use of the limited time and resources available to complete the study would be to identify Common Themes and formulate overall findings and recommendations (see Chapters 4 and 5). Given this situation, readers are cautioned against comparing the national and NASA priority scores for R&T Challenges from *different* panels to determine which is more important. The steering committee firmly believes that NASA should support research in all five R&T Areas, and the priorities identified in this report can be relied on to guide research planning within each of those areas.

REFERENCES

National Institute of Aerospace (NIA), National Strategy Team. 2005. Responding to the Call: Aviation Plan for American Leadership. Available online at <www.nianet.org/nianews/AviationPlan.php>.

National Research Council (NRC). 1992. Aeronautical Technologies for the Twenty-first Century. Washington, D.C.: National Academy Press. Available online at <http://books.nap.edu/catalog/2035.html>.

3

Research and Technology Challenges

The highest priority R&T Challenges for each R&T Area are listed and discussed below. The section for each Area includes a table showing the results of the quality function deployment (QFD) evaluation of R&T Challenges for that area. Each section also discusses general characteristics of high- and low-priority challenges in the relevant Area, R&T Thrusts that encompass multiple Challenges from a given area, and specific Challenges that rank high in national priority, but low in NASA priority. More detailed information for each Challenge appears in Appendixes A to E.

AERODYNAMICS AND AEROACOUSTICS

Introduction

Aerodynamics and aeroacoustics research is required to support development of advanced aeronautical systems. The scope of this R&T Area includes a wide range of fundamental fluid dynamic research ranging from low-speed, low-Reynolds-number flows to hypersonic, chemically reacting flows to aerodynamic issues associated with flight in alternative atmospheres. It does not include aerodynamic issues associated with ground transportation systems or fluid dynamic issues associated with hydrodynamic flows or the space environment.

The QFD process described in Chapter 2 was used to prioritize 19 R&T Challenges related to aerodynamics and aeroacoustics. Table 3-1 and Figure 3-1 show the results. The text that follows describes the 11 R&T Challenges that ranked highest in terms of NASA priority, the general characteristics of high- and low-priority Challenges, and the R&T Thrusts in this Area.[1] Further details on all the challenges, including the rationale for scoring, are found in Appendix A.

In terms of national priority, challenges A1, A2, A3, A6, and A7b all fall within a narrow range. Taking account of the weighting factors and scores, and noting that small changes in many of those elements can produce important changes in the final order, it should be concluded that these challenges are of roughly equal importance.

Top 11 R&T Challenges

A1 Integrated system performance through novel propulsion–airframe integration

Research into improved techniques for propulsion–airframe integration is required to enable greater aircraft flexibility and improve performance, especially as aircraft speeds increase. Improvements in the accuracy of predictions for three-dimensional (3-D) steady and unsteady interactions between external and internal aerodynamics and aeroacoustics would enable the design of advanced aeronautical systems, especially with systems of unconventional design. These interactions include the effects of steady and dynamic distortion on engine operations and the effects of hot, reacting exhaust flows on vehicle aerodynamics. These interactions are particularly important in the design of vertical and short takeoff and landing (V/STOL), extremely short takeoff and landing (ESTOL), supersonic, and hypersonic airplanes.[2]

---

[1]This chapter describes the top 10 R&T Challenges in each Area, except for the aeronautics and aeroacoustics Area. As shown in Figure 3-1, the NASA priority scores for Challenges A4a through A11 were relatively close, and there was a large difference in the scores for A11 and A12, so in this Area, unlike the remaining four, the top 11 R&T Challenges are described in the aeronautics and aeroacoustics Area.

[2]VTOL airplanes can take off and land vertically. This includes tiltrotors, the AV-8 Harrier, and the Joint Strike Fighter (JSF), for example. VTOL airplanes do not routinely take off or land vertically because of the range-payload penalty associated with the weight limitations of purely vertical operations. Rather, they use any available field length to develop some forward motion and wing lift during takeoff to increase the useful load (fuel plus payload). They tend to land vertically only at the end of the mission, when they are lighter, after burning fuel and/or dropping weapons.

STOL airplanes use high-lift systems to take off in less distance than

TABLE 3-1 Prioritization of R&T Challenges for Area A: Aerodynamics and Aerocoustics

		Strategic Objective							Why NASA?					
R&T Challenge		Capacity	Safety and Reliability	Efficiency and Performance	Energy and the Environment	Synergies with Security	Support to Space	National Priority	Supporting Infrastructure	Mission Alignment	Lack of Alternative Sponsors	Appropriate Level of Risk	Why NASA Composite Score	NASA Priority Score
	Weight	5	5	3	3	1	1		1/4 each					
A1	Integrated system performance through novel propulsion–airframe integration	9	3	9	9	9	9	**132**	3	9	3	9	**6.0**	**792**
A2	Aerodynamic performance improvement through transition, boundary layer, and separation control	9	3	9	9	3	3	**120**	3	9	3	9	**6.0**	**720**
A3	Novel aerodynamic configurations that enable high performance and/or flexible multimission aircraft	9	3	9	9	3	1	**118**	3	9	3	9	**6.0**	**708**
A4a	Aerodynamic designs and flow control schemes to reduce aircraft and rotor noise	9	1	3	9	3	1	**90**	3	9	3	9	**6.0**	**540**
A4b	Accuracy of prediction of aerodynamic performance of complex 3-D configurations, including improved boundary layer transition and turbulence models and associated design tools	3	3	9	3	3	3	**72**	9	9	3	9	**7.5**	**540**
A6	Aerodynamics robust to atmospheric disturbances and adverse weather conditions, including icing	9	9	3	1	9	1	**112**	3	9	3	3	**4.5**	**504**
A7a	Aerodynamic configurations to leverage advantages of formation flying	3	1	9	9	3	1	**78**	3	9	9	3	**6.0**	**468**
A7b	Accuracy of wake vortex prediction, and vortex detection and mitigation techniques	9	9	3	1	1	1	**104**	3	9	3	3	**4.5**	**468**
A9	Aerodynamic performance for V/STOL and ESTOL, including adequate control power	9	3	3	1	3	1	**76**	3	9	3	9	**6.0**	**456**
A10	Techniques for reducing/mitigating sonic boom through novel aircraft shaping	3	1	3	9	3	1	**60**	9	9	3	9	**7.5**	**450**
A11	Robust and efficient multidisciplinary design tools	3	3	9	9	3	3	**90**	3	9	3	3	**4.5**	**405**
A12	Accurate predictions of thermal balance and techniques for the reduction of heat transfer to hypersonic vehicles	1	1	3	1	9	9	**40**	9	9	3	9	**7.5**	**300**
A13	Low-speed takeoff and landing flight characteristics for access-to-space vehicles	1	3	1	1	3	9	**38**	3	9	9	9	**7.5**	**285**
A14	Efficient control authority of advanced configurations to permit robust operations at hypersonic speeds and for access-to-space vehicles	1	1	3	1	9	9	**40**	3	9	3	9	**6.0**	**240**
A15	Decelerator technology for planetary entry	1	1	1	1	3	9	**28**	3	9	9	9	**7.5**	**210**
A16	Low-Reynolds-number and unsteady aerodynamics for small UAVs	1	1	3	1	9	3	**34**	3	9	3	9	**6.0**	**204**
A17	Low-drag airship designs to enable long-duration stratospheric flight	1	3	1	3	9	1	**42**	3	3	3	9	**4.5**	**189**
A18	Prediction of communication capability through reentry trajectory and techniques to mitigate impact of communication blackouts	1	1	1	1	9	9	**34**	3	9	3	3	**4.5**	**153**
A19	Aircraft protective countermeasures based on a range of small deployed air vehicles	1	3	1	1	9	1	**36**	3	3	3	3	**3.0**	**108**

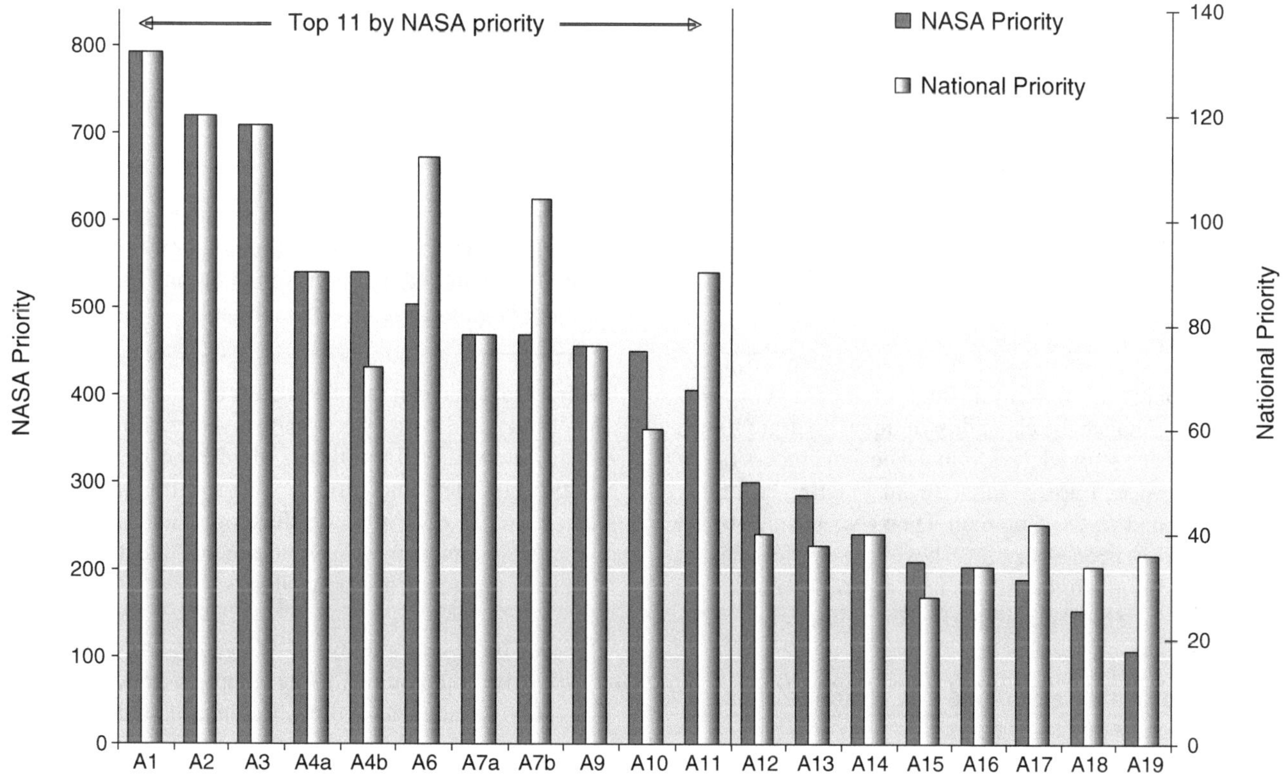

FIGURE 3-1 NASA and national priorities for Area A: aerodynamics and aeroacoustics.

Minimization of drag by propulsion–airframe integration will reduce fuel burn and CO_2 emissions.

A2 Aerodynamic performance improvement through transition, boundary layer, and separation control

Viscous drag at subsonic, supersonic, or hypersonic speeds may be reduced by controlling the onset of boundary layer transition using active control or passive 3-D design concepts. Direct reduction of skin friction drag is possible with extensive laminar flow, which can be achieved with a combination of vehicle shaping and flow control concepts. One example is natural laminar flow using reduced sweep and control of crossflow pressure gradients through shape optimization. A second example is boundary layer manipulation through suction, blowing, or distributed effectors. Related concepts may also be used to reduce separation at high lift and other conditions (e.g., buffet), which improves performance at high-lift conditions. In some conditions of flight, particularly at high lift, a turbulent boundary layer is needed. Active flow control techniques are emerging, including piezoelectric, voice-coil, dielectric barrier discharges, and surface electrical discharges. The potential advantages are clear, but implementation has been hampered by the lack of accurate and efficient methods for prediction (see Challenge A4b) and design and by the difficulty of conducting experiments that require high Reynolds numbers and are sensitive to disturbances such as free-stream turbulence and noise. Work on this Challenge should identify the most promising application domains, control approaches, and actuator concepts and develop efficient methods for design and experimental validation.

A3 Novel aerodynamic configurations that enable high performance and/or flexible multimission aircraft

Most classes of aircraft configuration have remained constant for many years (e.g., the tube and wing of a subsonic transport and the main rotor plus tail rotor of a helicopter). Novel aerodynamic configurations provide substantial opportunities for long-term breakthroughs in aircraft capabili-

conventional aircraft (typically a few thousand feet). Very few STOL aircraft can safely take off on runways shorter than 3,000 ft and none on runways less than 2,000 feet. (This class does not include ultralight aircraft, kit planes, etc. that can operate out of short fields due to their small size but do not have high-lift systems.)

ESTOL airplanes would be able to safely take off on runways of 2,000 ft. They would have high-lift systems and thrust-to-weight ratios that are higher than conventional aircraft but not as high as VTOL aircraft. ESTOL aircraft have not yet been developed for commercial or military operations.

V/STOL refers to both VTOL and STOL airplanes that convert to fixed-wing flight after takeoff; it does not include helicopters.

ties. A number of innovative concepts have been proposed in the past and pursued to differing levels. Examples include the blended wing body, canard rotor wing, oblique flying wing, and strut-braced wing. A sustained research program should be promoted to develop novel aircraft configurations, including further development of existing concepts where appropriate, with emphasis on achieving breakthroughs related to the high-priority Strategic Objectives.

Other R&T Challenges would also contribute to enabling novel aerodynamic configurations. Advances in flight mechanics and propulsion–airframe integration (R&T Challenge A1) are required to make advanced concept airplanes viable and robust. Flow control (R&T Challenge A2) could significantly enhance the capability of novel configurations, since it could be assumed a priori in the design process rather than added as an improvement to an existing airplane. Research related to the Common Theme of physics-based analysis tools is needed to move beyond empirical design tools.[3] In addition, flight testing is a critical element of a successful research program in novel configurations.

A4a Aerodynamic designs and flow control schemes to reduce aircraft and rotor noise

Many of today's airports now limit operations because of the noise emitted to the surrounding community. Future passenger growth at many airports will be limited if the noise levels emitted by the newer aircraft are not reduced further, thus adversely affecting capacity. Off-loading the main runway of regional jets by using ESTOL aircraft and rotorcraft, thus reducing congestion for larger passenger aircraft on the main runway, will dramatically increase capacity by allowing more takeoffs and landings at existing airports without increasing demand for runway usage (NRC, 2003; FAA, 2000). However, it will only be possible if these ESTOL aircraft and rotorcraft are quiet. Aerodynamic noise research should be pursued to (1) improve understanding of the underlying flow physics, (2) develop novel technologies, and (3) create improved and validated acoustic prediction and design tools. This research should include a balance of physics modeling, tool development, and experiments. Important physical phenomena that require research include cavity flows, unsteady flow–solid surface interactions, flow separation, rotor dynamic stall, and wake vortex dynamics. Novel needs include quiet, high-lift devices; technologies to enable steep, quiet, slow-approach trajectories; technologies to reduce the strength of vortices shed from the rotor blades and/or vortex/blade position control; integrated advanced control schemes for active rotorcraft noise reduction; and technologies to reduce rotor response to vortex-induced disturbances. Physics-based source noise prediction methods and improved computational aeroacoustic tools are key requirements. Design tools are needed both at the technology level and at the aircraft system level, with particular attention to integrated solutions for aerodynamic and operational issues.

A4b Accuracy of prediction of aerodynamic performance of complex 3-D configurations, including improved boundary layer transition and turbulence models and associated design tools

The aerospace industry lacks computational analysis and design tools that can rapidly and accurately predict complex flow behavior driven by boundary layer transition, flow separation, novel configurations, off-design operations, and multidisciplinary interactions. To meet this need, physics-based design tools must be developed and systematically validated in representative environments. Ideally, these tools should have the following attributes:

- Adaptive and intelligent self-generating grids that are easily implemented using simple computer-aided design surface instructions, minimal boundary condition definition, and desktop operation.
- Seamless applicability over the continuum of fluid flows (speed regimes, phase, periodicity) and reference frames.
- Ability to accurately predict transitional and separated flows, validated through experimentation.
- Ability to fully describe the state of the fluid at any point in the solution domain, with useful information on the surfaces.
- Inverse design capability.

The benefit of technologies developed by this Challenge would be enhanced by parallel development of multidisciplinary design tools to address complex nonlinear interactions, and parameter uncertainties and models, while still being computationally efficient (see Challenge A11).

A6 Aerodynamics robust to atmospheric disturbances and adverse weather conditions, including icing

Adverse weather conditions, including storms and icing conditions, significantly reduce the capacity and reliability of the air transportation system. Adverse weather also degrades system safety. This issue is of importance to both civil and military aviation. Research is needed to improve the ability to predict and monitor environmental conditions and develop aerodynamic designs and techniques that are robust to adverse conditions.

At present, wind-shear warning systems are built into commercial aircraft, icing hazards are handled by regulatory constraints on flight operations, and prediction techniques are largely empirical. Low-cost techniques to mea-

[3]"Physics-based" refers to the general use of scientific principles in the place of empirical data. It includes the use of principles from chemistry, biology, material science, etc.

sure upstream environmental conditions should be developed. Examples of promising techniques include microwave, lidar, and laser-acoustic measurement techniques. Efforts to miniaturize and reduce the cost of the measurement equipment should be supported. Techniques to predict and mitigate the impact of adverse environmental conditions on the aircraft operation should be improved. Required improvements include the development of models to predict the impact of multiphase, nonequilibrium situations encountered under icing conditions; validation of icing prediction capabilities to enable a reduction in the high cost of aircraft and helicopter icing certification; and models for the complex-flow, time-dependent, 3-D interactions encountered during wind shear or ambient turbulence on the aircraft flowfield.

A7a Aerodynamic configurations to leverage advantages of formation flying

Formation flight is currently used by military airplanes for a variety of operational reasons, although rarely for drag reduction. Recent breakthroughs in accurate navigation and control make possible extended precision formation flight at cruise and permit exploitation of favorable interference for vortex drag reduction. Although this phenomenon is well known, the magnitude of the potential savings is not widely appreciated. Three airplanes flying in formation and designed to best exploit these effects could reduce vortex drag by more than 50 percent in cruise, a greater reduction than that obtainable by extensive laminar flow control on the wing. This would mean roughly a 20 percent reduction in total drag under identical operating conditions. However, with less induced drag the optimum altitude increases, reducing viscous drag as well. The net result is almost a 30 percent reduction in total drag. Unlike the tight formations required for military applications, drag savings are possible even with longitudinal separations of several miles (Spalart, 1998), reducing safety concerns associated with formation flight.

Initial NASA work on autonomous formation flight has identified some of the technology requirements for achieving these savings, but considerable research remains in both control methodology and aerodynamic design to take most advantage of the concept. Applications to cargo airplanes, rotorcraft, and even supersonic flight are possible but have not been studied extensively. Aerodynamic challenges include vortex location prediction, sensing and control, and wing design for efficient high-lift cruise. Suggested work in this area would result in improved methods for predicting wake vortex evolution; design tools for evaluation and optimization of multiple interacting airplanes; and experimental validation, including flight testing (which is especially important for evaluating real atmospheric effects). The aerodynamic aspects of formation flying are related to R&T Challenges D1 and E2.

A7b Accuracy of wake vortex prediction, and vortex detection and mitigation techniques

Wingtip vortices produced by airplanes present a danger to following aircraft, so airplane designs and techniques that mitigate the strength of these vortices, techniques to locate and determine their strength, and techniques to predict their propagation and decay are important factors in minimizing aircraft separation and enhancing safety.[4] (Since aircraft lift is intimately tied to the production of circulation, these vortices cannot be completely eliminated.) Currently, aircraft separation standards are set by conservative estimates of the wake vortex trajectory (generally a sinking trajectory, but also affected by local weather conditions) and decay rate. Techniques to measure the characteristics of upstream wake vortices include lidar and laser-acoustic techniques, but these technologies are currently expensive (limiting their use to larger aircraft) and are less reliable than desired.

Research into techniques to predict the formation, trajectory, and decay of vortices needs to be performed. This includes development and validation of numerical methods to accurately predict the trajectory and dissipation of vortices, integration of local weather prediction techniques into existing larger-scale weather models, demonstration of low-cost techniques for locating and measuring the strength of wake vortices for both ground-based and aircraft-based applications, and investigation of airplane designs that mitigate the strength of wake vortices.

A9 Aerodynamic performance for V/STOL and ESTOL, including adequate control power

The development of ESTOL regional jets able to operate from 2,000 ft runways and taxiways and to cruise in existing air traffic corridors will significantly reduce congestion problems on the main runways of hub airports. V/STOL aircraft will be able to operate from taxiways and other paved areas at major airports, further relieving congestion. In responding to natural disasters and carrying out military operations, low-cost VTOL tactical transports would be able to operate from short, austere landing fields near the focus of attention (e.g., the location of injured civilians or troops, battle areas, and landslides).

Development of an efficient high-lift system is not the most important enabling technology for ESTOL airplanes. Conventional aerodynamic control surfaces become ineffective at the low landing and takeoff speeds of ESTOL airplanes (on the order of 65 knots). The challenge is to generate the forces needed for pitch trim and to control the aircraft at these slow speeds. It is also important to develop a thrust vectoring and reversing nozzle technology that not only provides the required lift but can also be integrated into a low-

[4]The scope of this Challenge does not include and would not directly apply to helicopter blade wakes.

drag configuration. (ESTOL airplanes require much more thrust than conventional or STOL airplanes, but not as much as VTOL airplanes.) In addition, wing design and fuselage shaping are needed to reduce cruise drag in the transonic regime for ESTOL regional jets.

An important task for research related to rotorcraft and VTOL airplanes is to improve hovering and cruise efficiency. Reductions in downward forces in near-hovering flight dramatically improve the payload capability of tiltrotor and powered-lift aircraft. Active control of large separation regions on these aircraft through blowing, zero-mass effectors, and integrated mechanical devices are promising methods for reducing download. Active twist control of the rotor also allows the rotorcraft to be designed to better match the hover and cruise design conditions, thereby improving efficiency. Active control of separation regions and smart design guided by high-fidelity codes will decrease cruise drag and improve the performance of VTOL airplanes.

A10 Techniques for reducing/mitigating sonic boom through novel aircraft shaping

Safe, efficient, cost-effective, environmentally acceptable supersonic flight over land remains elusive nearly 60 years after airplanes broke the sound barrier. The principal remaining problems are sonic boom mitigation, public acceptance, and sustained supersonic flight performance. Today, federal regulations prohibit civil supersonic flight over land. If this regulatory barrier can be overcome, it will probably stimulate investment that would overcome the other barriers and help usher in a new era of time-critical air travel. Building on the recent in-flight validation of NASA's theory of shaped sonic boom persistence, a robust and comprehensive plan of research for technology maturation and tool development should be pursued to determine if practical supersonic airplanes can be developed whose sonic boom is acceptable to the public (Pawlowski et al., 2005). Such a plan should comprise public sonic boom acceptability determination; community exposure testing; aircraft shaping techniques that result in a low-amplitude, acceptable acoustic signature with minimal performance impact; critical propulsion-airframe integration technologies commensurate with low-boom design; aircraft and acoustic scaling methodologies; sensitivities to off-design conditions under a variety of atmospheric conditions; rapid and inverse computational design tools that address multiple design constraints; systematic validation through ground and flight tests; and metrics to assess progress and guide continuation according to plan. This Challenge is closely tied to Challenge B8.

A11 Robust and efficient multidisciplinary design tools

Multidisciplinary design tools are pervasive in aeronautics. More effective multidisciplinary tools would likely shorten the design cycle time for conventional aircraft and facilitate the discovery of new highly integrated aircraft designs with better performance than conventional designs. The development of physics-based models for this design environment is addressed in R&T Challenge A4b. This Challenge is associated with the research required to efficiently and effectively integrate multidisciplinary design tools of varying fidelity and numerical complexity into a seamless design environment. Research is also needed on automated techniques for handling and propagating parameter uncertainties throughout the design to allow development of robust aircraft designs.

High-Priority R&T Challenges and Their Associated Thrusts

Some of the high-priority R&T Challenges significantly impact multiple Strategic Objectives; others are high priority because NASA possesses unique capabilities to address them. In particular, R&T Challenges that significantly improve capacity or safety and reliability scored high due to the relevant weightings. The principal factors affecting an increase in capacity relate to expanding the operational capabilities near airports, expanding flight capabilities under adverse weather conditions, and enabling an expansion of operation from smaller airports. The expansion of operations near airports will require research into noise reduction and aircraft wake physics. Expansion of operations under adverse weather conditions will require research associated with techniques to monitor and then mitigate adverse environmental conditions, including icing, wind shear, and free-stream turbulence. Expansion of operations from smaller airports involves research on shortened takeoffs and landings and the associated noise reduction.

The development of improved physical models and design tools for aerodynamic and aeroacoustic phenomena and techniques aimed at understanding and providing the option of controlling these phenomena rank high in the R&T Challenge prioritization. Mastery of these Challenges will enable significant advances in the performance and operability of aircraft through development of improved and possibly revolutionary designs and reduction of design margins associated with uncertainties.

The following four R&T Thrusts describe threads of commonality among the R&T Challenges within the aerodynamics and aeroacoustics Area.

Improved understanding and control of the fundamental physics of aerodynamic and aeroacoustic phenomena

Complex fluid dynamic processes often present barriers to improved aircraft performance, so a better understanding of these phenomena is required. These processes can occur across significant spatial and temporal scales and involve interactions with processes that come under the purview of other disciplines. With a deeper knowledge of the fundamental physical phenomena, effective techniques will likely evolve to control these processes, enabling improved aircraft performance.

Accurate and robust multidisciplinary design tools

Aeronautics is fundamentally multidisciplinary, so many aspects of aerodynamics and aeroacoustics are impacted by cross-discipline factors. Multidisciplinary aerodynamic and aeroacoustic design tools are needed that are accurate and robust yet cost-effective in terms of computing time and computational resources.

Sensing and responding to the external environment

Development of aircraft systems that respond dynamically to the local environment could significantly improve capacity and safety. With measurement techniques to sense the local environment ahead of an aircraft and allow it to respond accordingly, aircraft spacing can be reduced and operations in adverse weather can be expanded, with no degradation of safety.

Revolutionary aerodynamic configurations

Even though the basic design of civil aircraft has remained remarkably stable for many decades, it is not clear that the configuration has already been optimized. The steering committee believes that improved understanding and control of fluid dynamic phenomena will result in novel aircraft designs offering revolutionary advances in performance and operability in all mission areas.

Low-Priority R&T Challenges

No attempt was made to compile and assess all possible aerodynamic and aeroacoustic issues. All of the Challenges described above are relevant to fundamental aeronautics of civil aircraft. The aerodynamic Challenges that ranked low in the prioritization were largely research areas that support national or homeland security or the NASA space mission but minimally impact Strategic Objectives directly related to the performance of the air transportation system. Examples of these Challenges include hypersonic vehicle technologies, small UAVs, and stratospheric airships. These Challenges could play a vital role in NASA's space exploration mission and in matters of national and homeland security; however, they ranked low in terms of both national and NASA priority for this report, where the focus is on civil aeronautics.

Hypersonic technologies appear in Challenges throughout the prioritization list. Challenges associated with the development of a more complete understanding of hypersonic issues, such as transition, turbulence, and separation phenomena and the development of techniques to control these phenomena, are included in high-priority R&T Challenges that encompass multiple speed regimes. Challenges specific to hypersonic vehicles, such as low-speed handling characteristics, are rated much lower.

PROPULSION AND POWER

Introduction

This section describes key R&T Challenges and Thrusts associated with aircraft propulsion and electrical power generation that should be addressed via basic and applied research to advance national civil aeronautics capabilities. These advances will permit the U.S. aeronautics enterprise to bring highly competitive products to market and improve the national capacity to move people and goods quickly and affordably with minimal energy usage and environmental impact.

Historically, paradigm shifts in propulsion capability have enabled significant advances in aircraft performance. The replacement of water-cooled piston engines with radial, air-cooled engines enabled the great airframe advances of the first half of the 20th century, while those in the second half were greatly expedited by the gas turbine engine. The gas turbine will very likely continue to be the dominant means of propulsion for both civilian and military aircraft for the next half century. With oil prices at historic highs and increasingly stringent noise and emissions regulations, gas turbine designers face formidable obstacles to create more fuel efficient, cleaner, and quieter engines. Opportunities abound for significant advances, with current gas turbine performance still well below theoretical limits. For example, improvements in overall efficiency and, concomitantly, fuel economy, of more than 30 percent appear attainable (Koff, 2004) but will only occur with significant advances in high-temperature materials and rotating machinery aerodynamics. With advances in information technology, sensor miniaturization, and modeling, intelligent engines capable of self-diagnosis and adaptation, similar to those in the automotive realm, are in the offing. Advances in information technology are also driving electrical power demands for both flight systems and passenger needs—that is, entertainment and productivity. The desire for rapid yet affordable transcontinental and intercontinental travel will continue to motivate research into supersonic flight engines; it is difficult to imagine commercial aviation being restricted to subsonic flight regimes 50 years from now. Airbreathing engine technology also has the potential to contribute significantly to the development of reusable higher payload fraction, access-to-space vehicles. Technical progress will be greatly expedited by the use of validated, physics-based computational simulation tools, which will permit designers to optimize designs and greatly minimize the number of design cycles typical of empirical design-build-test-redesign approaches.

The QFD process described in Chapter 2 was used to prioritize 16 R&T Challenges related to the Area of propulsion and power. Table 3-2 and Figure 3-2 show the results. The text that follows describes the 10 R&T Challenges that ranked highest in terms of NASA priority, the general characteristics of high- and low-priority Challenges, and the R&T Thrusts in this Area. Further details on all Challenges, including the rationale for scoring, are found in Appendix B.

Top 10 R&T Challenges

B1a Quiet propulsion systems

Public concerns over the environmental impact of aircraft and airport operations—primarily noise and emissions—have prompted increasingly strict legal and regulatory requirements, which can severely constrain the ability of civil aviation to meet national and global needs for mobility, increased market access, and sustained economic growth. Aircraft noise concerns include takeoff and landing noise; taxi and engine run-up noise; flyovers at cruise altitude over very quiet areas; and sonic booms associated with supersonic flight.

Figure 3-3 shows how the impact of aviation noise on people living around airports has declined in the United States. It contrasts the growth of air travel with the reduction in the number of people exposed to 65-decibel (dB) day-night average sound level (DNL), which is what the federal government has defined as the "significant noise level." Since 1975, the number of persons exposed to significant noise levels has greatly declined, with the transition of commercial aircraft to quieter models even as air travel has grown dramatically. The availability of low-noise technologies, such as high-bypass-ratio engines, contributed significantly to this transition. Assuming the industry's continued recovery, and given the goal of doubling capacity over the next 10 to 35 years, the dramatic improvements in noise exposure in the last two decades are unlikely to persist. The environmental impact of aircraft noise is projected to remain roughly constant in the United States for the next several years and then increase as air travel growth outpaces expected technological and operational advancements (Waitz et al., 2004). Furthermore, the public currently reports considerable annoyance even when DNLs are below 65 dB. Regulatory actions to limit or reduce noise exposure will likely lead to even more stringent limits.

Future abatement efforts may need to reduce allowable noise levels to as low as 55 dB DNL in both the United States (NASA, 2003) and Europe (ACARE, 2001). Meeting future noise targets will be extremely challenging and will require continued fundamental research in noise phenomena and advanced propulsion technologies. The development of validated, physics-based noise prediction tools by NASA will greatly aid the development of quieter engines. Research is needed to reduce the noise of engine systems, including fan noise, jet noise, and core noise. Research should also encompass systems analysis; advanced concepts, such as adaptable chevrons; the community impact of aircraft noise; and improved metrics to quantify and mitigate these impacts.

B1b Ultraclean gas turbine combustors to reduce gaseous and particulate emissions in all flight segments

Emissions from aircraft constrain the growth of aviation due to their environmental impacts and potential human health consequences. For example, airports located in air quality nonattainment or maintenance areas increasingly find that air emissions add to the complexity, length, and uncertainty of the environmental review and approval of expansion projects (Akin et al., 2003).

Key pollutants of concern include oxides of nitrogen and sulfur (NO_x and SO_x), carbon monoxide (CO), unburned hydrocarbons (UHCs), hazardous air pollutants, and particulate matter (PM). In addition, emissions of CO_2 and water vapor (H_2O) in the upper troposphere and stratosphere are of concern because of their potential impact on Earth's climate (IPCC, 1999). Both CO_2 and H_2O are inherent combustion products of hydrocarbon fuels, and their emissions can only be reduced through improvements in overall cycle efficiency (see R&T Challenge B4)—or a change in fuels. Emissions of NO_x, CO, UHC, and PM from the combustor can be reduced through the development of ultraclean combustion approaches, a critical step to mitigate the environmental impacts of aviation.

Low NO_x emissions can be achieved with both rich- and lean-burning combustor designs. Lean combustion concepts have received substantial market penetration through their widespread implementation in land-based gas turbine applications over the last two decades. The key technical issues associated with these combustors concern unsteady combustion phenomena, including combustion instability, flame blow-off, flashback, and autoignition. Although combustors run lean overall, the majority of commercial aircraft engines run rich in the front end. The key issues associated with them are PM emissions and quench zone mixing (Lefebvre, 1999).

B3 Intelligent engines and mechanical power systems capable of self-diagnosis and reconfiguration between shop visits

In the future, advances in sensing, control, and information technology will lead to engines that are more sophisticated and more intelligent. Research thrusts should investigate how more intelligent systems can (1) improve engine health diagnostics and remedial actions in flight, (2) optimize the mission, and (3) use flight data to improve maintenance on the ground. For current engines, the focus will be very much on diagnostics. Better physics-based modeling will be essential. Development of better computational fluid dynamics (CFD) tools, life-prediction tools, and steady-state and dynamic performance checks will be keys to success. Reducing in-flight shutdowns by a factor of 3 and unscheduled engines removals and delays and cancellations by a factor of 5 should be achievable and would reduce maintenance costs by 50 percent. Requirements include (1) smaller sensors with better response and higher operating temperatures and (2) better materials with narrower property tolerances. This should increase disk and airfoil life by 50 percent.

Intelligent engine development will include active control of many engine components: combustor control to permit operation with leaner burners, leading to lower NO_x emissions; compressor active stall control to allow operation at higher pressure ratios, leading to higher fuel efficiency; and closed-loop clearance control to increase turbine efficiencies and extend on-wing life by 3 years.

B4 Improved propulsion system fuel economy

The fuel economy of gas turbine propulsion systems is a function of engine efficiency, propulsion-induced drag, and propulsion weight. Overall engine efficiency is the product of the efficiency of creating hot, high-pressure gases (thermal or cycle efficiency), the efficiency of transferring energy from the hot high-pressure gases to a more desirable form (transfer efficiency), and the efficiency of creating thrust from the engine fan and core flows (propulsion efficiency). The thermal efficiency for a gas turbine (Brayton cycle) is primarily a function of overall engine pressure ratio. That is, as long as the turbine can tolerate the inlet temperature corresponding to a given pressure ratio, the overall pressure ratio sets the efficiency of the cycle. Figure 3-4 illustrates very clearly that state-of-the-art gas turbines have not reached the theoretical limits of thermal efficiency. The technologies identified in the figure have the potential to improve the thermal efficiency of gas turbines, to significantly increase fuel economy, and to decrease the environmental impact of the air transportation system.

Transfer efficiency is determined by the component efficiencies of the fan and low-pressure turbine and the losses of the shaft bearings. High-efficiency, low-pressure turbines need high rotor speeds, but highly efficient fans require low rotor speeds. Therefore, engines with high transfer efficiency must have reduction gearboxes or other technologies that permit different rotor speeds for the fan and low-pressure turbine.

Propulsion efficiency is a function of the difference between the velocity of engine exhaust and the forward velocity of the aircraft. Increasing the mass flow of air through the system at slower speed improves propulsion efficiency and decreases noise. However, this increases the diameter of the engine, which increases friction and flow blockage. Since larger engines will also be heavier, the use of composites or other lightweight materials for construction of the large structural pieces of the turbofan will also be necessary.

As shown in Figure 3-4, improving thermal efficiency by 15 percent requires advances in several technologies: 3D aerodynamics, active flow control, cooled cooling air and a thermal management system, multiwalled cooling, and ceramic matrix composites (CMCs) and intermetallics. Over the long term, advances in all three efficiencies (thermal, transfer, and propulsion) should be able to improve fuel economy by 30 percent relative to the GE-90 for large commercial engines and 30 percent relative to T700/CT7 for small engines.

B5 Propulsion systems for short takeoff and vertical lift

The utilization of V/STOL airplanes and increased use of helicopters could greatly increase the capacity of civil aviation by allowing more takeoffs and landings at existing airports without increasing demand for runway usage (NRC, 2003). V/STOL airplanes include tilt-wing aircraft, tilt-rotor aircraft, vertical-lift fan aircraft, and blown-wing aircraft. Currently, the fuel economy of V/STOL propulsion systems is not on par with that of fixed-wing commercial airplanes. Propulsion systems for all new aircraft must also demonstrate extremely high levels of reliability. Propulsion systems for V/STOL aircraft are in an early state of development or do not exist for civil aircraft. In addition, engine-out strategies need to be developed and verified for certification.

This Challenge should support development of V/STOL and helicopter propulsion systems with fuel economy comparable to future small commercial aircraft—namely, 20 percent better than the CT7 family of engines that are currently in production for small conventional aircraft. Many of the same technologies that apply to large and small engines for conventional aircraft also apply to V/STOL propulsions systems. However, additional technologies such as high-efficiency, angled gearboxes; high-efficiency reduction gearboxes; large-bleed systems; thrust vectoring systems; noise reduction both inside and outside the aircraft; fan-tip-driven turbines; and high-power clutch systems will be required to put V/STOL airplanes into affordable, large-scale commercial service with minimal environmental impact.

There are three major technology efforts to be undertaken in support of V/STOL airplanes for civil aviation. The first and most important is to demonstrate an engine sized for most helicopters or for UAVs (roughly 3,000-shaft horsepower) that meets the fuel economy goals. The important characteristics of this demonstration engine are to achieve overall pressure ratios of 25:1 or 30:1 and turbine inlet temperatures of 2800°F. This will require some combination of the following technologies: (1) new compressor disk materials, (2) greatly improved turbine cooling configurations, (3) new turbine blade alloys and coatings, (4) component aerodynamics designed with the latest computational models, and (5) highly effective, low-pressure-drop dirt separation. Such an engine would benefit helicopters as well.

Second, the powertrain system of most V/STOL airplanes (as well as helicopters) will consist of shafting with speed reduction gearboxes, angled gearboxes, and perhaps

TABLE 3-2 Prioritization of R&T Challenges for Area B: Propulsion and Power

R&T Challenge		Strategic Objective: Capacity	Strategic Objective: Safety and Reliability	Strategic Objective: Efficiency and Performance	Strategic Objective: Energy and the Environment	Strategic Objective: Synergies with Security	Strategic Objective: Support to Space	National Priority	Why NASA?: Supporting Infrastructure	Why NASA?: Mission Alignment	Why NASA?: Lack of Alternative Sponsors	Why NASA?: Appropriate Level of Risk	Why NASA Composite Score	NASA Priority Score
	Weight	5	5	3	3	1	1		1/4 each	1/4 each	1/4 each	1/4 each		
B1a	Quiet propulsion systems	9	1	3	9	3	1	**90**	3	9	3	9	**6.0**	**540**
B1b	Ultraclean gas turbine combustors to reduce gaseous and particulate emissions in all flight segments	9	1	3	9	3	1	**90**	3	9	3	9	**6.0**	**540**
B3	Intelligent engines and mechanical power systems capable of self-diagnosis and reconfiguration between shop visits	3	9	3	3	3	1	**82**	3	9	3	9	**6.0**	**492**
B4	Improved propulsion system fuel economy	3	1	9	9	3	1	**78**	3	9	3	9	**6.0**	**468**
B5	Propulsion systems for short takeoff and vertical lift	9	1	3	3	3	1	**72**	3	9	3	9	**6.0**	**432**
B6a	Variable-cycle engines to expand the operating envelope	3	1	9	3	3	9	**68**	3	9	3	9	**6.0**	**408**
B6b	Integrated power and thermal management systems	3	1	9	3	3	9	**68**	3	9	3	9	**6.0**	**408**
B8	Propulsion systems for supersonic flight	3	1	3	1	9	9	**50**	9	9	3	9	**7.5**	**375**
B9	High-reliability, high-performance, and high-power-density aircraft electric power systems	1	3	9	3	3	3	**62**	1	9	3	9	**5.5**	**341**
B10	Combined-cycle hypersonic propulsion systems with mode transition	1	1	3	1	9	9	**40**	9	9	3	9	**7.5**	**300**
B11	Alternative fuels and additives for propulsion that could broaden fuel sources and/or lessen environmental impact	3	1	3	9	3	1	**60**	3	3	3	9	**4.5**	**270**
B12	Hypersonic hydrocarbon-fueled scramjet	1	1	3	1	9	9	**40**	9	3	3	9	**6.0**	**240**
B13	Improved propulsion system tolerance to weather, inlet distortion, wake ingestion, bird strike, and foreign object damage	3	9	3	1	3	1	**76**	3	3	3	3	**3.0**	**228**
B14	Propulsion approaches employing specific planetary atmospheres in thrust-producing chemical reactions	1	1	1	1	1	9	**26**	3	9	9	9	**7.5**	**195**
B15	Environmentally benign propulsion systems, structural components, and chemicals	1	1	1	9	3	1	**44**	3	3	3	3	**3.0**	**132**
B16	Reduced engine manufacturing and maintenance costs	3	3	3	3	3	1	**52**	3	1	1	3	**2.0**	**104**

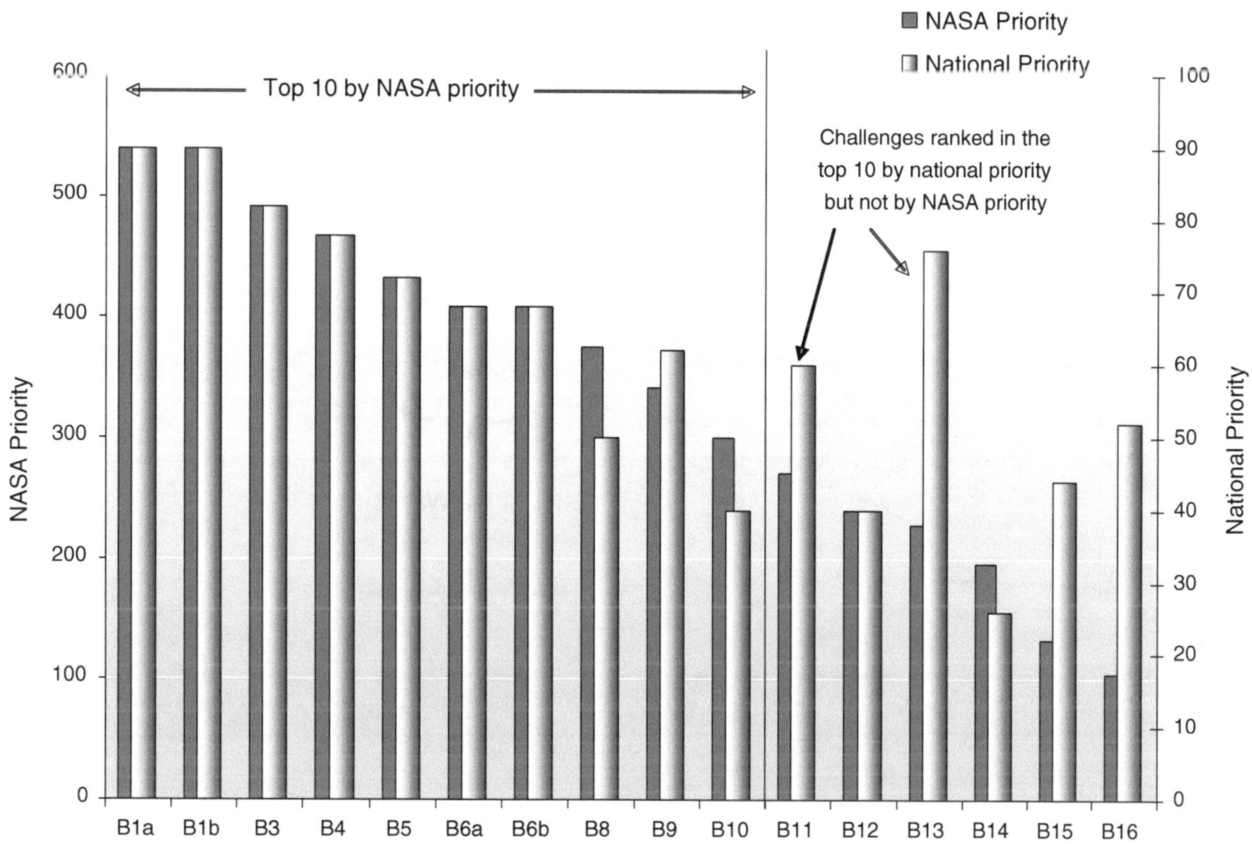

FIGURE 3-2 NASA and national priorities for Area B: propulsion and power.

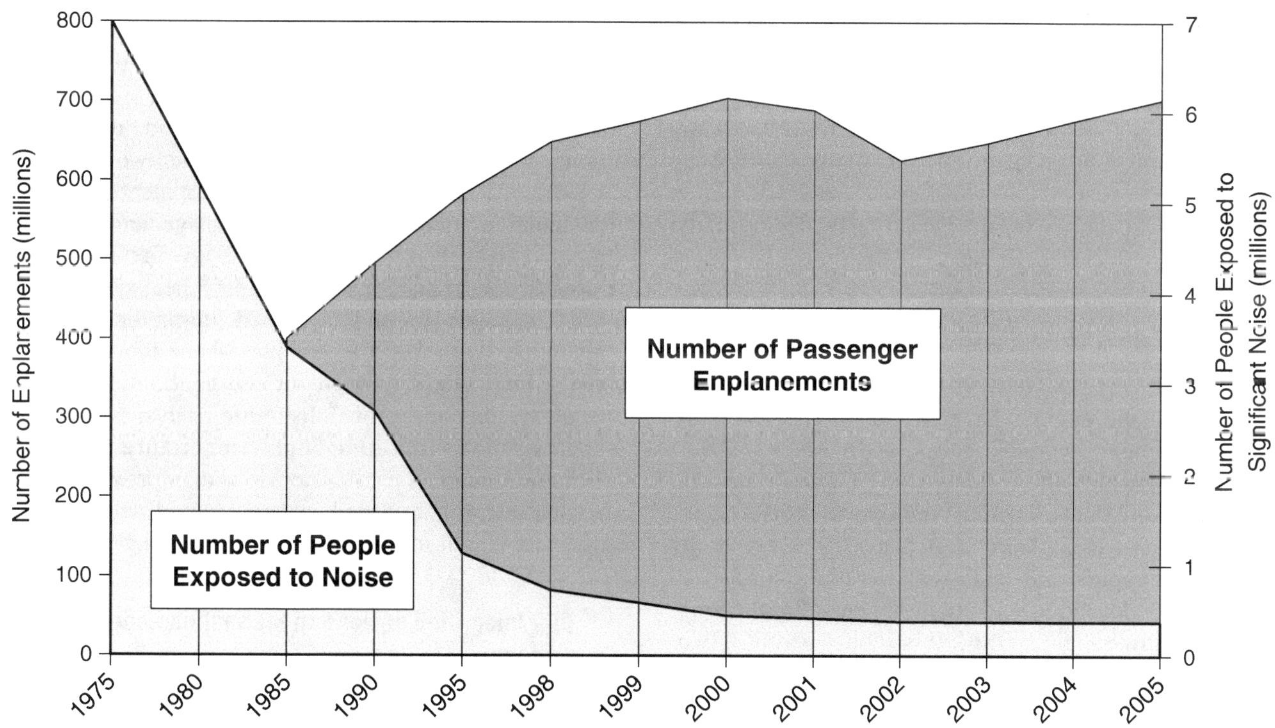

FIGURE 3-3 Actual and predicted exposure to significant noise (65-dB day-night average sound level) and enplanement trends for the United States, 1975-2005. SOURCE: C. Burleson, FAA, "Aviation environmental challenges," Presentation to Panel B, December 13, 2005.

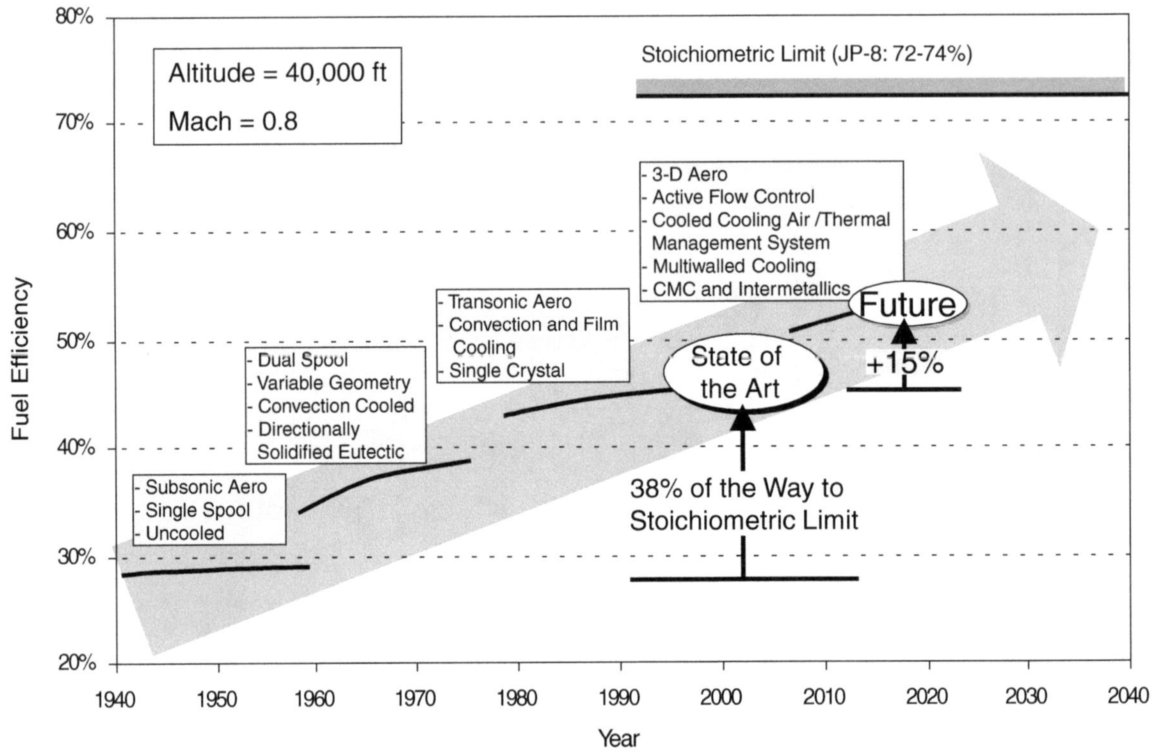

FIGURE 3-4 Considerable gas turbine fuel efficiency improvements are still possible. SOURCE: J. Stricker, Air Force Research Laboratory, Private communication to panel member D. Crow, February 2006.

clutch systems. The technology goal is to demonstrate highly reliable gearboxes with transfer efficiencies of about 99.8 percent and a power-to-weight ratio of 50 hp per pound. Reliable clutch operation would enable many new types of V/STOL aircraft.

Thirdly, engine-assisted wing lift, such as the blown wing, offers the simplest, most energy efficient short takeoff. Wing aerodynamics need to be developed, and the bleed or suction locations and quantities required need to be demonstrated for blown-wing V/STOL airplanes.

B6a Variable-cycle engines to expand the operating envelope

Variable-cycle engines have two or three flow paths through the engine, variable vanes, and variable exhaust nozzles, all of which allow them to vary engine bypass ratios and pressure ratios. They can improve the performance of both military and civil aircraft in many flight regimes by changing the bypass ratio and pressure ratio as a function of speed, altitude, and mission requirements. For the long-range Joint Strike Fighter (JSF), this should permit a twofold increase in rapid response radius, an eightfold increase in loiter capability, and a 30 percent reduction in gross weight. For a JSF follow-on aircraft, a 25 percent increase in lift and a 10-25 percent increase in range, depending on the mission, appear possible.

Variable-cycle engines have the potential to increase subsonic engine fuel economy. They also appear attractive for a supersonic commercial aircraft that has to accommodate stringent takeoff noise requirements and still achieve reasonable performance at supersonic speeds. For access to space, variable-cycle engines could provide a large reduction in payload costs as well as marked safety improvements.

This Challenge requires the development of numerous technologies: integrated thermal management approaches; reliable air-to-fuel heat exchangers; low-pressure-drop air-to-air heat exchangers; improved JP-8 heat sink capability; CMC technologies and associated life-prediction tools for operation above 2400°F; complex shape fabrication; high speed bearings; improved turbine cooling; better engine health predictions; probabilistic life analysis; in-flight data analysis; low-emission, high-temperature combustors; variable-geometry fan systems; and improved airframe-engine integration. This Challenge would benefit from the development of intelligent engines (Challenge B3).

B6b Integrated power and thermal management systems

Efficiency can be enhanced by integrating and optimizing, at the vehicle level, the traditionally severable airframe power and thermal management systems. "Integration" refers to physical, functional, and requirements integration of key propulsion and power system components, by combining them into fewer, multifunctional units all tied together in a more-electric architecture (see Challenge B9). Key components and functions include engine starting; electrical power generation, power conditioning, and routing; air cycle environmental control; avionics, fuel, and oil cooling; ventilation; flight control actuation; and overall vehicle and propulsion system thermal management, especially waste heat recovery and/or rejection. For example, engine start, auxiliary power, and environmental control systems may be combined into an airframe-mounted integrated power package that is physically coupled to the engine through power extraction and waste heat recovery. In this integrated approach, flight control systems are likely to be driven by electric or electrohydrostatic actuators, and thermal management is addressed in a seamless, system-level fashion. At the propulsion system level, electric power must be generated and integrated with airframe needs in the most efficient manner. This may be by a generator mounted on the shaft of the low-pressure turbine or, eventually, by fuel-cell-driven generators distributed within the airframe.

Today's modeling tools are derived from legacy approaches in which numerous component suppliers individually design, develop, and validate their product based on component-level requirements and specifications. New modeling and simulation infrastructures are necessary to use modeling tools in a system-level design framework, accommodating multiple platforms across multiple sites. A robust modeling framework is necessary to justify the system-level benefit of a given integrated component that may weigh or cost more than a traditional component or have different or enhanced functionality.

B8 Propulsion systems for supersonic flight

Commercially viable supersonic propulsion remains an elusive goal. Key issues include system performance and efficiency, the current ban on civil supersonic flight over the continental United States (14 CFR 91 ¶817), and Stage 4 noise standards.

Particularly for supersonic flight, propulsion systems development needs to be integrated with the design of the rest of the aircraft in a multidisciplinary effort to find an optimal trade-off between performance, efficiency, noise, emissions, and thermal management. Engine–airframe integration becomes more critical as the flight speed increases. This Challenge requires validated physics-based numerical simulation codes for component-level analysis and the improvement of multidisciplinary, system-level design tools for vehicle analysis.

Gas turbine research topics of interest include

- Variable-cycle engines optimized for both subsonic and supersonic flight with low specific fuel consumption, high thrust-to-weight ratios (T/Ws), and low noise.
- Lightweight, low-noise, efficient inlets and nozzles that also reduce wave drag and help to shape the sonic boom efficiently.
- Integrated airframe and propulsion controls to actively reduce vibration mode interactions between the engine and the plane (NIA, 2005).
- Noise and emissions data to validate models for sonic boom signature and determine its effect on humans (psychoacoustics), to assess the interaction of combustion products with ozone, and to help establish or confirm noise and emissions regulations.
- Electric actuation systems to eliminate the need for high-temperature hydraulic actuation systems.
- Active flow control to improve engine efficiency, reduce noise, and enable different airframe–propulsion integration concepts.
- Combustion process physics: modeling and experimental validation of injection, mixing, ignition, finite-rate kinetics, turbulence–chemistry interactions, and combustion instability to improve efficiency and life.
- Advanced materials and coatings (including high-temperature alloys for compressor and turbine disks) that meet requirements for operating temperature, service life, strength, and propulsion system noise.
- Alternative engine cycles for supersonic flight that might replace or enhance traditional gas turbines.

Many of these technologies are discussed in other R&T Challenges; much of the research proposed for subsonic engines will build a foundation for supersonic flight.

B9 High-reliability, high-performance, and high-power-density aircraft electric power systems

Future aircraft power systems must be able to meet the demands of more-electric aircraft (MEA). Future aircraft will progressively replace more and more mechanical and hydraulic systems with electrical systems, and electrical loads imposed by conventional systems will also continue to grow, to improve performance, convenience, and reliability. The higher power requirements of conventional loads are being driven by advances in avionics as well as by passenger entertainment and productivity needs. For example, the electric power demand on Boeing's 787 is nearly 1 MW, which is double that of the Boeing 777 and many times that of the first U.S.-built commercial jet, the Boeing 707 (Ames, 2005). The growth of new MEA loads is being driven by advances in the capabilities of electric actuators and controls, and it is being enabled by the development of more flexible and reliable aircraft generators. This Challenge can be met by improving key components and system-level technologies:

- Tenfold increase in power density for electric generators and motors suitable for aircraft use.

- Fivefold increase in energy and power density of suitable batteries and hybrid storage systems (e.g., the battery–ultracapacitor).
- An order of magnitude lighter optimized power system architectures (including, for example, a DC power bus, remotely controlled loads, and a wireless system control).
- Intelligent power management and distribution (PMAD) using advanced system models and wireless sensors or sensorless control technologies for graceful degradation and failsafe operation.
- Advanced analysis and simulation tools for multi converter power systems that can predict new modes of system dynamics and instability.

B10 Combined-cycle hypersonic propulsion systems with mode transition

The primary NASA hypersonics mission is for access to space in support of the Space Exploration Initiative and in placing and maintaining scientific payloads in low Earth orbit. A two-stage-to-orbit (TSTO) vehicle using a hydrogen-fueled, airbreathing first stage and a hydrogen-fueled rocket second stage could double the payload fraction to low Earth orbit relative to a two-stage, hydrogen-fueled rocket.[5] This would greatly reduce the cost of putting a payload into orbit. In addition, airbreathing hypersonic vehicles offer airplanelike operations, with increased safety and efficiency, more robust operation, and greater mission flexibility than rockets. A secondary mission for NASA hypersonics is to provide synergy with DoD programs in the development of missiles for time-critical targets; global strike and rapid resupply aircraft; and routine, on-demand access to space.

One combined-cycle hypersonic propulsion system under study for access to space is a turbine-based combined-cycle (TBCC) system. In order to design complex, combined-cycle hypersonic propulsion systems, experimentally validated, physics-based tools must be developed and refined because steady, full-enthalpy, clean air conditions cannot be reproduced in hypersonic ground test facilities. Experiments must be conducted on unit problems (e.g., jet injection into a supersonic stream) that contain the relevant flow physics but are amenable to simulation. Facility upgrades, such as for long-duration, high-temperature testing of engine materials and structures, should be completed in order to conduct the unit experiments under near-realistic flight conditions. Advanced diagnostics must be developed and used to obtain detailed databases in unit-problem experiments for complete validation of the computational tools, which can then be used for the vehicle design. Multiple-point validations are needed to verify that the tools produce results that can be extrapolated to conditions not available on the ground. Ultimately, flight testing must be conducted in order to obtain results under realistic operating conditions. Experiments should be flown on low-cost, suborbital rockets instead of expensive flight vehicles.

[5] P. Buckley, AFRL, "Payload mass fraction vs. staging velocity for TSTO vehicles to 51.7° orbit," Presentation to the DoD Technology Area Review and Assessment on March 29, 2004.

High-Priority R&T Challenges and Their Associated Thrusts

The rationale for the assignment of scores for each R&T Challenge is provided in Appendix B. In this section, the rationale for scoring will be discussed more generally. Table 3-2 shows that the top 10 R&T Challenges were all very relevant to NASA's mission, while those below the top 10 were less well aligned (with the exception of extraterrestrial planetary propulsion, which is clearly a NASA mission). In general, NASA has considerable infrastructure to support all the Challenges, with the exception of electric power systems, and NASA is particularly well equipped to conduct supersonic and hypersonic R&T. Other than propulsion in the atmospheres of extraterrestrial planets, industry, DoD, or, in a few cases, some other government agency will support R&T relevant to the high-priority Challenges. DoD, for example, has historically been a very strong supporter of V/STOL research. However, in the procurement-driven environment in which industry and the DoD live, time pressures often preclude achieving fundamental understanding, and empiricism must be resorted to when problems arise. Even though NASA may not be the only sponsor for some R&T, it can distinguish its research support by developing a fundamental understanding of phenomena, a strong commitment to physics-based modeling, and extensive validation of those models. All 10 high-priority Challenges entail moderate to high risk, which is the appropriate level for NASA R&T.

Not surprisingly, all of the top 10 Challenges involve gas turbine engines, with a strong focus on subsonic operations, the only flight regime currently supporting commercial capacity. V/STOL propulsion systems rank high for their potential to improve capacity, but this will not happen unless significant improvements are made in noise, fuel economy, and reliability. The top 10 Challenges will increase the efficiency of future aircraft, with greater levels of systems integration and optimization offering benefits not possible on aircraft designed component by component. Advances in information technology will lead to intelligent propulsion systems that invoke variability to optimize mission performance. These advances will increase demand for onboard electrical power, which will require electric power generation and distribution systems with more power and higher efficiency. Global air transportation is unlikely to be permanently confined to subsonic flight. Supersonic propulsion technologies will have strong synergies with DoD supersonic aircraft and space launch missions. In addition, many supersonic technologies will also be used to improve the performance of subsonic aircraft components and systems.

Combined-cycle hypersonic propulsion systems are expected to enable reusable launch vehicles with higher payload fractions and to benefit DoD as well.

The following four R&T thrusts describe threads of commonality among the R&T Challenges in the propulsion and power Area:

- High-temperature materials and structures.
- Validated physics-based modeling and simulation.
- Systems integration.
- Intelligent, adaptive systems.

Most of the Challenges in this area, regardless of rank, fall into one of these Thrusts, which are very important and will require significant investment of resources.

High-temperature materials and structures

Advanced materials are a key enabling technology for aeronautical and space vehicles and play a particularly critical role in propulsion systems. New developments in materials and processes for the production of these materials can deliver important improvements in performance, efficiency, safety, and reliability and can enable major advances in engine cycle design. In addition to developing materials with higher use temperatures, there is very significant payoff for high-temperature materials with (1) lower density or higher specific strength, (2) greater resistance to the combustion environment, (3) higher damage tolerance and predictable modes of degradation and failure, and (4) multifunctionality.

Significant NASA investment in materials is absolutely crucial for continued advances in subsonic, supersonic, and hypersonic propulsion and for continued U.S. leadership in advanced propulsion systems.

Gas turbines will continue to dominate civil aviation in the next few decades. Fuel costs, safety, and noise will drive major improvements in efficiency and reliability. Overall efficiency improvements will require higher pressure ratios for the overall cycle, higher turbine inlet temperatures, improvements in fan efficiency, and weight reduction in the large structural engine components. To achieve this, a number of materials developments must occur, including stronger compressor disk materials, higher temperature turbine disk and airfoil materials, and thermal barrier coating systems with higher temperature capability and increased reliability. For larger fan and structural components, low-density intermetallics and improved polymeric composites are needed. Over the past decade NASA has provided leadership and worked cooperatively with engine manufacturers in the development of advanced superalloy turbine disks and single-crystal airfoil alloys that will significantly improve the performance of the next generation of commercial engines. Continued support for research on airfoil and disk materials (including new processing approaches) with temperature capabilities 100°F to 200°F greater than current alloys is a high priority, since a broad exploration of new superalloys, refractory alloys, and intermetallics is beyond the scope and resources of any single engine manufacturer.

NASA has also contributed substantially to the fundamental knowledge base on oxidation of superalloys and coatings and the performance of bond coat/yttria-stabilized zirconia thermal barrier coating systems. Breakthroughs are needed in new ceramics and intermetallic bond coats for thermal barrier coating systems. New testing methodologies should be developed for these coatings to simulate engine environments, including the high thermal gradients that are characteristic of the turbine airfoil.

The development of intelligent engines will also require progress in life prediction, materials diagnostics, and multifunctional materials to enable computation-based life-prediction tools and complementary new approaches to in situ materials diagnostics.

Advances in supersonic and hypersonic propulsion will permit more efficient cross- and intercontinental travel and access to space, respectively. As Mach number increases, propulsion system temperatures escalate rapidly and oxidation becomes a major difficulty, particularly for air-breathing engines. The ceramics, CMCs, and high-temperature metallics (with active cooling) needed for these propulsion systems remain at low technology readiness levels. Materials systems in need of further development include carbon–carbon and carbon–silicon carbide composites, refractory alloys (rhenium-, niobium-, or molybdenum-based), and nickel alloys. Innovation in processing, joining, and close integration of materials with propulsion system design is essential. Significant progress in supersonic or hypersonic flight will require substantial investment in ultrahigh temperature ceramics, CMCs, and high-temperature metallics. No single U.S. industrial organization has the expertise to make the major breakthroughs in materials that are required.

Validated physics-based modeling and simulation

With the advances in computational speed, power, and affordability of the last two decades, aeronautics researchers have turned increasingly to computational simulation codes to model the complex physical and chemical conditions inherent in aircraft propulsion and power systems. Industry is appropriately enamored of the possibility of using computational simulation to reduce significantly both the cost and time of product development, to optimize system designs, and to increase reliability. Academic and government researchers also value the potential to attack more complex problems. Computational simulations generally employ a number of physics-based models within the governing conservation and state equations. Examples of models already in use include combustion–turbulence interactions, subgrid turbulence models in large eddy simulation (LES) codes, effects of unsteadiness in steady-state compressor codes, reduced-order chemical kinetic mechanisms, and droplet–

flow interactions. These physics-based models often contain adjustable parameters that are grossly calibrated to empirical data sets; the data sets themselves are often incomplete, particularly with regard to boundary conditions, prompting further untested assumptions to be incorporated. The entire codes themselves are often not validated in detail except for comparing their code predictions to input and output measurements. The codes often do not work well when the design space changes considerably, prompting more tweaking of the adjustable parameters. Nevertheless, within their applicable ranges, the computational simulation codes have enabled technical progress, as witnessed by the state of air craft propulsion today. Unfortunately, the applicable range limits themselves are often not well understood. NASA and its partners can greatly advance aircraft propulsion and power by developing and validating the constitutive physics-based models.

Physics-based models are readily assimilated by industry into their proprietary product system design codes. Research into physics-based models can be conducted jointly by NASA, industry, and academia since it is fundamental in nature, publishable, and shareable. It is work that takes time to mature, yet advances can readily be translated into practice as they occur. Validation involves the design of experimental facilities of appropriate scale and the use of advanced, nonperturbing diagnostics to measure parameters accurately in space and time to rigorously ascertain model fidelity. It is an iterative process culminating in submodels whose accuracy and range of applicability are well established.

Systems integration

This R&T Thrust is intended to support a clear trend in aeronautics design—namely, the movement toward aircraft system-level integration and optimization of traditionally separate airframe and engine subsystems. Improved systems integration will increase capacity by increasing operating flexibility, enabling the use of shorter runways (by improving the performance of powered lift or thrust vectoring systems), reducing end-user costs, and facilitating the design of commercial supersonic aircraft. Efficiency and safety will also be improved by more functional designs that are robust against adverse operational conditions (icing, wake ingestion, foreign object damage, and temperature extremes). "Integration" in this context refers to the physical, functional, and requirements integration of key propulsion and power components with each other, and with other systems, such as the airframe, the avionics, and the overall air transportation system. Optimization of key metrics (cost, weight, thrust, and fuel consumption) at the system rather than the component level is also included in this Thrust.

Integrated power and thermal subsystems were discussed under R&T Challenge B6b. A second example of the systems integration Thrust can be seen with the inlet and exhaust systems that will be required for innovative air platforms. Blended wing-body concepts have been proposed that use boundary-layer-ingesting engine inlets. This approach reduces the performance of the propulsion system but more than compensates for that loss with vehicle improvements in lift and drag. Similarly, although a variable-cycle engine for supersonic cruise might be heavier than a fixed-cycle engine of comparable thrust, it could also eliminate the need for heavy airframe-mounted, inlet-variable geometry, thereby increasing overall vehicle T/W. A higher degree of systems integration should be evident from the earliest design phases and may necessitate entirely new aircraft or engine architectures. New process modeling and simulation tools, along with business models, must also be developed to enable design and validation of integrated systems in a seamless, multiple-organization environment.

Intelligent, adaptive systems

The development of intelligent, adaptive systems technologies will be a key enabler for civil and military aeronautics and space. These technologies will permit (1) real-time, low-latency health monitoring systems; (2) optimization of the performance of current propulsion systems according to mission requirements and environmental conditions, including active control to enhance performance and avoid anomalous behavior; (3) new sets of tools for extended life and improved maintenance of commercial and military fleets; and (4) totally innovative systems for the future. These technologies involve engine and propulsion systems modeling; improved sensor capabilities; and innovative software for control logic that adjusts engine performance to enhance stability, improve distortion tolerance, minimize noise and emissions, and address deterioration issues in service.

Intelligent, adaptive systems are coming online, principally for the health monitoring of both commercial and military engine systems. This capability will anticipate and prevent failures, using control logic to reconfigure engine operation. This capability will improve time on wing, improve readiness, and reduce operating costs. Intelligent engine technologies are essential to the creation of variable-cycle engines. These engines use variable geometry to optimize performance for the mission takeoff, climb, cruise, descent, and landing.

Intelligent, adaptive technologies are needed to reduce fuel consumption and environmental impact by morphing the aircraft or engine to suit the needs of the moment—for example, a takeoff configuration to address noise requirements and a cruise configuration optimized for fuel burn and low NO_x emissions at altitude. Similar technologies will also be used to optimize supersonic and hypersonic engine configurations. For example, intelligent, adaptive engine technologies can be used to optimize a low-noise configuration for takeoff with high bypass ratios and then transition into supersonic or hypersonic configurations. These technologies also have direct application to space vehicles.

Intelligent, adaptive technologies need to be developed for current and future propulsion systems. With current systems, knowledge management should be developed in areas of software control to provide real-time assessment of the remaining life of critical engine components. In new systems, active control and variable geometry should be used to tailor propulsion flows to reduce sensitivity to inflow distortion, to enhance compressor stability, to control exhaust jet area and vector angle to reduce noise and emissions, and to enhance vehicle performance through powered lift.

Low-Priority R&T Challenges

R&T Challenges B11 and B13 ranked in the top 10 in terms of national priority but not by NASA priority. Challenge B11 (alternative fuels and additives for propulsion that could broaden fuel sources and/or lessen environmental impact) is clearly an important national priority. It was ranked lower as a NASA priority because DOE will need to take the lead in establishing the national infrastructure for an alternative fuel and because the combustion research needed to develop such a fuel will take much less time putting an alternative fuel infrastructure in place. Furthermore, aviation fuels are likely to have a first call on petroleum supplies should they become scarce, so that the use of alternative fuels for aviation is likely to follow their widespread use for ground-based applications, which would place less stringent demands on weight, volume, reliability, safety, and certification of new systems and technologies.

Challenge B13 (improved propulsion system tolerance to weather, inlet distortion, wake ingestion, bird strike, and foreign object damage) is ranked low in terms of NASA priority because the relevant technologies are more mature (and the attendant risk lower) and it is not as relevant to NASA's mission as the Challenges that scored in the top 10 by NASA priority.

MATERIALS AND STRUCTURES

Introduction

Advances in civil aeronautics materials and structures technologies are often the key enablers for new modalities of operation or regimes of flight. For example, improving jet engine efficiency requires continual introduction of new materials to allow the implementation of advanced aerodynamic concepts and higher operating temperatures to increase propulsion efficiency. A comprehensive multiphysics understanding of materials and structures enables innovative designs. New analysis techniques produce the next generation of design tools, which will allow revolutionary structural concepts to be accelerated into applications.

The assessment of R&T Challenges related to materials and structures was influenced by the globally competitive nature of the aerospace industry, particularly in the civilian aircraft market. New material and structural technologies that would help U.S. industry establish a clear advantage over its global competitors received high marks. R&T Challenges were also ranked bearing in mind global needs in aeronautics. Growth in demand for the movement of passengers and goods, especially against a backdrop of rapid economic development in Asia, calls for significantly increased capacity in the air transportation system. Similarly, environmental concerns related to fuel efficiency led to a focus on materials and structures Challenges for engine development and harvesting of energy from structural components and systems. Improvement in structural performance and efficiency was another key driver, and the assessment focused on design methods and tools required to facilitate such improvement. The changed climate for national and international security was another important factor.

The QFD process described in Chapter 2 was used to prioritize 20 R&T Challenges related to materials and structures. Table 3-3 and Figure 3-5 show the results. The text that follows describes the 10 R&T Challenges that ranked highest in terms of NASA priority, the general characteristics of high- and low-priority Challenges, and the R&T Thrusts in this Area. Further details on all Challenges, including the rationale for scoring, are found in Appendix C.

Top 10 R&T Challenges

C1 Integrated vehicle health management

Integrated vehicle health management (IVHM) refers to monitoring, assessing, and predicting the health[6] of aircraft materials and structures using networks of sophisticated onboard sensors. A fully integrated approach to IVHM relies on a multidisciplinary set of analysis, testing, and inspection tools, including miniaturized sensors and distributed electronics; sophisticated signal processing; data acquisition, integration, and database maintenance; artificial intelligence; damage science; and the mechanics of structures and their failure.

IVHM benefits all classes of aircraft, in all speed regimes and phases of flight. With a national fleet of aging aircraft and infrastructure in an industry with low profit margins, IVHM is increasingly important due to its ability to increase safety and reliability. It can also have a number of benefits for capacity. Decreasing the possibility of unexpected failure could speed the introduction of innovative material systems and structural concepts and enable the use of traditionally high-maintenance (and high-cost) systems, such as rotorcraft. More data would allow better understanding of the stresses experienced by a system, reducing the amount of overdesign motivated by uncertainty. In addition, aircraft could report the predicted lifetimes of their own parts and

[6]"Health" in this context implies either an absence of measurable material flaws or an ability to coordinate the growth rate of flaws with the safe life remaining for the element in question.

TABLE 3-3 Prioritization of R&T Challenges for Area C: Materials and Structures

	Strategic Objective							Why NASA?					
	Capacity	Safety and Reliability	Efficiency and Performance	Energy and the Environment	Synergies with Security	Support to Space	National Priority	Supporting Infrastructure	Mission Alignment	Lack of Alternative Sponsors	Appropriate Level of Risk	Why NASA Composite Score	NASA Priority Score
R&T Challenge **Weight**	**5**		**3**		**1**			**$^1/_4$ each**					
C1 Integrated vehicle health management	9	9	3	1	9	3	**114**	9	9	1	9	**7.0**	**798**
C2 Adaptive materials and morphing structures	9	3	9	3	9	3	**108**	9	9	1	9	**7.0**	**756**
C3 Multidisciplinary analysis, design, and optimization	9	3	9	1	3	3	**96**	9	9	3	9	**7.5**	**720**
C4 Next-generation polymers and composites	9	3	9	1	9	3	**102**	9	9	1	9	**7.0**	**714**
C5 Noise prediction and suppression	9	1	3	9	3	1	**90**	9	9	3	9	**7.5**	**677**
C6a Innovative high-temperature metals and environmental coatings	3	9	3	1	9	3	**84**	9	9	3	9	**7.5**	**630**
C6b Innovative load suppression, and vibration and aeromechanical stability control	3	9	3	1	9	3	**84**	9	9	3	9	**7.5**	**630**
C8 Structural innovations for high-speed rotorcraft	9	1	3	1	9	1	**72**	9	9	3	9	**7.5**	**540**
C9 High-temperature ceramics and coatings	3	1	9	3	3	9	**68**	9	9	3	9	**7.5**	**510**
C10 Multifunctional materials	3	3	9	3	9	9	**84**	3	9	3	9	**6.0**	**504**
C11 Novel coatings	3	9	3	3	1	1	**80**	3	9	3	9	**6.0**	**480**
C12 Innovations in structural joining	3	3	9	1	3	3	**66**	3	9	3	9	**6.0**	**396**
C13 Advanced airframe alloys	9	1	9	1	3	1	**84**	1	3	1	9	**3.5**	**294**
C14 Next-generation nondestructive evaluation	3	9	1	1	3	1	**70**	3	9	1	3	**4.0**	**280**
C15 Aircraft hardening	1	9	1	1	9	1	**66**	3	3	1	9	**4.0**	**264**
C16 Multiphysics and multiscale modeling and simulation	3	3	3	3	3	1	**52**	3	3	3	3	**3.0**	**156**
C17 Ultralight structures	3	1	3	1	3	3	**38**	3	9	1	3	**4.0**	**152**
C18 Advanced functional polymers	1	3	1	1	3	1	**30**	9	3	3	3	**4.5**	**135**
C19 Advanced engine nacelle structures	3	1	3	1	1	1	**34**	1	9	1	3	**3.5**	**119**
C20 Repairability of structures	3	3	3	1	1	1	**44**	3	3	1	3	**2.5**	**110**

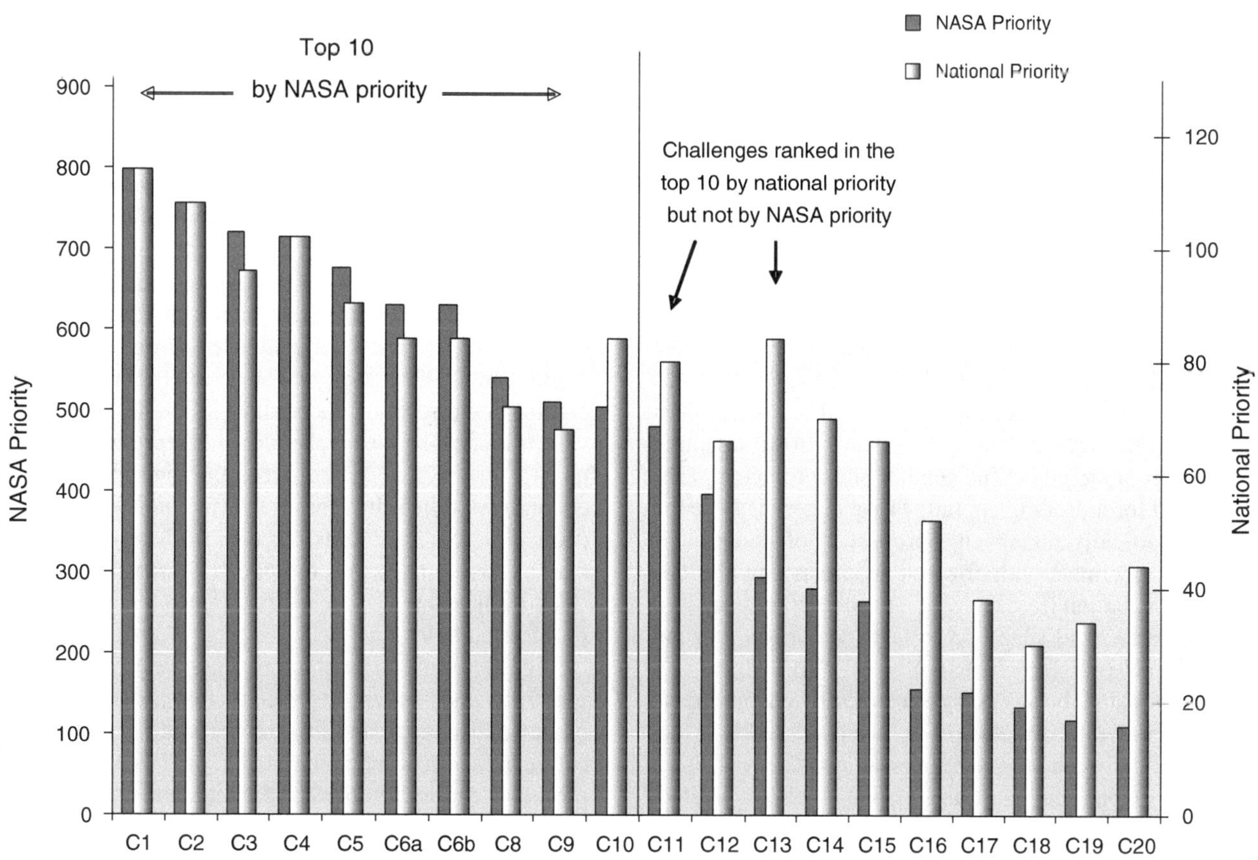

FIGURE 3-5 NASA and national priorities for Area C: materials and structures.

report the need for replacement parts, reducing operating costs and maintenance downtime. IVHM could quickly diagnose root problems, minimizing flight delays and increasing capacity (Powrie and Fisher, 1999; Simon, 2000). IVHM may also reduce vehicle operating cost and maintenance downtime and can speed the introduction of innovative material systems and structural concepts. Real-time onboard sensor systems that monitor the actual state of materials and structural components enable more efficient use of materials, including novel concepts.

There are two main features of the next generation of IVHM: (1) Sensor packages will be very small and exceedingly lightweight and (2) the reliance on humans to interpret the sensor output and assess the impact on structural integrity will be reduced or eliminated.

Three classes of IVHM systems warrant attention over the next decade, culminating in flight testing of full-scale IVHM systems that detect multisite damage. The first class includes fiber-optic sensor systems that use multiplexed fibers attached to or embedded within the structure, each with numerous sensing sites interrogated in turn by a single electro-optic module. The second class includes locally self-powered, wireless microelectomechanical sensors tiny enough that very large numbers of sensors become practical. Each sensor mote performs a point measurement, so many are used to effectively cover large areas. The third class includes discrete active and passive remotely powered sensor modules (e.g., by means of guided-wave ultrasonic or acoustic emission) that may be large compared to sensor motes but can interpret multimode vibrations or multiphysics parameters (temperature, stress, humidity, etc.) that propagate over relatively long distances within the key structural elements.

Successful application of IVHM also relies on continued research and refinement in fundamental structural mechanics and the mechanics of damage and failure for accurate interpretation of IVHM sensor data and to support autonomous decision making for damage recovery and mitigation.[7]

C2 Adaptive materials and morphing structures

Use of adaptive materials and morphing structures to change the aircraft shape (outer mold lines) and functions on demand represents a revolutionary approach for enabling optimal performance over a range of flight missions. Morphing wings that change their planform area by up to 50 percent and alter their sweep angle up to 50 degrees are emerging as a viable technology, and the benefits would be

[7]See R&T Challenges D4 and D5.

far more than the simple variable-sweep configurations of the past. However, the costs of incorporating such technology have not yet been evaluated.

During the past 2 years, two prototypes from the Defense Advanced Research Projects Agency (DARPA) Morphing Aircraft Structures program were successfully tested at transonic speeds in NASA Langley's Transonic Dynamics Tunnel at a scale sufficient to validate the concept. This design included concepts such as stretching skins, sliding skins, and seamless camber change. DARPA has also sponsored flight research of morphing technologies applied at the system level. These tests identified critical long-pole, component technologies that now limit the use of morphing technologies, particularly in heated supersonic flow. Adaptive materials have emerged as the number one component technology need for morphing aircraft. These materials have the ability to radically change the properties of component materials, to facilitate both effective load-carrying abilities and ease of actuation from one shape to another, as well as to change the structural shape, from large variations in wing area to seamless camber-changing.

Adaptive materials may be self-actuated by energy inputs such as light, heat, and electric or magnetic fields. They include heat-activated shape memory alloys like NiTiNOL; ceramics (e.g., lead zirconate titanate); photonically activated, lightweight, flexible shape memory polymers; electrically activated piezoelectrics; and magnetorheological fluids.

This Challenge requires development of commercial, high-speed, morphing airframe concepts, development of structural components such as stretching skins, and accelerated development of a special class of actuatable adaptive materials with lifetimes comparable to those of currently used materials. A fundamental task is to characterize the mechanical response of these inherently nonlinear materials, including hysteresis, fatigue, long-term behavior, and damage behaviors. Analysis and design tools that accurately predict these responses will open the door to even more applications of these revolutionary adaptive structural concepts, which could optimize performance and expand the flight envelope.

C3 Multidisciplinary analysis, design, and optimization

Methods for simulation-based, multidisciplinary design and optimization (MDO) are at the very core of a philosophy that moves away from the build-test-build approach, which has proven to be expensive and ineffective in exploring the aeronautical design space. MDO processes develop synergistic benefits by integrating people, analytical tools, experimentation, and information to design complex structural components and systems (Sobieszczanksi-Sobieski and Haftka, 1997). These approaches allow for development of optimal configurations, topologies, and dimensions for structural members and components to achieve design objectives, and they permit designers to examine the myriad what-if's that characterize sophisticated designs with interdisciplinary trade-offs.

After almost two decades of R&D, MDO processes for conventional designs have reached a high level of sophistication. In structural designs where the topology or outer mold lines are defined, analytical methods such as the structural finite-element technique, coupled with similar analytical tools for load assessment, provide a high level of success. However, for designs with a multiplicity of topologies, some of which are not well-defined, and for problems where a large number of design parameters and constraints must be considered in the early stages of the design process, MDO methodologies are still underdeveloped (Giesing and Barthelemy, 1998). Major effort must also be directed at including the effects of uncertainty in the design process, as well as increasing the level of detail in representing the structure. New ways of formulating problems that incorporate quantitative reliability measures to facilitate effective design decisions have been considered in this context. The extension of these approaches to large-scale structural and material design problems represents an entirely different level of problem complexity.

Significant new developments are required in both the platforms and the embedded tools that constitute the MDO process. Efficiency and effectiveness of the search process continue to be a problem, particularly in large-dimensionality problems and multimodal or disjointed search spaces. Current platforms are ill equipped to efficiently parse the vast amounts of data associated with the design process. There is a marked need for developing analysis modules for the search process to query in the design process. Such analysis modules must be based on the physics of the problem or on inferences derived from experimental data. While digital designs have enabled tight manufacturing tolerances and manufacturers can incorporate cost models, explicit mathematical modeling of manufacturing processes, repair, and environmental impact must be better integrated into the MDO process. These analysis tools must be developed at multiple levels of granularity and precision, to coincide with the appropriate stage of the design process. The numerical efficiency of these tools is paramount, and alternative paradigms that take advantage of a new generation of parallel computational hardware must be sought (Giesing and Barthelemy, 1998; Sobieszczanksi-Sobieski and Haftka, 1997). Uncertainty modeling in a data-lean environment, specifically for new concepts, continues to be an issue in this regard. There is a similar dearth of computationally efficient methods for reliability assessment, particularly in situations where uncertainty distributions do not conform to standard forms or where components or elements exhibit discrete behavior. The propagation of uncertainty in complex and highly coupled multidisciplinary systems needs to be modeled, and tools for design and optimization in a nondeterministic environment continue to be computationally intractable, especially when applied to design problems involving a large number of nondeterministic variables, parameters, and design constraints. Furthermore, the inclusion of risk and reliability analysis in the design process would yield a time-dependent description of risk associated with structural and material systems in service, facilitating decisions that enhance vehicle availability and reliability.

Use of commercial tools in optimization is not enough to advance the state of the art in MDO. Optimization is only one piece of the analysis, design, and optimization triad. It is the tightly integrated development of analysis and optimization tools that furthers the potential of MDO methods. In the aerospace arena, such expertise is unique to NASA. Additional gains can be realized with NASA working in close collaboration with researchers from academia and industry. A number of synergistic benefits could also be achieved by developing this aspect in concert with health-monitoring technologies (see R&T Challenge C1).

C4 Next-generation polymers and composites

Over the past 50 years, polymeric composites have revolutionized and improved the performance of aircraft structures. Future needs for enhanced structural performance, high-temperature capability, and durability can only be met by the next generation of polymer-based composites. Next-generation composites will take advantage of improved high-temperature polymeric matrices, new reinforcement materials, hybrid reinforcement approaches, improved joining technology, and science-based manufacturing with controlled 3-D placement of reinforcements. This Challenge includes development of tougher, higher-temperature adhesives for joining, innovative fillers to enhance performance, and new core materials for ultralightweight sandwich construction. It also includes development of repair techniques to restore structural integrity to damaged composite structures. The development of next-generation composites is dependent on three capabilities: multiscale modeling that links nano- and microstructure to structural composite response; science-based processing techniques that account for resin chemistry, cure kinetics, and flow physics to guide placement and distribution of the different reinforcement phases; and structural and mechanical testing to evaluate both the design and processing parameters. This next generation of composites will significantly improve structural efficiency, safety, and high-temperature performance; reduce data scatter; increase damage tolerance (e.g., delamination); and improve manufacturability (e.g., by eliminating hand lay-up). These composites will likely incorporate adaptive materials and multifunctional concepts, thus providing the enabling materials needed for visionary concepts in nacelle components, wing structures, and fuselage materials.

C5 Noise prediction and suppression

Local communities in this country and abroad are becoming extremely aggressive in passing stringent noise regulations, in order to substantially reduce the impact of aircraft noise. Takeoffs and landings at many airports have been restricted. The ability to reduce aircraft noise thus becomes an environmental as well as an operational constraint. Regulations passed recently in the European Union regarding noise inside commercial aircraft point to the need for cabin noise control as well as external noise control. There are a variety of promising materials and structures approaches that could be developed and validated in the next decade to substantially reduce both exterior and interior noise.

Noise is a multidisciplinary phenomenon. Effective noise control techniques must take into account multiple types of aerodynamic and acoustic excitations. Therefore, structural prediction tools must be integrated with computational aeroacoustic and fluid dynamic prediction tools for a fully coupled solution to the problem of structural noise. To validate these predictions, systematic tests should be carried out to measure noise signatures for a range of flight conditions in controlled environments such as anechoic wind tunnels. This should be followed by selective flight test of full-scale systems to measure noise signatures from the ground as well as inside the airframe.

Advanced materials for larger, stronger fan blades and higher-temperature turbine blades, together with the development of very-high-bypass-ratio engines, will be the biggest single factor in reducing external noise produced by jet aircraft. Advances in strong, lightweight composite nacelle structures, smart materials, and active structures would also reduce engine noise. Variable-geometry-chevron nozzles, which could be driven by the shape memory alloy NiTiNOL, have been demonstrated to reduce noise during takeoff and then reconfigure themselves to a more efficient shape for cruise (Calkins and Butler, 2004).

Major strides in noise suppression can also be achieved using advanced materials and active and passive structural techniques. Promising approaches include nanotechnology to enhance structural damping (noise absorption); morphing or tailored structures for laminar flow and noise source control; and multifunctional active composite structures with improved noise signature control, structural strength, health monitoring, and thermal insulation. The structural weight of additional materials or devices used for noise suppression is a key factor; with expanding advancements in smart structures technology and rapid miniaturization in data processing techniques, active noise control within the aircraft cabin appears more promising. Development of both sorts of noise suppression devices will be a key step to quieting current aircraft (interior and exterior) and could provide an impetus for an explosion of civil applications of rotorcraft and fuel-efficient prop-rotor aircraft.

C6a Innovative high-temperature metals and environmental coatings

Advanced high-temperature metallic turbine material systems (i.e., alloy substrates for the turbine blade, disk, and shroud, plus necessary environmental coatings) are critical to advancing the next generation of jet engines. These engines will power future subsonic and supersonic fixed-wing airplanes and rotorcraft, while enabling reduced operating costs and improved engine safety and reliability. Metallic material systems with higher operating temperatures will improve engine cycle efficiencies. Dramatic improvements in these materials are possible, but development has been retarded by the high cost of R&D given the current highly iterative nature of alloy design. For instance, intermetallic silicides may enable considerably higher operating temperatures than nickel superalloys; advanced disk alloys may greatly reduce creep and fatigue; and protective coatings with superior resistance to environmental degradation could significantly extend the service life of hot section components. But the length of time to develop these materials, often a decade or two, and the risk that success will not be achieved have been a huge disincentive to aggressive development.

The most difficult technical issue is the need to develop material systems that possess improved performance at higher temperatures while maintaining stability for tens of thousands of operating hours in an environment that is highly oxidative, corrosive, and erosive. However, strides are being made in materials modeling capability, driven by the success of new computational tools and ever-increasing desktop computer processing capability. The application of models to guide the advancement of these materials is just beginning, but it is becoming apparent that these tools can cut development time by half and focus alloy development on the most promising approaches, reducing development risk and cost (NRC, 2004). The drawbacks to new material development would be obviated by the ability to replace experiments with computer simulations, as is done with computational fluid dynamics. The goal of this Challenge is to provide the underlying technologies for material modeling tools that can predict properties of new high-temperature metallic materials and associated protective coatings. The effort would include generation of the necessary fundamental data, complemented by testing that simulates realistic jet engine operating conditions to validate the models. In concert with industry, these tools would then be applied to the development of innovative propulsion materials.

C6b Innovative load suppression, and vibration and aeromechanical stability control

This Challenge will minimize the impact of vibratory loads in aircraft using innovative passive and active techniques. It will also examine innovative techniques to increase aeromechanical stability margins in all flight modes.

Current aircraft use numerous passive devices to increase passenger comfort and to safeguard the functioning of key structural components and instruments. Some modern rotorcraft, such as the Sikorsky S-92 and Bell-Boeing V-22, have made successful use of active vibration control, as well. Additionally, the flight envelope is sometimes restricted due to low stability margins for some aircraft. The objective of this research is to couple advanced CFD methodology with comprehensive structural analysis, including multibody formulation, nonlinear structural and inertial couplings, and interactions between the flow and the structure to predict aeromechanical stability, vibratory loads, and vibration signatures at different stations in the airframe. To validate predictions, systematic tests in wind tunnels should be carried out using dynamically scaled and full-scale models to measure vibration loads and damping of different modes for a range of flight conditions. Selective full-scale flight tests should be carried out to measure vibratory loads and stability at level and maneuvering flight conditions. Innovative active and passive techniques should be developed to minimize vibration and increase stability margin. Finally, multidisciplinary optimization should be exploited to develop efficient, low-vibration, aeromechanically stable aircraft.

C8 Structural innovations for high-speed rotorcraft

One revolutionary vision is a next-generation, high-speed, high-lift rotorcraft that can cruise at over 250 knots, and that is "neighborly" quiet, runway independent, and economically competitive with a Boeing 737 aircraft (Johnson et al., 2006). Advances required to achieve such a vehicle include innovative rotor designs, active vibration and load control, variable-speed rotor technologies, active noise control, rotor morphing, lightweight and crash-absorbing airframe technologies, advanced composites with high damage tolerance, advanced transmission systems, diagnostics and prognostics of drive trains and rotor head systems, increased autonomy and maneuverability, and enhanced handling qualities. Reliable, comprehensive aeromechanics and technology tools must be developed and validated systematically, through dynamically scaled and full-scale tests in wind tunnels. This vision provides opportunities to incorporate many disruptive and nondisruptive technologies in rotorcraft design, with an enormous payoff in performance and life-cycle cost compared with existing helicopters.

C9 High-temperature ceramics and coatings

Advanced structural ceramics, including oxide-, carbide-, nitride- and boride-based systems, are characterized by high strength, stiffness, hardness, corrosion resistance, and durability. Such ceramics retain these properties at high temperatures, making them ideal for a wide range of demanding applications, including engine components for subsonic aircraft (combustor liners, exhaust-washed structures, high-temperature ducts, heat exchangers, and nacelle insulation) and airframe and propulsion systems for high-speed vehicles. The primary benefit of structural ceramic materials is the ability to withstand higher temperatures, which improves propulsion system efficiency, increases lifetime, enables higher operating speeds, and expands the margin of safety in airframe applications (NRC, 1998).

Oxide composites with operating temperatures as high as 1250°C and lifetimes of thousands of hours in highly oxidizing combustion or reentry environments are very suitable for some engine components, warm structures, and thermal management components.

Nonoxide composites made of silicon carbide reinforced either with carbon fibers or a combination of carbon fibers and silicon carbide fibers are capable of operating temperatures of 1300°C-2000°C for short times in highly oxidizing environments or for much longer times near the lower end of the thermal range when protected with environmental barrier coatings. Furthermore, because nonoxide fibers exhibit higher strength and better strength retention than oxide fibers, they are being widely researched for application in combustion environments as well as for hot structures of hypersonic and reentry vehicles.

Refractory metal (e.g., hafnium or zirconium) carbides and borides are capable of surviving thermal excursions up to 2000°C-2500°C for short times with little material recession, making them a strong candidate (in either monolithic form or as a composite matrix) for the sharp leading edges of hypersonic vehicles.

During the last 10 years, significant progress has been made in the processing, development, and demonstration of many ceramic systems for specific applications. Oxide composites deriving damage tolerance from highly porous matrices have been commercialized, and other systems with novel fiber coatings have been demonstrated in subscale testing for reentry vehicle thermal protection systems. Silicon carbide matrix processing approaches have advanced significantly, with systems produced by chemical vapor infiltration, melt infiltration, and preceramic polymer infiltration all having been demonstrated in subscale testing for jet or rocket engine components. NASA Glenn has led efforts to fabricate and test jet engine components such as exhaust nozzle liners, combustor liners, and turbine airfoils with silicon carbide matrix composites. Rocket nozzles fabricated from silicon carbide materials have been rig tested, and NASA Ames has demonstrated the ability to reproducibly fabricate refractory metal carbide and boride systems. Despite the above successes, component fabrication is not often taken much beyond the prototyping stage. Advancing the state of the art for high temperature ceramics suitable for aeronautical applications requires research in several key areas: fabrication and testing; modeling; and attachment methods.

Insufficient fabrication and testing experience deprives designers of confidence in the long-term behavior of these materials and in the design rules for translating material characteristics into component designs. Modeling tools to predict component life for these materials are inadequate. This causes inaccurate performance and cost assessments and further limits the use of ceramic materials. Since these materials are only considered for niche applications, no economy-of-scale cost savings can be anticipated. This could be alleviated through the development of better design tools, a more thorough understanding of the effects of process variations, and more efficient approaches to commercial fabrication.

Work is also needed to develop robust methods of attaching hot components to warm and cool structures as well as to develop textile approaches that can integrate complex component architectures with key features such as stiffeners, sensors, and cooling features.

C10 Multifunctional materials

Materials that possess multifunctional behavior combine electronic, magnetic, chemical, thermal, and mechanical properties at the macro, micro, or atomic level. These materials present unique opportunities for integrating communication, actuation, sensing, self-healing, and energy-harvesting functionalities into lightweight, load-bearing structures. Multifunctional materials enable a wide range of benefits, including improved aircraft telecommunications (wired, wireless, and optical); enhanced potential capabilities and flexibility for electronic and optoelectronic platforms, such as agile phased array and multifunctional radar systems; structural prognosis and nondestructive evaluation; self-sensing and self-repair; and local power generation through energy harvesting. The use of structural elements to provide new functions to aircraft platforms increases structural efficiency and enables new aircraft capabilities.

While the most research to date has been on materials with coupled electromechanical domains, a much broader vision is possible. Recent discoveries of electrochromic, magnetoelectric, and thermomechanical materials show substantial promise for future multifunctional materials.

High-Priority R&T Challenges and Their Associated Thrusts

High-priority R&T Challenges in the materials and structures Area had major relevance to at least one of three high-priority Strategic Objectives: capacity, safety and reliability, and efficiency and performance. Most of the Challenges in this Area were judged to have little or no relevance to energy and the environment, which is the fourth highest-priority Strategic Objective, although one was judged to have major relevance.

The most highly ranked R&T Challenges are those that could radically change the way new aircraft are designed, manufactured, and maintained. The key to success for all of these Challenges is interdisciplinary collaboration. Such a strategy will derive full benefit from NASA's extensive infrastructure in (1) materials development and characterization and (2) structural analysis, optimization, and testing. For example, the highest priority materials and structures Challenge is integrated vehicle health management (IVHM). Success will require a multidisciplinary set of analysis, testing, and inspection tools, including miniaturized sensors and distributed electronics; sophisticated signal processing; data acquisition, integration, and database maintenance; artificial intelligence; damage science; and the mechanics of structures and their failure. Multiple aspects of materials science, structural design, and aeronautics are brought to bear on the problem of assuring vehicle health. IVHM holds the promise of reducing vehicle cost, weight, and maintenance downtime as well as speeding the introduction of new material systems and structural concepts. In addition, IVHM has the strongest influence on both of the most highly weighted Strategic Objectives: improved capacity and enhanced safety and reliability.

Four R&T Thrusts describe threads of commonality among the R&T Challenges within the materials and structures Area. Most of Challenges in this area, regardless of rank, fall into one of these Thrusts, described below.

Visionary materials and structures concepts

Visionary aeronautical concepts often depend on new materials that possess unprecedented properties or behavior and allow aircraft designers to consider new flight regimes, aircraft configurations, and operational paradigms, such as high-speed rotorcraft and morphing aircraft with the ability to change their outer mold lines. Visionary concepts also hinge on innovative structural components, often made possible with newly developed materials, alloys, and coatings. This R&T Thrust is fundamental to NASA's aeronautics mission of enabling revolutionary concepts and innovative designs. New structural concepts take advantage of emerging analytical design tools and advanced structural materials, and they promise to significantly reduce the weight of structures while maintaining structural integrity and improving efficiency. New material concepts include next-generation polymers, metals, and composites, as well as advanced functional polymers and adaptive and multifunctional materials and coatings.

Comprehensive multilevel predictive methodologies for design and analysis

The second R&T Thrust for materials and structures technology is developing and understanding the multiscale and multiphysics behavior of aircraft materials and structures in a comprehensive manner and then bringing together previously separate design and analysis methodologies and tools, starting from the initial design concept all the way through to operation. This thrust moves beyond the realm of existing MDO techniques to incorporate key aspects of risk-based design and reliability. With new tools that allow systematic inclusion of the effects of uncertainty, whether in loading, material behavior, or mission requirements, this process would yield a rational approach for quantifying the risk associated with a certain design and allow for meaningful trade-off studies to be performed among competing design concepts. These tool sets would revolutionize aircraft design by reducing weight, noise, and vibration and by increasing stability control. NASA could have a unique role in developing and benchmarking such tools at the precompetitive stage, prior to their adoption by industry.

Novel technologies for improved structural efficiency and safety

Many of the R&T Challenges for materials and structures relate to improving structural efficiency and safety. Some, like ultralight structures and advanced joining, reduce weight directly. Alternatively, functionality can be added to a structural component or to the materials in the component to increase overall design efficiency. This Thrust includes adaptive structures that change shape and functions on demand, allowing efficient, multipoint adaptability for optimal performance. Also included are materials that perform dual roles in systems by virtue of their ability to serve as structural elements and to generate power, manage thermal loads, or impart some other additional functionality. IVHM, nondestructive evaluation (NDE), and vibration control use advances in sensing and actuation technology to reduce requirements on the structures themselves. This Thrust also includes aircraft hardening (increasing survivability of an aircraft in the event of an explosion or biological or chemical threat), which should enable new levels of safety.

Materials and structures for extreme environments

Materials and structures suitable for use in extreme environments are relevant to many R&T Challenges. Extreme environments are characterized by very high temperatures, chemical reactions, and erosive and/or corrosive conditions. They can be found in engine interiors and in the supersonic and hypersonic speed regimes. Expanding the operational envelope of each class of structural materials (polymers, metals, ceramics, and composite systems) would increase efficiency and safety margins for airframe and engine materials and structures.

Low-Priority R&T Challenges

The QFD process identifies areas of high national and NASA priority. In general, these two scores were highly correlated. However, as seen in Figure 3-5, two R&T Challenges in the top 10 by national priority did not make the top 10 by NASA priorities: novel coatings and advanced airframe alloys.

C11 Novel coatings

Novel coatings had only middling scores when it came to supporting infrastructure available at NASA and lack of alternative sponsors, primarily because these coatings were, for the most part, either underdeveloped or well developed.

Underdeveloped coatings (including self-sensing, acoustically active, and functionally graded coatings) are still in the very fundamental research stage. Rather than focus on them directly, NASA should establish partnerships with universities to develop these coatings. The well-developed coatings (including superhydrophobic and ice shredding coatings) already have significant commercial potential and should be handled by industry.

C13 Advanced airframe alloys

Airframe alloys had low scores in supporting infrastructure and lack of alternative sponsors. Industry has significant interest, resources, facilities, and expertise to address this Challenge, whereas NASA's own capability has eroded owing to the retirement of expert personnel. Several new and promising metallic materials can be used for critical structural applications. NASA can make a significant contribution to developing these alloys by collaborating with universities doing fundamental multiscale physics research necessary to gain a fundamental understanding of their material behavior. This knowledge could then be leveraged with industry's efforts to enable the design of materials for specific properties. This will dramatically shorten the development cycle for new alloys by focusing limited resources on the most promising candidates.

Other low-priority challenges

Some materials and structures R&T Challenges scored low because they did not fit within the decadal time frame that is the scope of this survey, they were already being pursued by industry or other government agencies, or they were not viewed as major contributors to civil aviation. Many of these Challenges address emerging technologies, and relevant research will yield useful products, but their utility to and impact on commercial aviation is either limited or unknown. Most intriguing is advanced functional polymers (Challenge C18), which includes self-healing polymers for passive repair of damage, reversible liquid crystal adhesives, light-harvesting polymers for collecting solar energy, superabsorbent polymers for flame retardation, and mechanochromic polymers that can change color in response to damage. A logical role for NASA in this dynamic and diverse area would be to vet new materials and devices and develop criteria for long-term materials investment.

R&T Challenges such as innovations in structural joining (Challenge C12), advanced airframe alloys (Challenge C13), aircraft hardening (Challenge C15), ultralight structures (Challenge C17), advanced engine nacelle structures (Challenge C19), and repairability of structures (Challenge C20) are also worthy of note since they could improve current civil aircraft design and also apply to military aircraft. In some cases, these Challenges are regarded as natural candidates for company investment, not cutting-edge NASA efforts. Joining, repair, and lightweight design aspects of airframes are also addressed in the highly ranked multidisciplinary analysis and the new composites tasks. Ultralightweight structural concepts benefit from the integration of advanced composites, adaptive materials, multifunctional materials, and multidisciplinary structural optimization. Thus, even though these topics received low rankings on their own, they will be addressed in a broader, integrated context in the more highly ranked tasks.

Similarly, advances in NDE are synergistic and closely allied with IVHM efforts. NDE provides input to prognosis and life-prediction systems, and its development fits well with NASA's mission in terms of aviation safety. NDE facilitates the development and insertion of new materials and structures and processes by ensuring that they are manufactured according to specifications and behave as designed when they are put into service. However, NDE is also an active area of research in DoD and in nonaerospace industries. For the next generation of NDE, the data-acquisition hardware is of less interest than the new tools for interpreting the multiphysics NDE measurement data. NASA should work in collaboration with academia to develop analysis tools for automating the interpretation of the multiphysics NDE measurement data. This role would provide NASA a unique focus, different from the work currently sponsored by others.

Multiphysics and multiscale modeling and simulation (Challenge C16), while important to the efficient design of future materials and structures, has many contributors from several federal agencies as well as academia. It lacks a clear, focused set of objectives and a clear tie to NASA's mission. The long time frame associated with R&D in multiphysics, multiscale modeling makes this Challenge more suited to academia.

DYNAMICS, NAVIGATION, AND CONTROL, AND AVIONICS

Introduction

In this report, dynamics is defined as the motion characteristics of the aircraft due to the forces that act on it. Navigation generally refers to determination of the aircraft's state (i.e., its position, velocity, and attitude rates) in three dimensions at a particular time. Control is closely linked to guidance; together they refer to determination of the aircraft's future state and the processes for reaching and staying on a specified trajectory. Finally, avionics consists of aviation electronics, both onboard and off-, which implements navigation, guidance, control, surveillance, communications, and other functions. Avionics includes the development, production, and use of aviation electronics, including both hardware and software.

The QFD process described in Chapter 2 was used to prioritize 14 R&T Challenges related to dynamics, navigation, and control, and avionics. Table 3-4 and Figure 3-6 show the re-

TABLE 3-4 Prioritization of R&T Challenges for Area D: Dynamics, Navigation, and Control, and Avionics

		Strategic Objective							Why NASA?					
		Capacity	Safety and Reliability	Efficiency and Performance	Energy and the Environment	Synergies with Security	Support to Space	National Priority	Supporting Infrastructure	Mission Alignment	Lack of Alternative Sponsors	Appropriate Level of Risk	Why NASA Composite Score	NASA Priority Score
R&T Challenge	**Weight**	**5**		**3**		**1**			**$^1/_4$ each**					
D1	Advanced guidance systems	9	9	9	3	3	3	**132**	9	9	3	9	**7.5**	**990**
D2	Distributed decision making, decision making under uncertainty, and flight-path planning and prediction	9	9	9	3	3	3	**132**	3	9	3	9	**6.0**	**792**
D3	Aerodynamics and vehicle dynamics via closed-loop flow control	1	9	9	3	3	3	**92**	9	9	3	9	**7.5**	**690**
D4	Intelligent and adaptive flight control techniques	3	9	9	3	3	9	**108**	3	9	3	9	**6.0**	**648**
D5	Fault-tolerant and integrated vehicle health management systems	3	9	3	1	3	9	**84**	9	9	3	9	**7.5**	**630**
D6	Improved onboard weather systems and tools	9	9	3	1	1	1	**104**	9	9	3	3	**6.0**	**624**
D7	Advanced communication, navigation, and surveillance technology	9	9	9	3	3	3	**132**	3	9	3	3	**4.5**	**594**
D8	Human–machine integration	3	9	9	1	3	3	**96**	3	9	3	9	**6.0**	**576**
D9	Synthetic and enhanced vision systems	3	9	3	1	1	3	**76**	9	9	3	3	**6.0**	**456**
D10	Safe operation of unmanned air vehicles in the national airspace	3	9	3	1	9	1	**82**	3	9	3	3	**4.5**	**369**
D11	Secure network-centric avionics architectures and systems to provide low-cost, efficient, fault-tolerant, onboard communications systems for data link and data transfer	9	9	9	1	9	3	**132**	3	3	1	3	**2.5**	**330**
D12	Smaller, lighter, and less expensive avionics	1	3	9	3	3	9	**68**	3	3	3	3	**3.0**	**204**
D13	More efficient certification processes for complex systems	3	9	9	1	1	3	**94**	3	1	1	3	**2.0**	**188**
D14	Design, development, and upgrade processes for complex, software-intensive systems, including tools for design, development, and validation and verification	3	9	3	1	1	3	**76**	1	3	1	1	**1.5**	**114**

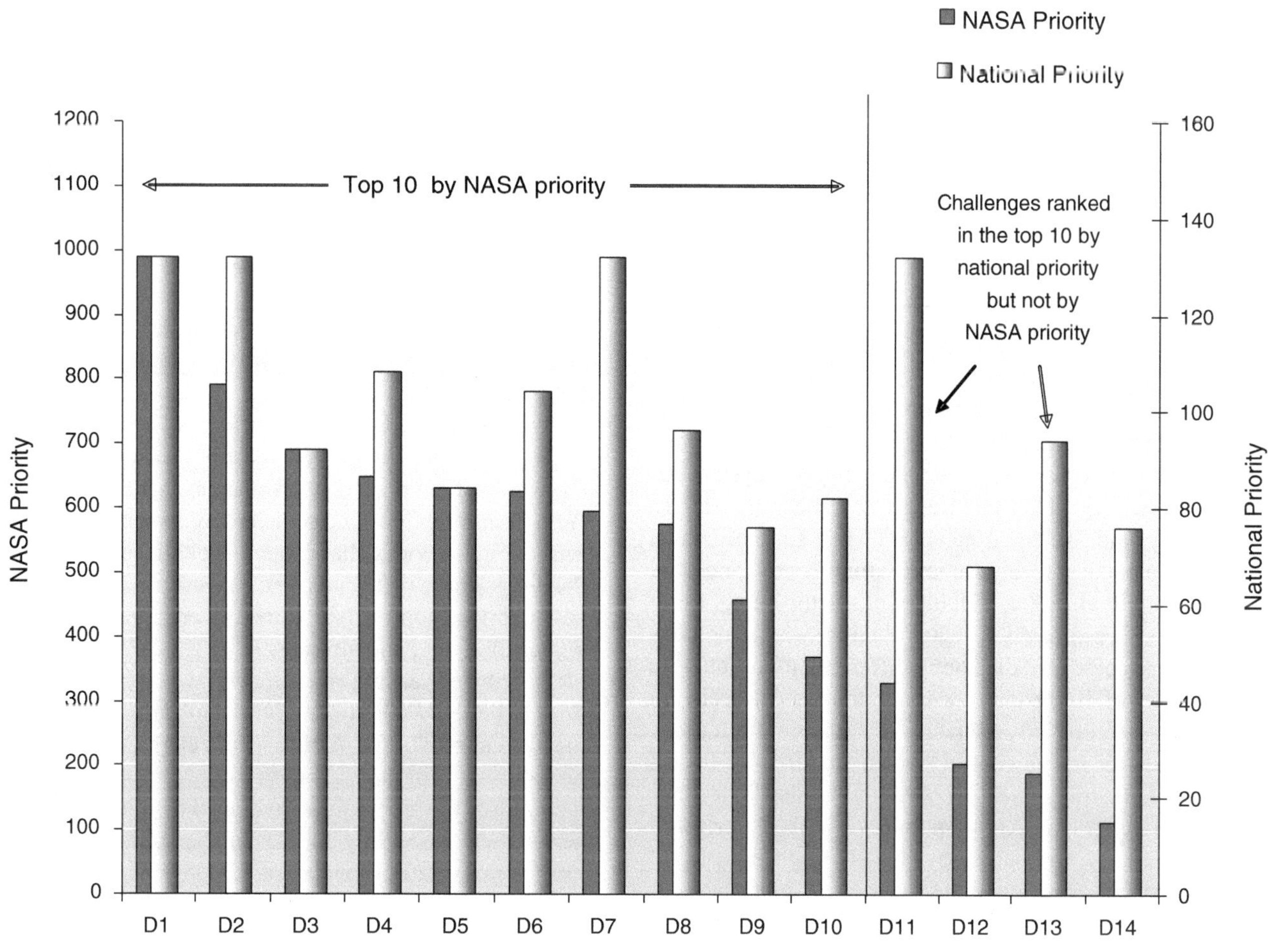

FIGURE 3-6 NASA and national priorities for Area D: dynamics, navigation, and control, and avionics.

sults. This section describes the 10 R&T Challenges that ranked highest in terms of NASA priority, the general characteristics of high- and low-priority Challenges, and the R&T Thrusts in this Area. Further details on all the Challenges, including the rationale for the scoring, are found in Appendix D.

Top 10 R&T Challenges

D1 Advanced guidance systems

Advanced guidance systems consist of subsystems and processes (hardware and software) assembled for the purpose of providing an aircraft, spacecraft, or other dynamic system with desired state trajectories. These trajectories can be defined using either discrete or continuous data and can include information such as current velocity, acceleration, time of arrival, and desired position. The determination of the desired trajectory usually takes into account mission-dependent constraints, which can include obstacles (such as terrain, wake vortices, or other aircraft), hazards (such as weather), coordination with other aircraft (such as cooperative and multiaircraft guidance, formation flight, or swarming), and regulatory constraints (such as airspace class restrictions) (Doebbler et al., 2005).

State-of-the-art guidance systems enable aircraft to follow waypoints, perform automatic obstacle avoidance, and fly in formation with other aircraft (Schierman et al., 2004). Additional research is needed to develop guidance algorithms and mature them into flight-ready systems,[8] to develop improved reconfigurable and adaptive guidance systems, and to develop advanced guidance systems for UAVs. One concern, for example, is the need to develop improved technologies to avoid controlled flight into terrain, particularly in the case of all-weather operation of advanced rotorcraft. Some important research is inhibited by the limited number of programs and facilities capable of implementing and flying these systems on real aircraft. Also, certification and regulatory issues must be resolved so that the air trans-

[8]R. Duren, associate professor, Baylor University, "Avionics research challenges," Presentation to Panel D on November 15, 2005.

portation system can take advantage of the full capabilities of current and future guidance systems for piloted aircraft and UAVs.

Advanced guidance systems have the potential to greatly improve the capacity, safety, and efficiency of the air transportation system. In addition, they can enhance the performance of many existing and future military systems.

D2 Distributed decision making, decision making under uncertainty, and flight path planning and prediction

Improving the decision-making process used by pilots and aircraft systems, when coupled with improvements in flight-path planning and prediction, has been theorized as an effective approach to improving air transportation system capacity and safety. This Challenge has the potential to significantly improve the timeliness of real-time decisions to alter flight paths in the dynamic environment of congested airspace (Ding et al., 2004; Helbing et al., forthcoming; Rong et al., 2002). Coordinated decision making, which includes the direct exchange of data among different aircraft and the deconfliction of flight paths without the need to rely on ground-based controllers, addresses the inherent limitations of centralized air traffic control systems in terms of uncertainty and fault tolerance. A coordinated, distributed approach to decision making increases air transportation system reliability and safety by distributing control and mission management capabilities among multiple agents. It also allows for rapid response to changing dynamics and minimizes vulnerability to system failures.

Automated systems can help improve decision making and flight path planning. Levels of automation ranging from "pilot aid" (that is, systems that advise pilots to take specific action) to "fully autonomous" are achievable but have not yet been developed to the point where they can support high levels of automation for civil aircraft. Until now, coordinated distributed algorithms for constraint reasoning (for example, to optimize flight paths) have not been applied to the air transportation system because implementation with such a complex system would require aircraft to exchange a large number of messages, which raises substantive communications, bandwidth, and man–machine interface issues.

This Challenge should address the needs of a wide variety of conventional and unconventional aircraft types, including those with no distributed decision-making capability. Aircraft types of interest include commercial airliners, general aviation aircraft, civil helicopters, military aircraft, and UAVs.

This Challenge also has the potential to be of great benefit when applied to complex, nonaviation systems that operate in dynamically changing environments and require high-quality, real-time decision making.

D3 Aerodynamics and vehicle dynamics via closed-loop flow control

Closed-loop flow control appears to offer tremendous promise in improving aerodynamic performance. For example, active flow control approaches should allow the airfoil lift:drag (L/D) to remain high over large changes in angle of attack.[9] Flow control R&T could also be used to develop a spoiler-aileron to replace complex and heavy control surfaces and to reduce or eliminate turbulent flow over aircraft surfaces to reduce skin-friction drag. These applications could lead to new aircraft configurations (Chavez and Schmidt, 1994).

The mechanization of flow control systems may require a large number of distributed sensors measuring pressure or shear stress over the wing and changes in the boundary layer. Actuation might be accomplished by morphing the wing or introducing devices that induce sucking or blowing along the wing. These distributed sensors and actuators are coordinated so that control is obtained over large flight regimes, angles of attack, and attitudes.

Distributed sensing and actuation would also permit structures to be self-aware for health monitoring, thereby increasing system reliability. Airframe and engine structures could be monitored for changes in behavior.

Some of the techniques developed by this Challenge may also advance modeling and design capabilities applicable to morphing aircraft (Tandale et al., 2005; Valasek et al., 2005). Heretofore, aircraft have generally been fixed-frame structures. Morphing aircraft would be designed with distributed actuation and controls and with mechanization as an inherent property. They would lead to new capabilities and concepts in aircraft design. Examples include (1) biomorphic aircraft, such as ornithopters, that could maneuver robustly in complex environments and (2) hunter-killer aircraft that change shape to optimize performance for different tasks (e.g., surveillance, reconnaissance, and ground attack). Morphing technology might also enable aircraft capable of perching.

D4 Intelligent and adaptive flight control techniques

The missions and capabilities of future aircraft, both manned and unmanned, will be more multifunctional than those of the current generation of specialized aircraft. Achieving aggressive performance targets in range, payload, reliability, safety, noise, and emissions will require a total system that is integrated to a far higher level than existing aircraft. R&D for military aircraft has been able to push the

[9]The flow over the specially shaped GLAS II airfoil remains naturally separated at the rear of its upper surface over a wide range of angles of incidence; in the absence of active control, its L/D does not exceed 25. At an incidence angle of 10 degrees, its L/D is nearly 500 (Glauert, 1945; Glauert et al., 1948).

technological envelope associated with intelligent and adaptive flight control techniques farther than R&D for civil aircraft because of different safety limits. In the far term, as it advances, application of military technology to civil aircraft may be possible.

The vehicle management systems (VMS) paradigm offers the most promising path to realizing goals related to this Challenge. VMS takes a top-down systems approach to specifying, designing, and validating the aircraft as a single system with highly integrated inner and outer loops. It thereby unifies the traditionally separate fields of propulsion control, flight control, structural control, noise control, emissions control, and health monitoring. The current state of the art in VMS uses traditional feedback control, consisting of measurements of vehicle states such as airspeed, altitude, angle of attack, and linear and angular acceleration (Jaw and Garg, 2005). By incorporating an online learning capability to cope with new and unforeseen events and situations and nonlinear adaptive control, in which the controller self-tunes to maintain stability and tracking in the presence of disturbances and changing vehicle parameters, an intelligent and adaptive VMS can be developed with the promise of significant advances in capability, safety, and supportability (Tandale and Valasek, 2003).

Significant advances in the state of the art are required to develop an intelligent and adaptive VMS. Current nonlinear adaptive control approaches assume that (1) sensor information is reliable and (2) known nonlinearities can be modeled as slowly varying parameters that affect the system linearly. However, advanced actuators for flow control and structural control will have characteristics that are much more nonlinear than those of conventional control actuators. Control laws and control actuator allocation are currently treated as separate problems, such that optimization of the integrated control law is difficult or impossible. Finally, the problem of multiple correlated, simultaneous failures remains unsolved. Approaches that use analytic redundancy to finding failed sensors generally assume that aircraft dynamics have not changed, while adaptive or reconfigurable control approaches assume that sensor information is reliable. On an affordable aircraft with limited or no sensor redundancy, it is difficult or impossible to tell the difference between a degraded sensor and damage to the aircraft that changes the way it flies.

D5 Fault-tolerant and integrated vehicle health management systems

Development of IVHM system technologies is key to the acceptance of the automation needed in the transformation of the air transportation system. The technology provides an increased capability to accurately discover and assess system faults and reconfigure or recover from them. Although highly integrated, health management aspects consist of related components: fault detection and isolation, recovery and reconfiguration, and condition-based maintenance (CBM). In addition, modeling plays an important role in the development of these functions (Garg, 2005; Litt et al., 2005; Tandale and Valasek, 2006).

Fault detection, isolation, recovery, and reconfiguration

Fault detection, isolation, recovery, and reconfiguration involve processes and approaches that enable robust detection of faults from measured or estimated error residuals and isolation of faults with minimal latency in the presence of noise and environmental effects during aircraft operation. Fault detection, isolation, recovery, and reconfiguration are platform specific and should cover all flight regimes and mission types. Recovery and reconfiguration systems are developed with regard to the possibilities of faults, the nature of the latency of the fault detection and isolation system, and the controls available for recovery and reconfiguration. Redundancy management strategies for avionics and the airframe directly influence options for recovery and reconfiguration.

Condition-based maintenance

CBM involves maintenance processes and capabilities derived from real-time assessment of aircraft system conditions obtained by software from embedded and redundant sensors. The combination of software and sensors can create important communications and bandwidth problems. More robust diagnostics and prognostics are needed to achieve the goal of CBM, which is to perform maintenance only on evidence of need to prevent a failure from reducing aircraft availability. In addition, CBM includes processes that couple real-time assessment of system and component performance with ground- and air-based logistics to improve aircraft system readiness and maintenance practices. CBM is a form of proactive equipment maintenance that forecasts incipient failures. CBM also aims to ensure safety, equipment reliability, and reduction of total ownership cost. Fault tolerance is achieved when CBM is married to decision strategies for safe and reliable operation of manned and unmanned aircraft.

Modeling

Physics-based models of sensors, actuators, avionics, components, and vehicle flight dynamics contribute to the development of methods for forecasting aircraft system performance and, thereby, help uncover faults. In addition, these models can be used for examining architectures and control strategies to reconfigure systems and ensure safety and reliability.

An aircraft is a very complex system. While individual fault-tolerant functions can be set up for each subsystem, the value of fault-tolerant designs is maximized when the system is modeled as a whole, since the behavior of each subsystem can influence that of other subsystems. The advantage of working with a total system model lies in the ability to discover a fault through its effects on other parts of the system before the fault is discovered in the individual subsystem itself. One primary thrust of fault-tolerant technology development is to identify system models that characterize the behavior of systems properly without developing an overly detailed and unnecessary representation. In other words, an optimum system is not a collection of optimized subsystems.

To advance the state of the art in fault-tolerant aircraft systems, fundamental R&T is required in the three topics above to develop a more robust image of the state, or health, of an aircraft in the presence of uncertainty. With a better model of itself the aircraft can trace back system anomalies through the multitude of discrete state and mode changes to isolate aberrant behavior. Fault-tolerant systems combine simple rule-based reasoning, state charts, model-free monitoring of cross-correlations among state variables, and model-based representations of aircraft subsystems. Together, these models form a hybrid system model. Advances in computing resource technology have allowed hybrid system models to run in real time.

Fault-tolerant aircraft systems, coupled with CBM, may improve aircraft safety and reduce aircraft life-cycle maintenance and ownership costs. Critical research tasks include developing (1) robust and reliable hardware and software tools for monitoring components, detecting faults, and identifying anomalies; (2) prognosis analysis tools for predicting the remaining life of key components; (3) approaches for recovering from detected faults, including reconfiguration of the flight control system for in-flight failures of manned and unmanned aircraft; and (4) low-cost, lightweight, wireless, self-powered sensors with greater memory and processing capability.

D6 Improved onboard weather systems and tools

Pilots—and the avionics software that provides in-flight four-dimensional trajectory replanning and commands to the pilot or autopilot—require additional weather information to minimize the impact of weather on the control of flight in high-density traffic. Basic research is needed to determine the most cost-effective way of integrating real-time weather information into four-dimensional integrated control of flight. This information might include information from data links with ground sites and other aircraft and weather video from ground stations and satellites (Bokadia and Valasek, 2001; Lampton and Valasek, 2005, 2006).

Other aircraft could provide information about geospatial position, wind, icing conditions, turbulence, lightning, and precipitation, as well as imagery from radars and other sensors. Data links with the ground could provide actual and forecast information on winds at different flight levels, pressure, icing potential, precipitation, ground-level temperatures, weather fronts, severe weather, airport surface conditions, and other information from significant meteorological information reports (SIGMETs); pilot reports (PIREPs); meteorological aviation reports (METARs), terminal area forecasts (TAFs), imagery from satellites and radars, and so on.

D7 Advanced communication, navigation, and surveillance (CNS) technology

The capacity of the air transportation system is dependent on minimum spacing requirements for safe operation. Minimum spacing depends on many factors, including the capability of each aircraft to precisely fly a predetermined, geospatially time-referenced flight path.

Advanced, integrated, accurate, secure, and reliable CNS capabilities are required for network-centric operations, which can increase capacity in very high density airspace. Each aircraft may be considered a node in a network-centric, distributed, fault-tolerant ATM system. Communications between nodes (aircraft to aircraft, aircraft to ground, and aircraft to satellite to ground) must be highly reliable. (For example, the probability of a missed or incorrect message should be less than 10^{-7} per flight hour, depending on the consequence of the fault.) Safe, secure, accurate, and certifiable CNS technologies that provide required capabilities are needed.

More precision aircraft navigation, coupled with the precise six-dimensional[10] guidance algorithms used in advanced flight management systems, will enable reduced spacing between aircraft operating en route and in the terminal airspace. CNS system functions must be tightly coupled in terms of information integrity, and they should allow pilots to operate cooperatively with ground systems without controllers continuously in the control loop. The CNS should transmit navigation, guidance, and other sensor data to other aircraft and ground operation centers via multichannel data links while, at essentially the same time, they receive similar information about other aircraft, the weather, airport conditions, etc. This information can prevent accidents by revealing the current and future status of other aircraft, weather phenomena, terrain, buildings, and vehicles on the ground at airports. This Challenge should also increase the affordability of onboard avionics to encourage aircraft owners and operators to procure more capable avionics. This Challenge encompasses the following CNS issues:

- Communications issues.
- Fault-tolerant network connectivity and security.
- Dynamic network control and reconfiguration.
- Quality of service.

[10]The six dimensions refer to three position coordinates and three velocity vectors to define aircraft location, speed, and direction of motion.

- Spectrum allocation and usage.
- Adequate communication bandwidth.
- Required communications capability as a function of geospatial location and phase of flight.
- Navigation issues.
  —High-precision, six-dimensional estimate of aircraft state as a function of time.
  —Integration of satellite navigation with other navigation modes.
  —Navigation system capability, including reliability and quality of input signals.
  —Functional integration of navigation system with guidance and flight control systems to ensure high-integrity, integrated control of flight during automatic and manual modes.
- Surveillance issues.
  —Capability of data links to provide accurate time-referenced data from navigation systems, guidance systems, and other sensors when interrogated by external systems or periodic broadcast.
  —Handling of multiple, simultaneous interrogations using multiple channels to provide high integrity, secure data.
  —Processing and reacting to incoming data about other aircraft, hazardous weather, etc.
  —Continuous improvement in situational awareness through advanced sensors, communication links, and human–system interfaces.

D8 Human–machine integration

The ever-increasing demand for air transportation, combined with the rapid pace of technological change, poses significant challenges for effective integration of humans and automation. For the foreseeable future, humans will continue to play a central role in key decision-making tasks that directly influence the efficiency and safety of civil aviation. As technology evolves, it may be anticipated that the role of humans and the nature of their task will change accordingly. In order to maintain or improve on existing standards of performance and safety, it is critical that the allocation of functions between humans and automation and the design of the human–machine interface be optimized based on a solid foundation of scientific principles that reflect our best understanding of human sensory, perceptual, and cognitive processes. Human–machine integration should remain an important element of NASA research directed toward civil aeronautics applications.[11] However, the emphasis should be shifted from development and testing of specific input and output devices toward more fundamental research involving modern instruments that measure brain physiology. Research should also include voice command and recognition technology, coupled with increased machine contextual understanding, to reduce workload. This will help define the future role of humans in complex, highly automated systems. Key research topics include human–machine integration methods, tools, and integration technologies for vehicle applications.

D9 Synthetic and enhanced vision systems

Synthetic and enhanced vision systems provide an out-the-window view of terrain, obstacles, and traffic. These systems can also be used as flight crew interfaces for flight trajectory and planning operations (Kelly et al., 2005). The synthetic vision systems that use databases to generate terrain and obstacles require high-fidelity, high-integrity information and a self-healing capability. Enhanced vision systems use forward-looking sensors such as infrared, radar, and laser ranging to allow the flight crew to visualize the real world when visibility is hindered. Currently, vision systems are limited by weather, human factors issues, and other issues. New sensors and improved sensor fusion are needed.

A combined synthetic and enhanced vision system has future potential as a navigation, approach, and landing sensor. The ability to "see" the airport in poor weather has the potential to reduce the likelihood of a go-around. Information fusion that exploits the capabilities of sensors and compensates for their deficiencies is needed, and the immature state of this art represents the most difficult obstacle to achieving these benefits.

Synthetic and enhanced vision systems are also intended to aid airport surface operations in poor weather, reducing runway occupancy and taxiing errors and reducing gate-to-gate travel time. Research topics of interest are as follows:

- Database integrity and quality.
- Information fusion.
- Object detection and avoidance.
- Human–machine interface issues.
- Verification of accuracy, fault tolerance, and reliability.

D10 Safe operation of unmanned air vehicles in the national airspace

The use of UAVs for a variety of civil applications (e.g., farming, communications relays, border monitoring, power line and pipeline monitoring, and firefighting) will continue to increase. Flight operations of military UAVs in civil airspace are also expected to increase. To facilitate these operations, UAVs should be integrated into the air transportation system. This requires them to be at least as safe as manned aircraft.

Most UAV technologies, capabilities, and processes are shared with manned aircraft and require research in several key topics, including the following four:

[11] J. Vagners, professor emeritus, aeronautics and astronautics, University of Washington, Presentation to Panel D on November 15, 2005.

- *Aircraft.* Automation, system upgrade issues, and communications systems, all of which are distinct from those for manned aircraft.
- *Human–machine interaction.* Function allocation, human interface design, situational awareness, training, and required level of proficiency in the remote operation of the aircraft.
- *Maintenance and support.* In matters where UAVs differ distinctly from traditional aircraft.
- *Flight operations.* Sense- or see-and-avoid issues, person-to-person interfaces between operators and controllers, assurance of positive control of the aircraft (especially with highly automated UAVs that are not directly controlled by ground-based operators in real time), and automated contingency management.

High-Priority R&T Challenges and Their Associated Thrusts

R&T Challenges that significantly impact multiple strategic objectives or for which NASA possesses unique capabilities ranked high on the technology prioritization list. All of the top 10 Challenges received the maximum score for relevance to safety and reliability, and seven of the top 10 also received the maximum score for relevance to capacity and/or efficiency and performance. None of the top 10 received the maximum score for energy and the environment, one received the maximum score for relevance to national and homeland security, and two received maximum scores for support to space. Most of the Challenges, regardless of rank, fall into one of five R&T Thrusts that describe threads of commonality among the R&T Challenges within the dynamics, navigation, and control, and avionics Area. These thrusts are discussed next.

Increased integration

Avionics systems are becoming more integrated within individual aircraft, and the control of aircraft flights is more tightly integrated in the air transportation system as a whole. The future air transportation system will see increased integration and information sharing among components of the air transportation system—including individual commercial, business, and general aviation aircraft; ATM facilities; and operation centers for passenger airlines, air cargo operators, and the military. Capacity increases can be achieved, for example, by reducing separation between aircraft, but this could threaten safety. An individual aircraft requires information on the relative position of other aircraft and ground hazards, which may be fixed (terrain and buildings) or moving (aircraft on taxiways and runways). Functional as well as information integration will be needed. All of this will require fault-tolerant, integrated, secure, reliable, flight-critical communications.

Multifunctional, highly integrated guidance and control

The missions and capabilities of future aircraft, both manned and unmanned, will be more multifunctional than the current generation of specialized aircraft. Achieving aggressive performance targets in range, payload, reliability, safety, and emissions will require aircraft to be much more integrated than existing aircraft systems.

Distributed decision making and control

Most scenarios for the future air transportation system envision increased distribution of decision making. Currently, most decision making is centralized and ground-based, with air traffic controllers responsible for coordinating the movement of aircraft in the air and the FAA center at Herndon, Virginia, responsible for national flow control. While centralization has advantages in terms of ensuring safety, it is inflexible and it limits decision making by the pilots and airlines. It also has inherent limits—for example, the mental limits of individual human controllers—that contribute to capacity problems in the system.

Technological advances such as advanced communication systems, satellite navigation, and sophisticated decision-making technologies could further distribute decision making and control among various airspace systems and move these tasks from the ground to the air. These changes, however, will require much greater complexity and functionality in the airborne systems, and important questions must be answered before these changes can become a reality. Relevant questions include how to ensure safety in such a distributed environment, how to provide reliable and efficient communications, how to develop and implement sophisticated decision-making algorithms, and how to define and implement appropriate human–machine interactions.

Intelligent use of automation

Used intelligently, automation has the potential to greatly enhance the safety and efficiency of civil aviation. The rapid evolution of technologies for sensing, processing, and communicating information enables designers to consider new systems with unprecedented levels of automation. The current trend toward increased automation is introducing fundamental, qualitative changes in human roles and tasks. In some cases, these changes have assigned humans tasks for which they are ill-equipped, such as monitoring highly automated processes.

NASA has substantial facilities and expertise that could be applied to this Thrust. Historically, these resources have primarily been used to develop and demonstrate specific system- and subsystem-level solutions to particular operational or safety problems. Transitioning these point designs to practical applications has been problematic. A more productive use of NASA's considerable capabilities would be for NASA to become a provider of basic research products, enabling technologies, and system engineering tools to support system development by industry and certification by the FAA. There is a compelling need for a focused program of research that will yield practical, validated technologies, processes, and tools to support effective human–machine integration in civil aviation.

Revolutionary vs. evolutionary approaches

Aeronautics research can use revolutionary and/or evolutionary approaches. A revolutionary approach allows the researcher to look for the best solution to a problem (assuming "best" can be properly defined) using the latest technology available without concern for the technology that is currently fielded. The evolutionary approach looks for a solution that is a derivative of or an incremental improvement to a current system.

NASA plays an important role in aeronautics with its capability to do revolutionary research. By following a revolutionary approach to ATM and avionics, for example, NASA can set a goal for the end state—a picture of what the system might look like in 10, 15, or 25 years. NASA can use its modeling and simulation expertise and its proof-of-concept flight demonstration capabilities to predict the system efficiencies for the end state. It can set the long-term vision for aeronautics and the air transportation system. This role requires communication and interaction with the FAA, DHS, DoD, and other members of the Joint Planning and Development Office that is defining the nature of the Next Generation Air Transportation System and the research program necessary to make it a reality. The FAA, which must implement modifications to the system in an evolutionary manner, can develop its system roadmap in part by using the NASA end state.

Low-Priority R&T Challenges

Four R&T Challenges were not in the top 10. Three of the four (D12, smaller, lighter, and less expensive avionics; D13, more efficient certification processes; and D14, design, development, and upgrade processes for complex, software-intensive systems) had a significant impact on only one or two Strategic Objectives. For example, Challenge D14 is very relevant to the safety and reliability Strategic Objective but had only a minimal or modest impact on the other five Objectives.

The fourth R&T Challenge that did not make the top 10 was D11, secure, network-centric avionics architecture and systems. This Challenge has a significant effect on four of the Objectives, and it tied for the highest score in terms of national priority. However, it ranked low in terms of NASA priority because it scored worse than all of the top 10 Challenges in terms of alignment with the NASA mission and the availability of alternative sponsors.

Challenges D13 and D14 also scored in the top 10 in terms of national priority but not in terms of NASA priority. Challenge D13 is important because all new technologies must be certified before they can be put in service. D14 is important because many new systems will be software intensive. However, these Challenges scored low in terms of "Why NASA?" because (1) other organizations in government, industry, and academia are already working on relevant technologies and (2) NASA has relatively little infrastructure or expertise to contribute. However, because Challenges D11, D13, and D14 scored high as a national priority, some part of the national civil aeronautics effort should support relevant R&T.

INTELLIGENT AND AUTONOMOUS SYSTEMS, OPERATIONS AND DECISION MAKING, HUMAN INTEGRATED SYSTEMS, AND NETWORKING AND COMMUNICATIONS

Introduction

Aeronautics research encompasses much more than airframes and engines. For many years NASA has been in the forefront of discovering how human beings interface with aviation hardware. NASA has also been a leader in development of autonomous systems and communications interfaces. Accordingly, R&T Challenges in the Area of intelligent and autonomous systems, operations and decision making, human integrated systems, and networking and communications focus on issues associated with the air transportation system of today and tomorrow as a complex interactive system; issues associated with the performance of systems in individual aircraft are addressed in the preceding sections.

The Next Generation Air Transportation System (NGATS) Joint Planning and Development Office (JPDO) is striving to achieve eight key capabilities (NGATS JPDO, 2005, pp. 7-9):

- Network-enabled information access, which will give decision makers throughout the air transportation system quick access to the critical information they need in normal and emergency conditions.
- Performance-based services, which will maximize the performance of all categories of aircraft.
- Weather assimilated into decision making, which will take advantage of improved probabilistic weather information.
- Layered, adaptive security, which will be more efficient, more effective, and less intrusive.
- Broad-area precision navigation, which will allow pilots to make precision landings at airports that do not have control towers, radar, or an instrument landing system (ILS).
- Aircraft trajectory-based operations, which will include automatic, continuous analysis of trajectories to increase capacity and assure safe separation of aircraft.

- Equivalent visual operations, which will allow pilots and controllers to see the same picture, enabling controllers to delegate some tasks to pilots.
- Super-density operations, which will use advanced capabilities, including detection and avoidance of hazardous wake vortices, to enable closely spaced and converging approaches in the air as well as more efficient airport ground operations.

This study did not assess and does not necessarily endorse the above set of capabilities. However, many of the capabilities would be supported by R&T Challenges in this Area. These Challenges also encompass the basic and applied research necessary to establish a proper balance between automated and human-centric system configurations and operational concepts in the air transportation system. The scope of research in this Area includes new NASA R&T Thrusts as well as the expansion of existing technology programs. One Area of particular near-term interest is the incorporation of autonomous and semiautonomous aircraft into the national (and global) air transportation system.

The QFD process described in Chapter 2 was used to prioritize 20 R&T Challenges related to intelligent and autonomous systems, operations and decision making, human integrated systems, and networking and communications. Table 3-5 and Figure 3-7 show the results. The text that follows describes the 10 R&T Challenges that ranked highest in terms of NASA priority, the general characteristics of high- and low-priority Challenges, and the R&T Thrusts in this Area. Further details on all Challenges, including the rationale for scoring, are found in Appendix E.

As shown in Table 3-5, many of the Challenges in this Area ranked high because they would enhance the performance of the air transportation system as a whole, bringing about noteworthy improvements related to many of the air transportation system strategic objectives (capacity, safety and reliability, etc.). As shown Figure 3-7, the top 10 Challenges fall into three groups:

- R&T Challenge E1 stands alone, with a NASA priority score of 936.
- R&T Challenges E2 and E3 stand together with scores of 780 and 744.
- R&T Challenges E4 to E8c stand together with scores of 624 to 576.

The difference in scores and rankings of the R&T Challenges in each of the last two groups is not significant.

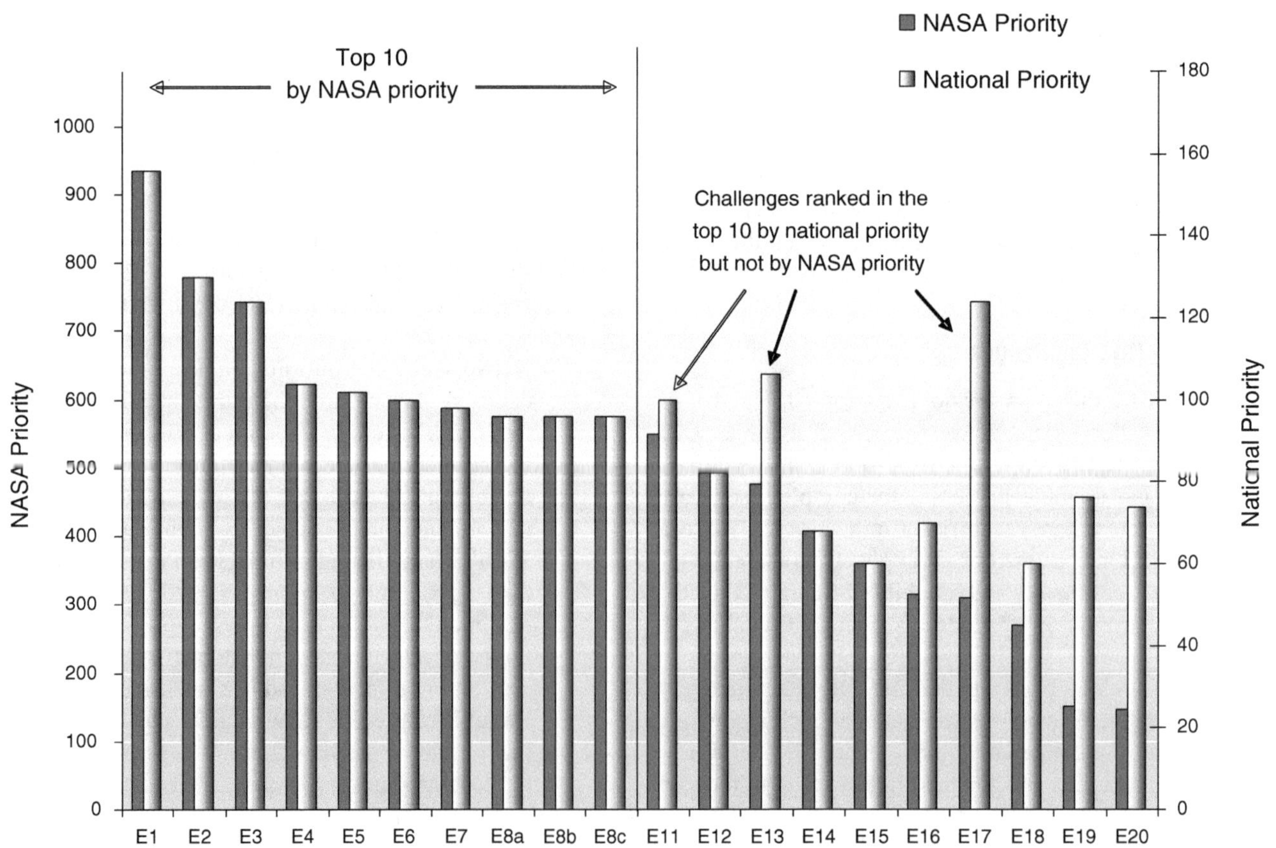

FIGURE 3-7 NASA and national priorities for Area E: intelligent and autonomous systems, operations and decision making, human integrated systems, and networking and communications

TABLE 3-5 Prioritization of R&T Challenges for Area E: Intelligent and Autonomous Systems, Operations and Decision Making, Human Integrated Systems, and Networking and Communications

R&T Challenge		Strategic Objective: Capacity	Strategic Objective: Safety and Reliability	Strategic Objective: Efficiency and Performance	Strategic Objective: Energy and the Environment	Strategic Objective: Synergies with Security	Strategic Objective: Support to Space	National Priority	Why NASA?: Supporting Infrastructure	Why NASA?: Mission Alignment	Why NASA?: Lack of Alternative Sponsors	Why NASA?: Appropriate Level of Risk	Why NASA Composite Score	NASA Priority Score
	Weight	5	5	3	3	1	1		1/4 each	1/4 each	1/4 each	1/4 each		
E1	Methodologies, tools, and simulation and modeling capabilities to design and evaluate complex interactive systems	9	9	9	9	9	3	**156**	3	9	3	9	**6.0**	**936**
E2	New concepts and methods of separating, spacing, and sequencing aircraft	9	9	9	3	3	1	**130**	3	9	3	9	**6.0**	**780**
E3	Appropriate roles of humans and automated systems for separation assurance, including the feasibility and merits of highly automated separation assurance systems	9	9	9	1	3	1	**124**	3	9	3	9	**6.0**	**744**
E4	Affordable new sensors, system technologies, and procedures to improve the prediction and measurement of wake turbulence	9	9	3	1	1	1	**104**	3	9	3	9	**6.0**	**624**
E5	Interfaces that ensure effective information sharing and coordination among ground-based and airborne human and machine agents	3	9	9	1	9	3	**102**	3	9	3	9	**6.0**	**612**
E6	Vulnerability analysis as an integral element in the architecture design and simulations of the air transportation system	3	9	9	1	9	1	**100**	3	9	3	9	**6.0**	**600**
E7	Adaptive ATM techniques to minimize the impact of weather by taking better advantage of improved probabilistic forecasts	9	3	9	3	1	1	**98**	3	9	3	9	**6.0**	**588**
E8a	Transparent and collaborative decision support systems	3	9	9	1	3	3	**96**	3	9	3	9	**6.0**	**576**
E8b	Using operational and maintenance data to assess leading indicators of safety	3	9	9	1	3	3	**96**	3	9	3	9	**6.0**	**576**
E8c	Interfaces and procedures that support human operators in effective task and attention management	3	9	9	1	3	3	**96**	3	9	3	9	**6.0**	**576**
E11	Automated systems and dynamic strategies to facilitate allocation of airspace and airport resources	9	3	9	3	3	1	**100**	3	9	1	9	**5.5**	**550**
E12	Autonomous flight monitoring of manned and unmanned aircraft	3	9	3	1	9	1	**82**	3	9	3	9	**6.0**	**492**
E13	Feasibility of deploying an affordable broad-area, precision-navigation capability compatible with international standards	9	9	3	1	3	1	**106**	3	3	3	9	**4.5**	**477**
E14	Advanced spacecraft weather imagery and aircraft data for more accurate forecasts	3	3	9	3	1	1	**68**	3	9	3	9	**6.0**	**408**
E15	Technologies to enable refuse-to-crash and emergency autoland systems	1	9	1	1	3	1	**60**	3	9	3	9	**6.0**	**360**
E16	Appropriate metrics to facilitate analysis and design of the current and future air transportation system and operating concepts	3	3	9	3	3	1	**70**	3	9	3	3	**4.5**	**315**
E17	Change management techniques applicable to the U.S. air transportation system	9	9	9	1	3	1	**124**	1	3	3	3	**2.5**	**310**
E18	Certifiable information-sharing protocols that enable exchange of contextual information and coordination of intent and activity among automated systems	3	1	9	1	9	1	**60**	3	9	3	3	**4.5**	**270**
E19	Provably correct protocols for fault-tolerant aviation communications systems	3	9	3	1	3	1	**76**	3	3	1	1	**2.0**	**152**
E20	Comprehensive models and standards for designing and certifying aviation networking and communications systems	3	9	3	1	1	1	**74**	3	3	1	1	**2.0**	**148**

Top 10 R&T Challenges

E1 Methodologies, tools, and simulation and modeling capabilities to design and evaluate complex interactive systems

The U.S. air transportation system is a complex interactive system whose behavior is difficult to simulate with currently available models. Methodologies, tools, and simulation and modeling capabilities suited for the design and integration of complex interactive systems are needed to understand the air transportation system as an integrated, adaptive, distributed system that includes aircraft, ATM facilities, and airports, each with its own complex systems, all of which interact with one another, the environment, and human operators. Simulations and models for complex interactive systems are needed to accurately estimate system performance, to properly allocate resources, and to select appropriate design parameters. Additionally, the large number of possible future system designs requires models that can be reconfigured to model a wide range of design parameters.

E2 New concepts and methods of separating, spacing, and sequencing aircraft

Expected growth in the demand for air transportation will require efficient, denser en route and terminal area operations. This necessitates procedures that reduce minimum spacing requirements during all phases of flight and in all weather conditions, through an integrated approach that leverages a suite of emerging technologies such as required navigation performance and automatic dependent surveillance broadcast (ADS-B). The objective of this Challenge is to efficiently accommodate a large number and wide range of aircraft, including UAVs, through spacing and sequencing based on aircraft type and equipment rather than a common worst-case standard. Several concepts of operation should be systematically compared in terms of their technological, business, and human factors issues as well as their impact on capacity, safety, and the environment. This Challenge will study reduced separation operations within the context of existing ATM protocols and revolutionary paradigms that could significantly increase capacity, although the latter would involve a much more complicated transition process.

Integration of UAVs into the air transportation system will require procedures that can safely manage aircraft with diverse performance characteristics and highly automated onboard flight management systems (Sabatini, 2006). Safe, high-capacity operations in a complex future airspace environment will require fundamental research into alternative ATM paradigms such as simultaneous noninterfering operations (Xue and Atkins, 2006) in which general aviation, rotorcraft, and UAV traffic are threaded through airspace unused by commercial air traffic. As onboard automation and cooperative control algorithms are matured (McLain and Beard, 2005), UAV traffic might also be efficiently managed using formations of UAVs that are coordinated locally but treated as a single entity by air traffic controllers and pilots of nearby aircraft.

E3 Appropriate roles of humans and automated systems for separation assurance, including the feasibility and merits of highly automated separation assurance systems

Air traffic control is currently a labor-intensive process. FAA controllers—aided by radar, weather displays, and procedures—maintain traffic flow and assure separation by communicating instructions to aircraft in their sector of responsibility. Limitations to this traditional paradigm are, in some areas, constraining the capacity of the air transportation system. For example, the FAA required airlines serving the Chicago O'Hare airport to reduce some of their flights during 2005 because of congestion-related delays. A recent study of en route sector congestion suggested that capacity could be increased by a factor of two or more while maintaining existing spacing, by developing new systems that merge human and computer decision making and automate time-critical separation assurance tasks (Andrews et al., 2005).

Initiatives to reduce aircraft separation by providing automated advisories to air traffic controllers and flight crews have not lived up to expectations, because of controller workload concerns, institutional resistance, and other factors. The advent of UAVs has caused additional concern because it may not be feasible for UAVs with human-in-the-loop collision avoidance schemes to act in time to prevent midair collisions. This has led to interest in determining whether automating aircraft separation, whereby the controller is neither in the loop nor responsible for separation, is feasible and desirable. However, changing the role of the controller from tactical separation to traffic flow management and trusting automated systems to manage the tactical separation of aircraft would require resolution of major human factors, safety, and institutional issues (Wickens et al., 1998; Woods and Hollnagel, 2006). Collisions could occur if a UAV fails to respond or the automated traffic separation system fails and if human intervention is not effective. This Challenge would determine the appropriate roles of humans and automated systems to assure separation in high-density airspace during nominal and off-nominal operations. As part of this challenge, NASA should assess the feasibility and merits of highly automated separation assurance systems.

E4 Affordable new sensors, system technologies, and procedures to improve the prediction and measurement of wake turbulence

Existing wake vortex separation standards reduce system capacity during takeoff and landing operations and instrument approaches. Encounters with a wake vortex are also a growing concern in en route Reduced Vertical Separation Minima (RVSM) airspace (Reynolds and Hansman, 2001).[12]

Current research by the FAA and NASA is focused on procedural enhancements that take advantage of wake transport by winds (Mundra, 2001). For example, the capacity of San Francisco International Airport is expected to improve by using this approach to enable arrivals on both closely spaced parallel runways during low-visibility weather. However, the relaxation of in-trail wake separation standards awaits improved measurement and prediction of wake behavior.

Existing sensors and models do not adequately characterize wake decay phenomena, especially at typical final approach altitudes. Improved sensors, including coherent pulsed lidars, capable of directly measuring wake rotational momentum, are needed to support phenomenological studies and enable more accurate predictions of wake magnitude and decay in various atmospheric conditions. Those predictions, combined with models of aircraft upset risk, should allow reduced wake separation standards without degrading safety.

R&T Challenge A10 will conduct research to improve techniques for predicting and measuring the formation, trajectory, and decay of vortices, including methods to accurately predict wingtip vortex formation and define changes in aircraft design to mitigate the strength of the vortices. This Challenge would complement that work by developing affordable new sensors, system technologies, and procedures to improve prediction and measurement of wake strength, location, motion, and aircraft upset risk in terminal and en route airspace. Together, Challenges A10 and E4 will enable safe flight with reduced in-trail wake separation.

E5 Interfaces that ensure effective information sharing and coordination among ground-based and airborne human and machine agents

The potential for sharing a wide range of information within the air transportation system raises additional questions about how multiple agents (pilots, controllers, other system users, and automated system elements) can coordinate and share information given their disparate viewpoints and contexts. For information sharing to be effective, information must be provided to the right agents, at the right time, and in a fashion that facilitates accurate interpretation regardless of the source of the information. Some of the shared information may be factual (e.g., aircraft position, speed, heading, altitude, and flight plan), while some of it may be less tangible (e.g., potential responses to disruptions). The information elements will also likely vary in their timeliness and accuracy, and access to some information will be restricted for security and business reasons. Developing appropriate interfaces (in terms of information-sharing protocols, as well as display and visualization technology) is a nontrivial challenge, because agents can be easily overwhelmed by too much information or by the need to translate and analyze the information relative to their own situation and goals (Woods et al., 2002). Interfaces for human agents, in particular, will need to include methods for visualizing and interpreting operational situations to facilitate effective judgments and decisions. In addition, information sharing and decision-making processes will often be conducted collaboratively by multiple agents. Therefore, they will require knowledge of both individual human cognition and of collaborative work among agents with potentially conflicting goals and different representations of the immediate situation (Brennan, 1998; Olson et al., 2001). Information-sharing protocols become exceptionally critical during crises, such as 9/11, when control of the national airspace was transferred to the military. Communications and decision-making protocols were fragmented. Research related to this Challenge must be coordinated with DoD and DHS to avoid a recurrence of such problems. The Challenge should also capitalize on technologies pioneered in the telecommunications industry that would facilitate the transfer of diverse information through dynamically reconfigured networks using thousands of disparate nodes.

E6 Vulnerability analysis as an integral element in the architecture design and simulations of the air transportation system

More than three-fourths of air transportation system delays are weather related (Meyer, 2005). Snow or thunderstorms at major hub airports often significantly reduce overall system capacity and efficiency. Abnormal en route winds cause unexpected peaking and depeaking at arrival gateways. En route convective weather causes disruptive and unpredictable rerouting, precipitating en route delays and reducing capacity and efficiency. Disruptions can also be caused by natural disasters (such as volcanoes, hurricanes, tornadoes, and wildfires), electronic attacks (such as power outages, hurricanes, GPS spoofing, spurious communication messages, and hacking into navigation aids), and physical attacks (such as destruction of control facilities and radars). The effects of these disruptions may be local, regional, or national. In all cases, system capacity and efficiency are directly affected, and, more important, the safety of the air transportation system may be compromised by an inadequate response.

Airlines use a variety of techniques to respond to such disruptions. Some reduce schedule to preposition aircraft for the recovery, when the weather abates; others try to fly their full schedule, hoping that the recovery will take care of itself.

[12]Reduced Vertical Separation Minima apply to the airspace from flight levels 290 to 410 (which is equivalent to altitudes of approximately 29,000 feet to 41,000 feet) and create twice as many usable flight levels, decreasing the vertical separation between aircraft from 2,000 feet to 1,000 feet. While increasing capacity, this also could exacerbate the effects of wake turbulence.

Assessing vulnerabilities and risks should be the first step in reducing the likelihood and consequences of unplanned system disruptions (Volpe, 2003, p. 4). System safety impacts of disruptions should be evaluated early in the development cycle of new ATM system architectures, operating concepts, and system components. An agile ATM design should include provisions to counter or recover from system disruptions, and the design of the overall air transportation system should be evaluated by research and simulation to develop both system design concepts and/or operational procedures. In addition, quantitative analyses should be used to assess the safety impact of system architecture options. This Challenge would introduce vulnerability analyses as an integral element in the architecture design and simulations of the air transportation system to reduce the likelihood that the system will experience major system disruptions, to mitigate the severity of specific system disruptions, and to facilitate recovery from system disruptions. The result would be an air transportation system that is self-diagnosing and self-healing.

E7 Adaptive ATM techniques to minimize the impact of weather by taking better advantage of improved probabilistic forecasts

Adaptive traffic flow management methods are needed to take advantage of recent improvements in automated aviation weather forecasts. About 70 percent of aviation delay is due to operationally significant weather, including thunderstorms, low ceilings and visibilities, high winds, and turbulence. Exploitation of weather data collected from ground sensors and satellites using advanced image processing and machine intelligence has enabled significant improvements in aviation weather forecasts. One- to two-hour storm motion products are now being routinely displayed in key airport and en route air traffic facilities and in airline dispatch centers. Included are automatically updated estimates of the forecast accuracy, expressed as a probability (Robinson et al., 2004). This information is beginning to be used by air traffic managers and dispatchers, but only manually (Wolfson et al., 2004).

Algorithms are needed that automatically translate the weather forecasts into actionable traffic flow recommendations, with the goal of fully incorporating the weather data into air traffic automation designs. A few examples of automation that translate probabilistic weather forecasts into traffic flow recommendations have been developed, and FAA air traffic managers have shown they can reduce delays. For example, the LaGuardia Airport traffic flow managers are using storm motion forecast tools, such as the Route Advisory Planning Tool, to automatically identify safe departure routes (Evans, 2006). However, many automation systems are not incorporating the new weather information into their designs. This Challenge would demonstrate the use of automated weather forecasts in making traffic flow decisions and determine where this capability is cost beneficial.

E8a Transparent and collaborative decision support systems

Air traffic operations are enhanced by effective decision support systems that assist pilots, controllers, traffic flow managers, and airline personnel in tasks such as routing, flight planning, scheduling, and traffic separation. These decision support systems contribute to safe and efficient operations by using technology to enhance human capabilities and collaborate with the operator, as opposed to fully automated systems, which use technology rather than an operator to perform tasks. Collaborative decision support systems are most effective when the operators understand the basis for and limitations in the system's reasoning process and can judge the appropriateness of system-generated recommendations. Similarly, the system's recommendations should take into account operators' knowledge and intentions as well as the context in which they operate. Support for reciprocal information sharing and mutual understanding of intentions and actions—a process called grounding—is critical to avoid breakdowns in human–machine collaboration and overall system performance (Sorkin et al., 1988; Lee and Moray, 1994; Smith et al., 2001; McGuirl and Sarter, 2006). This Challenge will identify the type of information to be shared between human operators and automated decision support systems and develop candidate designs for these systems.

E8b Using operational and maintenance data to assess leading indicators of safety

Safety analysis is often a reactive, ad hoc process made difficult, in part, by the very high level of safety required of air transportation in the United States. Few unambiguous data points (accidents) are available for analysis, the number of data points continues to decrease because of the success of ongoing safety efforts, and accidents that do occur are increasingly the result of a complex chain of unlikely circumstances, each of them benign (Leiden et al., 2001). While human error is often cited as a major safety concern, successful human performance is also a major (and underreported) contributor to system safety. Thus, a particular concern for safety analysis is the human contribution to safety, especially when predicting the safety impact of dramatic changes to the role of human operators and increased reliance on automation. Likewise, safety analysis must consider individual aircraft as well as systemwide safety, which involves complex interactions among many agents. Using a common set of safety metrics (see R&T Challenge E16), this Challenge would develop methods both for monitoring the current system through ongoing analysis of operational and maintenance data and for predicting potential safety problems associated with proposed changes to the air transportation system.

E8c Interfaces and procedures that support human operators in effective task and attention management

The expected growth in air transportation demand will likely require operators to perform a wider range of tasks and to collaborate more closely with one another and with modern technologies. Pilots may begin to play a more active role in traffic separation or spacing and will need to coordinate their activities and intentions with other pilots and controllers. They will need to interact and exchange information, often interrupting each other and creating new tasks for one another. In general, more information will need to be distributed in a timely manner, task sets will increase, interruptions will become more likely, and the tolerance for delayed action or intervention will probably be reduced. It will be critical to ensure that operators are supported in properly scheduling and prioritizing their tasks, to improve attention management and avoid errors caused by unnecessary task switching, unnecessary interruptions, or inappropriate dismissals of demands (i.e., the failure to switch attention when appropriate and necessary) (Woods, 1995; McFarlane and Latorella, 2002; Ho et al., 2004).

High-Priority R&T Challenges and Their Associated Thrusts

Only one R&T Challenge in this Area (E1) had major relevance and impact on the energy and environment Strategic Objective. The other nine high-priority R&T Challenges in this Area each had major relevance and impact on at least two of the other three highly weighted strategic objectives (capacity, safety and reliability, and efficiency and performance).

In some cases, the R&T Challenges in this Area would have direct operational impact, for example, by developing new concepts and methods of separating, spacing, and sequencing aircraft to increase capacity and safety in all flight conditions (E2). In other cases, the benefits would be less direct, for example, by developing more capable methodologies, tools, and simulation and modeling capabilities to design and evaluate complex interactive systems (E1) and determining appropriate roles of humans and automated systems for separation assurance in high-density airspace during nominal and off-nominal operations (E3). Although the results of research in these areas would take longer to produce operational benefits, the research is essential, it is appropriate for NASA to include the research in a 10-year plan, and NASA involvement in the research is necessary to ensure that this research moves forward and can be readily applied to the air transportation system.

The ultimate objective of NASA's aeronautics research as it relates to intelligent and autonomous systems, operations and decision making, human integrated systems, and networking and communications is to provide the fundamental capabilities required for an adaptive and robust air transportation system that meets the nation's goals for economic growth, public well-being, and national security. Because the air transportation system is a complex interactive system, the linkages among its component systems are just as important as the component systems themselves. The committee identified four R&T Thrusts that describe threads of commonality among the Challenges in the intelligent and autonomous systems, operations and decision making, human integrated systems, and networking and communications Area:

- Decision making, negotiation, collaboration, information sharing, and allocation of airspace resources.
- Aircraft separation, spacing, and sequencing.
- Simulation, modeling, and analysis of complex, adaptive distributed systems.
- Wake and weather sensing, modeling and prediction, and other enabling air transportation technologies.

Each of these thrusts is discussed below. As shown in Figure 3-8, the first two Thrusts would lead directly to improvements in air transportation system operations, and the last two Thrusts would provide enabling technologies and capabilities that support the first two.

Decision making, negotiation, collaboration, information sharing, and allocation of airspace resources

Mechanisms must be constructed to facilitate and structure the interactions of all air transportation system components—businesses, organizations, individual humans, technologies—such that the emergent system performance is adequate. To do so requires foundational research into several topics, including the appropriate roles of automation; methods of supporting effective decision making and task management by individual humans, by automated systems, and by humans and automation working together; and information sharing, negotiation, and coordination within and between organizations. One outcome of these functions of particular importance to the design of the Next Generation Air Transportation System is allocation of airspace and air port resources, insofar as the method used to allocate these resources can determine how demand relates to capacity and whether airports and flights experience outcomes such as delays or denial of service. The following R&T Challenges are most closely related to this Thrust:

E5, Interfaces that ensure effective information sharing and coordination among ground-based and airborne human and machine agents.

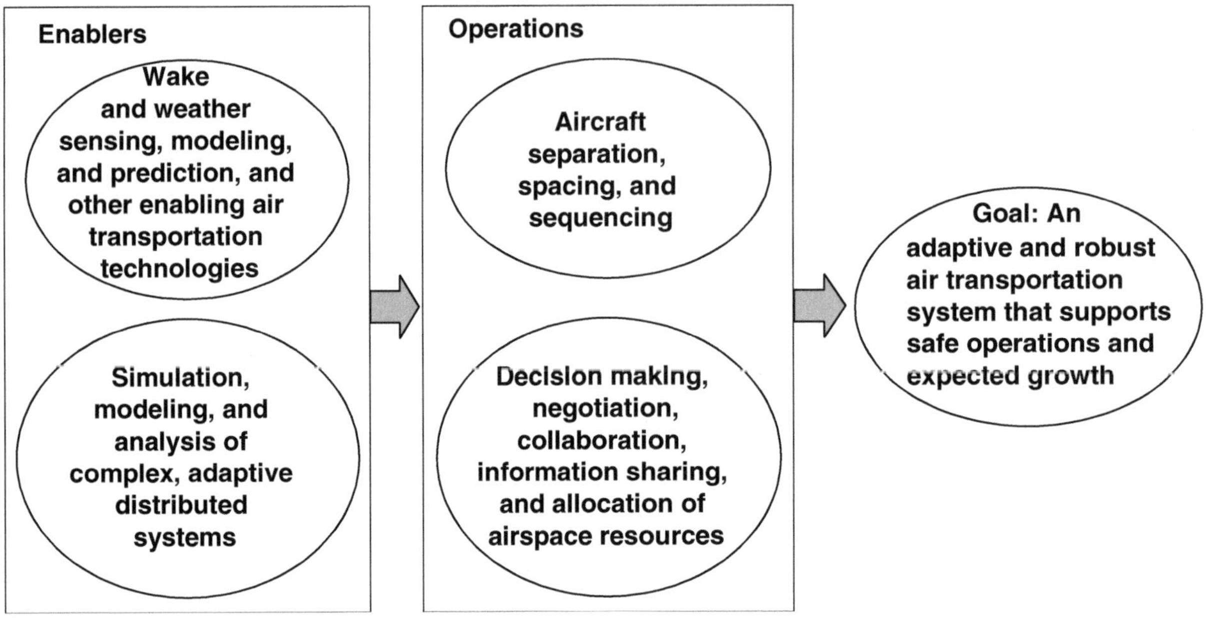

FIGURE 3-8 R&T Thrusts related to Area E: intelligent and autonomous systems, operations and decision making, human integrated systems, and networking and communications.

E8a, Transparent and collaborative decision support systems.

E8c, Interfaces and procedures that support human operators in effective task and attention management.

E11, Automated systems and dynamic strategies to facilitate allocation of airspace and airport resources.

E18, Certifiable information-sharing protocols that enable exchange of contextual information and coordination of intent and activity among automated systems.

Aircraft separation, spacing, and sequencing

The high-density airspace of the future will require effective management of aircraft separation, spacing, and sequencing in all flight conditions. The future air transportation system must develop improved models and novel operational concepts to support capacity growth and accommodate scheduled and unscheduled operations in airspace shared by manned and unmanned aircraft without compromising safety. Of particular importance to this research are separation assurance methods and understanding the appropriate roles of humans and automation in high-capacity airspace. Research in several R&T Challenges is needed to provide critical inputs, including accurate wake and weather forecasting as well as flight monitoring capabilities. The following Challenges are most closely related to this Thrust:

E2, New concepts and methods of separating, spacing, and sequencing aircraft.

E3, Appropriate roles of humans and automated systems for separation assurance, including the feasibility and merits of highly automated separation assurance systems.

Simulation, modeling, and analysis of complex, adaptive distributed systems

Design of the Next Generation Air Transportation System is a tremendous engineering challenge. This network of safety-critical, complex interactive systems will be vast in scope and involve multiple disparate organizations with separate objectives and capabilities. Examining the challenges facing this development, the committee found that individual technologies and systems will contribute to system performance only indirectly through their influence on how the larger system is collectively operated by the many user organizations; thus, system operations must also be a focus of research and development. Understanding the complexities of these operations requires new design tools and methodologies. Metrics are also important, because system performance metrics have a direct impact on the design of the system: Parameters that are not measured—or are measured incorrectly or incompletely—will not be fully considered or accounted for in the final design. Thus, it is critically important that the appropriate metrics be identified and incorporated into system analysis and design tools and processes. However, there is no comprehensive, widely held set of metrics to analyze and design the current and future air transportation system. Because many issues (e.g., economic, efficiency, safety, environment) must be addressed simultaneously, the problem cannot be decomposed into isolated examinations of capacity, safety, technology, human factors, etc. New simulation tools are needed with extensive predictive capabilities. Likewise, new analysis methods are required to integrate safety and vulnerability assessments into the design process. Coordinating all of these insights requires understanding of complex, adaptive distributed systems with the unique characteristics of air transportation, including the need to simulate and predict the behavior of radically different system configurations.

The following R&T Challenges are most closely related to this Thrust:

E1, Methodologies, tools, and simulation and modeling capabilities to design and evaluate complex interactive systems.

E6, Vulnerability analysis as an integral element in the architecture design and simulations of the air transportation system.

E8b, Using operational and maintenance data to assess leading indicators of safety.

E16, Appropriate metrics to facilitate analysis and design of the current and future air transportation system and operating concepts.

E17, Change management techniques applicable to the U.S. air transportation system.

E20, Comprehensive models and standards for designing and certifying aviation networking and communications systems.

Wake and weather sensing, modeling and prediction, and other enabling air transportation technologies

Several critical technologies warrant fundamental research by NASA for their likely value as enablers of many possible operational concepts, and because some are still in a nascent state. Of particular importance are the sensing, modeling, and prediction of aircraft wakes and hazardous weather. Other enabling technologies center on the creation of enhanced automation capabilities for safety (e.g., "refuse-to-crash") and for capacity (e.g., agents for negotiating resource allocation) beyond those that the research community currently knows how to develop and certify as robust in a wide range of conditions. Additional technologies focus on further enhancements to the communication and navigation needs of air traffic management functions. The following R&T Challenges are most closely related to this Thrust:

E4, Affordable new sensors, system technologies, and procedures to improve the prediction and measurement of wake turbulence.

E7, Adaptive ATM techniques to minimize the impact of weather by taking better advantage of improved probabilistic forecasts.

E12, Autonomous flight monitoring of manned and unmanned aircraft.

E13, Feasibility of deploying an affordable broad-area, precision-navigation capability compatible with international standards.

E14, Advanced spacecraft weather imagery and aircraft data for more accurate forecasts.

E15, Technologies to enable refuse-to-crash and emergency autoland systems.

E19, Provably correct protocols for fault-tolerant aviation communications systems.

Low-Priority R&T Challenges

Seven of the 10 R&T Challenges that did not rank in the top 10 by NASA priority also ranked low in national priority. Which is to say, they had substantial impact on only one of the highly weighted strategic objectives (capacity, safety and reliability, efficiency and performance, or energy and the environment). The other three R&T Challenges (E11, E13, and E17) would have substantial impact on two of the highly weighted strategic objectives, but they ranked low in terms of NASA priority because of low Why NASA? scores. Because these Challenges rank high in national priority, it is important that some part of the national civil aeronautics R&T effort (by NASA, other government agencies, industry, or academia) support work to overcome them.

E17 Change management techniques applicable to the U.S. air transportation system

The current ATM airspace architecture and associated procedures are antiquated and so reliant on interim fixes that there is significant resistance to additional changes. A novel and consistent approach is needed to manage changes to the ATM system and to overcome barriers and organizational inertia within the FAA and other stakeholders. The FAA should support research related to this Challenge, though it may take external pressure (e.g., from the Office of Management and Budget, the Government Accountability Office, and the White House) to prompt action. It may be appropriate for NASA to also conduct research related to this Challenge, but NASA would first need to increase its relevant capabilities.

E11 Automated systems and dynamic strategies to facilitate allocation of airspace and airport resources

The current allocation of airspace and airport resources (e.g., airport departure and arrival slots) at major airports is heavily biased toward airline operations. The competition for airspace and airport resources will be exacerbated by growth in commercial and private air travel, including the introduction of very light jets. Future air transportation systems would benefit from built-in automatic response systems driven by software agents that quickly negotiate and make decisions regarding, for example, real-time allocation of airspace and landing slots among aircraft with diverse size and performance characteristics, while considering the needs of all stakeholders, air transportation system efficiency, and energy conservation (Cramton et al., 2002). This Challenge ranked low in NASA priority because the FAA already is funding some related research and is best suited to address this Challenge.

E13 Feasibility of deploying an affordable broad-area, precision-navigation capability compatible with international standards

Many small airports cannot operate when visibility is restricted because they do not have the equipment (e.g., a Category I, Category II, or Category III instrument landing system) necessary for a safe approach and landing.[13] The number of ILS frequencies available in large metropolitan areas, where multiple runways require precision approach and landing capabilities, is limited. Increased access to these airports and runways would increase efficiency significantly, particularly for some segments of the aviation industry (e.g., feeder airlines, business aircraft, and air taxis). While satellite navigation systems are currently deployed by the United States and others, they do not provide sufficient coverage, accuracy, and reliability for aviation requirements to replace or substitute for ground-based aids, particularly as regards precision landing guidance provided by ILS (Shively and Hsaio, 2005). The objective here is to design, develop, and deploy a space-based navigation system augmentation that complements GPS, Galileo, and the Global Navigation Satellite System (GLONASS), so that guidance equivalent to Category IIIc ILS is universally available and no additional ground-based capability (e.g., pseudolites) needs to be installed. Deployment of this capability would open up any airport or temporary landing area for all-weather operations with no need for expensive and time-consuming construction; facilitate reconfiguration of approach paths; and improve safety. It would allow many existing runways and small airports not equipped with ILS to operate in low visibility conditions and would support emergency operations and homeland security. It would have the added benefit of allowing the FAA to remove thousands of existing en route and terminal navigation aids, such as very high frequency (VHF) omnidirectional range (VOR) equipment, distance measurement equipment (DME), tactical air navigation (TACAN) systems, ILSs, and nondirectional beacons (NDB). Decommissioning these systems would eliminate associated maintenance and operating costs. This Challenge ranked low in NASA priority primarily because the feasibility of deploying an affordable, broad-area precision navigation capability will be determined by technical, economic, and regulatory issues. The technical feasibility issues are well aligned with NASA's aeronautics mission, but economic feasibility issues are better handled by industry, and regulatory feasibility issues are better handled by the FAA.

[13]An ILS is a ground-based precision approach system that provides course and altitude guidance to pilots as they prepare to land. ILS systems are rated according to their capabilities:

A Category I system can provide guidance regarding course and glide slope down to an altitude of 200 feet with a runway visual range of not less than 1,800 (or 2,400 feet depending on runway lighting and configuration).

A Category II system can provide guidance regarding course and glide slope down to an altitude of 100 feet with a runway visual range of not less than 1,200 feet.

A Category IIIa system can provide guidance regarding course and glide slope all the way to touchdown as long as the pilot has some external visual reference during the final phase of landing and the runway visual range is not less than 700 feet.

A Category IIIb system can provide guidance regarding course and glide slope all the way to touchdown even without any external visual references, as long as the runway visual range is not less than 150 feet (for taxi operations after landing).

A Category IIIc system can provide guidance regarding course and glide slope all the way to touchdown and during taxi operations without any external visual references and with zero visibility.

REFERENCES

Advisory Council for Aeronautical Research in Europe (ACARE), 2001. European Aeronautics—A Vision for 2020. Available at http://www.acare4europe.com/docs/Vision%2020.pdf.

Akin, Gump, Strauss, Hauer, and Feld, LLP. 2003. Airport Air Quality Guide—A Guide to Air Quality Issues Under the Clean Air Act.

Ames, B. 2005. Power electronics drive next-generation vehicles. Military and Aerospace Electronics, July 1. Available online at <http://mae.pennnet.com/Articles/Article_Display.cfm?ARTICLE_ID=232549&p=32& pc=ENL/>.

Andrews, J., H. Erzberger, and J. Welch. 2005. Safety Analysis for Advanced Separation Concepts. 6th USA/Europe Seminar on Air Traffic Management Research and Development, Baltimore, Md., June 27-30. Available online at <www.eurocontrol.int/eec/public/standard_page/EEC_News_2005_2_Seminar_Best.html>.

Bokadia, S., and J. Valasek. 2001. Severe Weather Avoidance Using Informed Heuristic Search. AIAA-2001-4232, AIAA Guidance, Navigation, and Control Conference, Montreal, Canada, August 6-9.

Brennan, S.E. 1998. The grounding problem in conversations with and through computers. In S.R. Fussel and R.J. Kreuz (eds.), Social and Cognitive Psychological Approaches to Interpersonal Communication. Hillsdale, N.J.: Lawrence Erlbaum.

Calkins, F.T., and G.W. Butler. 2004. Subsonic Jet Noise Reduction Variable Geometry Chevron. AIAA 2004-190, 42nd AIAA Aerospace Sciences Meeting and Exhibit, Reno, Nev., January 5-8.

Chavez, F.R., and D.K. Schmidt. 1994. Analytical aeropropulsive-aeroelastic hypersonic-vehicle model with dynamic analysis. Journal of Guidance, Control, and Dynamics 17(6): 1308-1319.

Code of Federal Regulations, Title 14. Part 91.

Cramton, P., L.M. Ausubel, and P. Milgrom. 2002. Comments on Alternative Policy Options for Managing Capacity and Mitigating Congestion and Delay at LaGuardia Airport before the U.S. Department of Transportation, Washington, D.C., Dockets FAA-2001-9852 and FAA-2001-9854, June 20. Bethesda, Md.: Market Design, Inc. Available online at <www.cramton.umd.edu/papers2000-2004/mdi-comment-to-faa-on-managing-capacity-at-lga.pdf>.

Ding, Y., J. Rong, and J. Valasek. 2004. Feasibility Analysis of Aircraft Landing Scheduling for Non-Controlled Airports. AIAA-2004-5241, Proceedings of the AIAA Guidance, Navigation, and Control Conference, Providence, R.I., August 16-19.

Doebbler, J., P. Gesting, and J. Valasek. 2005. Real-Time Path Planning and Terrain Obstacle Avoidance for Aircraft. AIAA-2005-5825, Proceedings of the AIAA Guidance, Navigation, and Control Conference, San Francisco, Calif., August 15-18.

Evans, J.E. 2006. Implications of a Successful Benefits Demonstration for Integrated Weather/Air Traffic Management (WX/ATM) System Development and Testing. 12th American Meteorological Society Conference on Aviation, Range and Aerospace Meteorology, Atlanta, Ga., January 29-February 2.

Federal Aviation Administration (FAA). 2000. Federal Aviation Administration Office of System Capacity, Aviation Capacity Enhancement Plan. Washington, D.C.: FAA. Available online at <www.faa.gov/ats/asc/00ACE.html>.

Garg, S. 2005. NASA Glenn Research in Controls and Diagnostics for Intelligent Aerospace Propulsion Systems. NASA/TM—2005-214036. Cleveland, Ohio: Glenn Research Center.

Giesing, J., and J.M. Barthelemy. 1998. A Summary of Industry MDO Applications and Needs. AIAA-1998-4737, 7th AIAA/USAF/NASA/ISSMO Symposium on Multidisciplinary Analysis and Optimization, St. Louis, Mo., September 2-4. Available online at <endo.sandia.gov/AIAA_MDOTC/sponsored/summary_paper.pdf>.

Glauert, M.B. 1945. The design of suction aerofoils with very large C_L-range. Aeronautics Research Council Reports and Memoranda 2111, National Research Establishment, November.

Glauert, M.B., W.S. Walker, W.G. Raymer, and N. Gregory. 1948. Wind-tunnel tests on a thick suction aerofoil with a single slot. Aeronautics Research Council Reports and Memoranda 2646, National Research Establishment, October.

Helbing, K., T. Spaeth, and J. Valasek. Forthcoming. Improving aircraft sequencing and separation at a small aircraft transportation system airport. Journal of Aircraft.

Ho, C-Y., M. Nikolic, M. Waters, and N.B. Sarter. 2004. Not now: Supporting interruption management by indicating the modality and urgency of pending tasks. Human Factors 46(3): 399-409.

Intergovernmental Panel on Climate Change (IPCC). 1999. Special Report: Aviation and the Global Atmosphere. Geneva: World Meteorological Organization.

Jaw, L.C., and S. Garg. 2005. Propulsion Control Technology Development in the United States: A Historical Perspective. NASA/TM-2005-213978, October.

Johnson, W., G. Yamauchi, and M. Watts. 2006. Design and Technology Requirements for Civil Heavy Lift Rotorcraft. Proceedings of the American Helicopter Society Vertical Lift Aircraft Design Conference, San Francisco, Calif., January 18-20.

Kelly, W., J. Valasek, D.Wilt, J. Deaton, K. Alter, and R. Davis. 2005. The Design and Evaluation of a Traffic Situation Display for a SATS Self Controlled Area. Proceedings of the 24th Digital Avionics Systems Conference, Washington, D.C., October 30-November 3.

Koff, B.L. 2004. Gas turbine technology evolution: A designer's perspective. Journal of Propulsion and Power 20: 577-595.

Lampton, A., and J. Valasek. 2005. Prediction of Icing Effects on the Stability and Control of Light Airplanes. AIAA-2005-6219, Proceedings of the AIAA Atmospheric Flight Mechanics Conference, San Francisco, Calif., August 15-18.

Lampton, A., and J. Valasek. 2006. Prediction of Icing Effects on the Lateral/Directional Stability and Control of Light Airplanes. AIAA-2006-6487, Proceedings of the AIAA Atmospheric Flight Mechanics Conference, Keystone, Colo., August 21-24.

Lee, J., and N. Moray. 1994. Trust, self-confidence, and operators' adaptation to automation. International Journal of Human-Computer Studies 40: 153-184.

Lefebvre, A. 1999. Gas Turbine Combustion, Philadelphia, Pa.: Taylor and Francis.

Leiden, K., J. Keller, and J. French. 2001. Context of Human Error in Commercial Aviation. Boulder, Colo.: Micro Analysis and Design, Inc. Available online at <http://human-factors.arc.nasa.gov/ihi/hcsl/HPM_pubs/AvHumanError.pdf>.

Litt, J.S., D.L. Simon, S. Garg, T. Guo, C. Mercer, R. Millar, A. Behbahani, A. Bajwa, and D.T. Jensen. 2005. A Survey of Intelligent Control and Health Management Technologies for Aircraft Propulsion Systems. NASA/TM-2005-213622, ARL-TR-3413, May.

McFarlane, D.C., and K.A. Latorella. 2002. The scope and importance of human interruption in human-computer interaction design. Human-Computer Interaction 17(1): 1-61.

McGuirl, J., and N.B. Sarter. 2006. Supporting Trust Calibration and the Effective Use of Decision Aids by Presenting Dynamic System Confidence Information. In press.

McLain, T., and R. Beard. 2005. Coordination variables, coordination functions, and cooperative timing missions. AIAA Journal of Guidance, Control, and Dynamics 28(1): 150-161.

Meyer, D. 2005. Corridor Integrated Weather System. Presentation to the National Business Aviation Association Friends/Partners in Aviation Weather Forum, Orlando, Fla., November 11. Available online at <www.ral.ucar.edu/general/fpaw2005/4%20_Segment/meyer.pdf>.

Mundra, A.D. 2001. An Assessment of the Potential for Wake Vortex-Related Enhancements in NAS [national airspace system]. McLean, Va.: MITRE Center for Advanced Aviation System Development.

National Aeronautics and Space Administration (NASA). 2003. Aerospace Technology Enterprise Strategy, NASA Report NP-2003-01-298-HQ.

National Institute of Aerospace (NIA). 2005. Responding to the Call: Aviation Plan for American Leadership, Available online at <www.nianet.org/nianews/AviationPlan.php>.

National Research Council (NRC). 1998. Ceramic Fibers and Coatings: Advanced Materials for the 21st Century. Washington, D.C.: National Academy Press. Available online at <http://fermat.nap.edu/catalog/6042.html>.

NRC. 2003. Securing the Future of U.S. Air Transportation: A System in Peril. Washington, D.C.: The National Academies Press. Available online at <http://fermat.nap.edu/catalog/10815.html>.

NRC. 2004. Accelerating Technology Transition: Bridging the Valley of Death for Materials and Processes in Defense Systems. Washington, D.C.: The National Academies Press. Available online at <http://fermat.nap.edu/catalog/11108.html>.

Next Generation Air Transportation System Joint Planning and Development Office (NGATS JPDO). 2005. 2005 Progress Report to the Next Generation Air Transportation System Integrated Plan. Washington, D.C.: JPDO. Available online at <www.jpdo.aero/site_content/pdf/ngats-np_progress-report-2005.pdf>.

Olson, G.M., T.W. Malone, and J.B. Smith. 2001. Coordination Theory and Collaboration Technology. Hillsdale, N.J.: Lawrence Erlbaum.

Pawlowski, J., D. Graham, C. Boccadoro, P. Coen, and D. Maglieri. 2005. Origins and Overview of the Shaped Sonic Boom Program. AIAA-2005-0005, 43rd AIAA Aerospace Sciences Meeting and Exhibit, Reno, Nev., January 10-13.

Powrie, H.E.G., and C.E. Fisher. 1999. Engine health monitoring: Towards total prognostics. Proceedings of the 1999 IEEE Aerospace Conference 3: 11-20.

Reynolds, T.G., and R.J. Hansman. 2001. Analysis of Aircraft Separation Minima Using a Surveillance State Vector Approach. Updated version of paper from 3rd International Air Traffic Management R&D Seminar (ATM2000), Naples, Italy, June 13-16, 2000. Available online at <http://icat-server.mit.edu/library/fullRecord.cgi?idDoc=183>.

Robinson, M., J. Evans, B. Crowe, D. Klingle-Wilson, and S. Allan. 2004. Corridor Integrated Weather System Operational Benefits 2002-2003: Initial Estimates of Convective Weather Delay Reduction. Project Report ATC-313. Lexington, Mass.: MIT Lincoln Laboratory.

Rong, J., S. Geng, J. Valasek, and T. Ioerger. 2002. Air Traffic Conflict Negotiation and Resolution Using An Onboard Multi-Agent System. DASC-345, 21st Digital Avionics Systems Conference (DASC) on Air Traffic Management for Commercial and Military Systems, Irvine, Calif., October 22.

Sabatini, N.A. 2006. Testimony of the FAA Associate Administrator for Aviation Safety before the House Committee on Transportation and Infrastructure, Subcommittee on Aviation on Unmanned Aircraft Activities, March 29, 2006. Available online at <www.faa.gov/news/news_story.cfm?contentKey=4029>.

Schierman, J.D., D.G. Ward, J.R. Hull, N. Gandhi, M.W. Oppenheimer, and D.B. Doman. 2004. An approach to integrated adaptive guidance and control with flight test results. Journal of Guidance, Control and Dynamics 27(6): 975-988.

Shively C.A., and T.T. Hsaio. 2005. Error and availability analysis of category IIIb LAAS augmented by radar altimeter. Navigation 52(3).

Simon, D.L. 2000. An Overview of the NASA Aviation Safety Program Propulsion Health Monitoring Element. 36th Joint Propulsion Conference and Exhibit, Huntsville, Ala., July 16-19.

Smith, P.J., E. McCoy, and J. Orasanu. 2001. Distributed cooperative problem-solving in the air traffic management system. In G. Klein and E. Salas (eds.), Naturalistic Decision Making. Hillsdale, N.J.: Lawrence Erlbaum.

Sobieszczanski-Sobieski, J., and R.T. Haftka. 1997. Multidisciplinary aerospace design optimization: Survey of recent developments. Structural Optimization 14(1): 1-23.

Sorkin, R.D., B.H. Kantowitz, and S.C. Kantowitz. 1988. Likelihood alarm displays. Human Factors 30(4): 445-459.

Spalart, P.R. 1998. Airplane trailing vortices, Annual Review of Fluid Mechanics 30: 107-138.

Tandale, M.D., and J. Valasek. 2003. Structured Adaptive Model Inversion Control to Simultaneously Handle Actuator Failure and Actuator Saturation. AIAA-2003-5325, AIAA Guidance, Navigation, and Control Conference, Austin, Tex., August 11-14.

Tandale, M.D., J. Valasek, J. Doebbler, and A.J. Meade. 2005. Improved Adaptive-Reinforcement Learning Control for Morphing Unmanned Air Vehicles. AIAA-2005-7159, Proceedings of the AIAA Infotech@ Aerospace Conference, Arlington, Va., September 26-29.

Tandale, M.D., and J. Valasek. 2006. Fault tolerant structured model inversion control. Journal of Guidance, Control, and Dynamics 29(3).

Valasek, J., M.D. Tandale, and J. Rong. 2005. A reinforcement learning-adaptive control architecture for morphing. Journal of Aerospace Computing, Information, and Communication 2(5): 174-195.

Volpe National Transportation Systems Center. 2003. Risk assessment and prioritization. Volpe Journal 2003: Transportation and Security. Available online at <www.volpe.dot.gov/infosrc/journal/2003/index.html>.

Waitz, I., J. Townsend, J. Cutcher-Gershenfeld, E. Greitzer, and J. Kerrebrock. 2004. Aviation and the Environment; A National Vision Statement, Framework for Goals and Recommended Actions. Report of the Partnership for AiR Transportation Noise and Emissions Reduction (PARTNER) to the United States Congress, December. Available online at <http://partner.aero>.

Wickens, C.D., A.S. Mavor, R. Parasuraman, and J.P. McGee, eds. 1998. The Future of Air Traffic Control: Human Operators and Automation. Washington, D.C.: National Academy Press. Available online at <http://fermat.nap.edu/catalog/6018.html>.

Wolfson, M.M., B.E. Forman, K.T. Calden, W.J. Dupree, R.J. Johnson, R.A. Boldi, C.A. Wilson, P.E. Bieringer, E.B. Mann, and J.P. Morgan. 2004. Tactical 0-2 Hour Convective Weather Forecasts for the FAA. Proceedings of the 11th Conference on Aviation, Range and Aerospace Meteorology, Hyannis, Mass., October 4-8.

Woods, D.D. 1995. The alarm problem and directed attention in dynamic fault management. Ergonomics 38(11): 2371-2393.

Woods, D.D., and E. Hollnagel. 2006. Joint Cognitive Systems: Patterns in Cognitive Systems Engineering. New York, N.Y.: Taylor and Francis.

Woods, D.D., E.S. Patterson, and E.M. Roth. 2002. Can we ever escape from data overload? A cognitive systems diagnosis. Cognition, Technology, and Work 4(1): 22-36.

Xue, M., and E. Atkins. 2006. Noise-minimum runway-independent aircraft approach design for Baltimore-Washington International Airport. AIAA Journal of Aircraft 43(1): 39-51.

4

Common Themes and Key Barriers

COMMON THEMES

Chapter 3 describes R&T Thrusts, which are threads of commonality among the R&T Challenges identified by each panel within its own R&T Area. The steering committee also identified threads of commonality among the R&T Thrusts and the R&T Challenges from different R&T Areas and called them Common Themes:

- Physics-based analysis tools
- Multidisciplinary design tools
- Advanced configurations
- Intelligent and adaptive systems
- Complex interactive systems

These Themes are not an end in themselves; they are a means to an end. Each Theme describes enabling approaches that will contribute to overcoming multiple Challenges. Exploiting the synergies identified in each Common Theme will enable NASA's aeronautics program to make the most efficient use of available resources.

Physics-Based Analysis Tools

Description

Physics-based analysis tools attempt to predict the behavior of physical and/or chemical phenomena by solving the fundamental governing conservation, constitutive, and state equations together with appropriate closure equations based on first-principle physical models. In other words, "physics-based" refers to the general use of scientific principles in place of empirical data. The tools can be hierarchal in space and time, and the lowest order model (e.g., zero-dimensional steady state and two-dimensional unsteady state) that predicts a phenomenon to the accuracy desired should be employed. This Theme also includes models derived from other branches of science, such as chemistry, biology, materials science, computer and information science, and cognitive science, though many are not strictly physics-based. For complex problems such as three-dimensional, unsteady, heterogeneous flows, computational simulations that provide numerical solutions must generally be used.

Background

This Theme is particularly applicable to three R&T Areas examined in this study: aerodynamics and aeroacoustics, propulsion and power, and materials and structures. Physics-based analysis tools offer the opportunity to decrease significantly the use of empirical approaches in aeronautics R&T. Empiricism, as defined here, refers to the generation of information through cut-and-try experimentation and testing. It is not inherently bad as long as the results are integrated with models, lead to knowledge and understanding, and are not widely extrapolated beyond the ranges of the test parameters. To a great degree, enlightened empiricism was responsible for many of the aeronautical advances of the previous century. Empiricism, however, can be expensive and time consuming and may not lead to a fundamental understanding of phenomena. From a national perspective, empirical modeling and design can be an inefficient use of resources and may lead to compromised, nonoptimal designs that rely on unnecessarily large design margins.

An important benefit of advances in physics-based analysis tools is the new technology and systems frontiers they open. New concepts often emerge from a greater understanding of the underlying physics offered by new analytical capabilities. In these cases, experimentation might never lead to the level of insight offered by even relatively simplistic physics-based tools. An example of this is sonic-boom mitigation technology. It is highly unlikely that any practical amount of experimentation will lead to a design for a low-sonic-boom aircraft. The development of linear and nonlinear physics-based analysis tools is necessary to mature this technology.

With advances in computational speed, computing power, and the affordability of digital processors in the last two decades, aeronautics researchers in industry, academia, and the government have turned increasingly to computational simulations to model complex physical and chemical phenomena. Industry is motivated by the possibility of using computational simulations to reduce the cost and time of product development, while increasing product reliability. Academic and government researchers also value the ability to attack more complex problems. These computational simulations generally employ a number of physics-based models within the governing conservation and state equations. Examples include models that describe droplet behavior and interactions, particulate matter formation, turbulence, turbulence–chemistry interactions, boundary-layer growth and transition, fracture, crack propagation, and material phase boundaries. A lack of fundamental understanding often requires these models to contain adjustable parameters that are grossly calibrated to empirical data sets. These data sets are often incomplete, which means that untested assumptions must be incorporated in the models. The computational simulations are generally not validated in spatial detail except for comparison of code predictions to input and output measurements. Additionally, the adjustable parameters are tweaked to match predictions with measurements. It is not uncommon to find that the codes do not extrapolate well when the design space changes considerably, prompting more tweaking of the adjustable parameters. Also, when results are presented, details are usually omitted in connection with the use of boundary conditions or how adjustable parameters were set, making it harder for independent researchers to reproduce the results. Thus to a large extent, empiricism has transitioned from the physical to the computational realm, but it persists. Nevertheless, within their applicable ranges, computational simulations have enabled technical progress, as witnessed by the state of aeronautics today. Unfortunately, limits on the use of simulations are often not well understood.

Suggested approach

NASA and its academic and industrial partners can make very significant contributions in developing and validating physics-based analysis tools. These are readily assimilated by industry into their proprietary product design codes. NASA, industry, and academia can jointly participate in research into physics-based analysis tools because it is fundamental in nature, publishable, and sharable. This research will take time to mature, yet advances can readily be translated into practice as they occur. Furthermore, given the budget- and schedule-driven nature of the aerospace business, this is the type of work that industry can no longer afford to pursue. Developing physics-based tools whose accuracy and range of applicability limits are well established is a lengthy, iterative process. Validation requires well-designed experiments to elucidate the underlying physics as well as experimental facilities of appropriate scale and advanced, nonperturbing diagnostics to perform detailed, spatially and temporally resolved measurement of parameters.

Benefits of synergy

Advances in physics-based analysis tools would help address R&T Challenges in all of the R&T Areas. For example, turbulence modeling is a key element in the accurate prediction of mixing, which is very important in many aspects of aerodynamics (A2), aeroacoustics (A4a, B1a), and combustion processes (B1b). Accurate predictions of flow separation are a prerequisite to the successful design of both nonreacting (A2, A4a, A4b) and reacting (B1a, B1b, B5, B8) flow devices. Mathematical models of material properties and reactions are essential for structural response (A4a, C4, C10). Droplet–droplet and droplet–flow interactions are important processes in predicting both icing (A6) and combustion behavior (B1b, B8, B10). Modeling flow–structure interactions accurately is an important element of aeroelasticity and noise generation (A4a, B1a, C5, C6b). The development of higher-temperature alloys is key to improving propulsion system fuel efficiency (B4, C6a). Many of the computational science issues associated with large, complex computational simulations, such as automated grid generation, parallelizing codes, and error propagation analyses, are common elements across several R&T Areas.

Relevant R&T Challenges

The following R&T Challenges would benefit significantly from using physics-based analysis tools: A1, B1a, B1b, E1, A2, E2, D3, A4a, A4b, B4, C4, D4, E4, B5, C5, A6, B6a, C6a, C6b, A7b, B8, A9, B10, and C10.

Multidisciplinary Design Tools

Description

Discipline-specific design tools, including optimization and inverse design, have improved the performance of airfoils, wings, structures, control systems, and propulsion systems for many years, and they are now critical parts of the design process. The next step in the design of more complex systems involves more than just combining these disciplinary tools or gluing together discipline-specific analyses and optimization. New multidisciplinary tools are needed to integrate high-fidelity analyses with efficient design methods and to accommodate uncertainty, multiple objectives, and large-scale systems. Research in efficient methods for including large numbers of design variables (e.g., adjoint methods, multifidelity models, and surrogate models), probabilistic design methods, and tools for distributed, complex systems is particularly important to the development of future aeronautical systems.

Background

Methods for simulation-based multidisciplinary design and optimization (MDO) are at the very core of a philosophy that moves away from empirical methods that have proven to be expensive and often have not met expectations in exploring the aeronautical design space. MDO processes bring together people, analytical tools, experimentation, and information to design complex structural components and systems.

The design of aeronautical systems requires a system-level, multidisciplinary approach to assessing potential costs, benefits, and risks. Design tools that couple a small number of disciplines in a restricted design space have reached a level of maturity and fidelity that make them important parts of the design process. For example, aeroelastic design tools are now within reach that couple computational fluid dynamics, multibody dynamics, and finite-element analyses for full aircraft configurations. In structural designs where the topology or outer mold lines are defined, analytical methods such as the structural finite-element technique, coupled with similar analytical tools for load assessment, promise success. More recently, high-fidelity multidisciplinary design tools have begun to incorporate a broader range of disciplines and are starting to be used earlier in the design process.

However, for designs with a multiplicity of topologies, some of which are not well defined, and for problems where a large number of design parameters and constraints must be considered in the early stages of the design process, the multidisciplinary design process is still underdeveloped. One of the major limitations of past efforts to create MDO tools has been a low level of fidelity, driven by a lack of physics-based models that are efficient and appropriate for system-level design. In addition, MDO tools have often lacked flexibility and have been developed and applied for very specific applications.

Suggested approach

Significant new developments are required in both the design strategies and the embedded tools that constitute the multidisciplinary design process. Key issues associated with next-generation multidisciplinary design tools include fidelity, computational efficiency, and the ability to handle uncertainty.

Efficiency and effectiveness continue to be a problem, particularly in large-dimensionality problems and multimodal or disjointed search spaces. Most current approaches to design are ill-equipped to deal with the vast amounts of data associated with the design process. High numerical efficiency is paramount for multidisciplinary design tools, and alternative paradigms that take advantage of a new generation of parallel computational hardware must be sought. Furthermore, not all methods are ideal for all problems. The goal of this Theme is not to generate one perfect, all-encompassing algorithm but to use the most efficient and effective method or combination of methods for each problem. Proper algorithm selection in itself is an important research topic.

Uncertainty modeling in a data-lean environment, which is often the case with new concepts, continues to be an issue, particularly in situations where uncertainty distributions do not conform to standard forms or where components or elements exhibit discrete behavior. The propagation of uncertainty in complex and highly coupled multidisciplinary systems needs to be modeled, and tools for design and optimization in a nondeterministic environment continue to be computationally intractable, especially when applied to design problems involving a large number of nondeterministic variables, parameters, and design constraints.

Methods for distributed design (where the design team is geographically dispersed) and for the design of large-scale distributed systems have achieved some success but have been restricted to special types of problems. Continued development of more general, scalable approaches to this problem is also critical for the design of complex systems.

The practical resolution of these issues will require fundamental research efforts in the development of design-oriented, physics-based models; new design methodologies that can seamlessly manage models of multiple fidelities for the various components of the system; methods to increase the computational efficiency of tools; methods to handle complex interactions with high accuracy; and methods to manage uncertainty in the design process.

Benefits of synergy

Multidisciplinary design processes develop synergistic benefits by integrating people, analytical tools, experimentation, and information to design complex components and systems. Their importance is reflected in the relevant R&T Challenges listed below.

Relevant R&T Challenges

Many Challenges in each R&T Area rely on improved multidisciplinary design tools. These include Challenges A1, A2, A3, A4a, B1a, B1b, B4, B8, C6b, D3, and D4. In addition, many R&T Challenges identify multidisciplinary design tools and design under uncertainty as core technologies, including A11, C3, D2, and E1.

Advanced Configurations

Description

Advanced configurations embody innovative, outside-the-box approaches to better meet the strategic objectives outlined in this report. They serve as technologies in themselves when they represent advancements in system-level definitions beyond those possible with conventional design tools, methods, and expertise. Examples of advanced configuration technologies include revolutionary aircraft concepts and advanced structural designs.

Background

Integration of innovative technologies into advanced configurations has long been a part of U.S. aeronautical development. For example, the Bell X-1A demonstrated supersonic flight, thus pushing the envelope beyond what was once thought to be an impassable barrier. Other advanced system configurations, such as the X-15, X-29, and X-35, have demonstrated multiple advanced technologies. Other examples, such as the Gossamer-Condor, Voyager, and Helios aircraft, demonstrated advanced vehicle configurations that were groundbreaking innovations and went beyond validated analytical methodologies.

Suggested approach

Creativity and good ideas have been at the center of revolutionary advances in aeronautics. One of NASA's roles is to foster the implementation of innovative solutions to challenging technological barriers. The development of innovative concepts needs to include freedom to innovate as well as physics and engineering checks. It is implicit that available design tools (from component-level, physics-based tools to MDO models) and empirical knowledge will be used to screen concepts before new tools and models are created.

Innovation is not possible, however, without tolerance for failure. The progression from technology identification, maturation, and demonstration to implementation is rarely linear. Advanced configurations, regardless of "success," form the basis for validating component-level and system-level physics-based models and MDO approaches. For example, the development programs for the SR-71 and NASA's XV-15 tilt-rotor had technology problems, but both produced functional aircraft. Even today, technical gaps persist in modeling the high-speed flight aerodynamics and combustion processes of the SR-71. However, by having a good balance between innovation, tolerance of failure, sound technical knowledge and judgment, and engineering analysis, the SR-71 came to fruition and became the fastest and highest-flying production aircraft ever built. The XV-15 demonstrated V/STOL capabilities, and programs such as the V-22 Osprey and the Bell/Agusta BA609 civil tilt-rotor aircraft have greatly benefited from and advanced the concepts demonstrated by the XV-15. Design and testing of advanced configurations should continue to have an important presence in civil aeronautics R&T.

Another aspect of innovation on advanced configurations is the process of integration itself. Oftentimes, outside-the-box thinking is needed to seamlessly integrate technologies that have been optimized individually but not yet integrated into a system. How to best integrate different technologies is a topic common to many R&T Challenges.

Benefits of synergy

Many synergies arise when developing advanced system configurations that integrate diverse technologies. For example, there is a direct synergy between advances in variable-cycle engines and the development of supersonic aircraft. Similarly, research on sonic boom mitigation is integral to the design of engines and propulsion systems for supersonic civil aircraft. In addition, for hypersonic vehicles (e.g., scramjet), the propulsion system cannot be designed separately from the rest of the vehicle. In this case, technologies that support the engine and vehicle often mature hand-in-hand as the systems are integrated. Many advanced configurations would also benefit from new sets of active control techniques and smart components (engines, materials, structures) that can self-diagnose and repair. Moreover, from the operational point of view, advanced (and even current) configurations benefit from a change in paradigm in the way that guidance, control, and real-time weather information is shared and used by pilots, controllers, and air traffic managers.

Relevant R&T Challenges

The following R&T Challenges are closely related to advanced configurations: A1, C1, C2, E2, A3, B3, D3, A4a, C4, D4, B5, C5, D5, B6a, B6b, C6a, E6, A7a, E7, B8, C8, A9, A10, B9, B10, and C10.

Intelligent and Adaptive Systems

Description

When an emerging detailed knowledge of physical phenomena is combined with the development of miniaturized sensors, compact actuators, and powerful computational capabilities, the potential exists to develop intelligent and adaptive systems with significantly improved performance and robustness. This Common Theme encompasses aircraft-level R&T Challenges aimed at sensing the operational environment, actively responding to that environment, and learning from the resulting interactions. Examples include (1) flow control techniques for improving aerodynamic performance, reducing noise, increasing maneuverability, and making aircraft robust to atmospheric disturbances and adverse weather conditions and (2) methods for improving the interaction of humans with aircraft systems. This Theme also involves technologies aiming at development of smart engines and mechanical power systems, adaptive materials and morphing structures, load suppression, and vibration and aeromechanical stability control.

The development of innovative classes of aircraft and complex systems will be facilitated by techniques to overcome the design and operational constraints and the physical limits of current systems. Each of the R&T Challenges related to this Theme involves the measurement of physical characteristics of a system in an effort to develop responsive and flexible schemes to improve system performance, robustness, efficiency, and safety.

Background

While some technologies encompassed by this Theme are relatively immature, significant performance improvements can be expected through development and execution of a coordinated research plan. Many promising flow control techniques have already been developed, such as microfluidic injectors, piezoelectric synthetic jets, voice-coil actuators, dielectric barrier discharges, and surface plasma discharges. These techniques have shown promise in the laboratory with limited, scaled flight testing. Adaptive materials with the ability to radically change their properties are also being explored, with the goal of affecting load-carrying capability and allowing large variations in wing area or shape. In the past 2 years, prototypes from DARPA's Morphing Aircraft Structures program have been demonstrated at transonic speeds in the Transonic Dynamics Tunnel at NASA Langley. Significant advancements in sensing techniques have also been realized with respect to miniaturization, frequency response, and allowable environmental operating conditions. Techniques currently exist to measure both surface and in-stream properties useful for adaptive control techniques. Finally, basic control techniques are available, but research is needed in the flight control laws for systems with a large number of highly distributed sensors and actuators, nonlinear adaptive control techniques, and adaptive techniques compatible with the failure of distributed sensors and actuators.

Suggested approach

To fully realize the benefits of the research within this Theme, cross-disciplinary teams will be required. Coordination across the R&T Challenges should be pursued to leverage promising developments in overlapping technologies. Efforts aimed at improving aircraft performance will require people with detailed knowledge in the following areas:

- The fundamental physical processes being controlled.
- Novel actuator designs, including material and structural response and electronics.
- Innovative sensing techniques.
- Information technology.
- Control theory.

The cross-disciplinary teams should interact frequently with designers and operators of current systems to clearly understand evolving constraints of existing systems. In addition, control of one parameter may have unexpected consequences for other parameters. These trade-offs must be identified and understood before they can be addressed.

Benefits of synergy

Integration of the R&T Challenges within the Common Theme on intelligent and adaptive systems would facilitate the cross-pollination of ideas and techniques. For example, flow control actuators developed for improving external aerodynamics may well find application in propulsion systems, while adaptive materials and structures developed for morphing aircraft may find application in noise reduction efforts. With this research conducted as an integrated Theme, rapid and effective implementation of advancements could be realized across historically disparate domains.

Synergies also exist between this Theme (which focuses on aircraft R&T) and the Common Theme on complex interactive systems (which focuses on the air transportation system as a whole). Intelligent and adaptive systems developed for use on aircraft potentially provide information useful in the operation of larger, more complex air transportation systems. For example, sensors incorporated into an aircraft to detect icing may well provide information useful to the ATM system.

Relevant R&T Challenges

The following R&T Challenges are closely related to intelligent and adaptive systems: E1, C1, A2, C2, D2, E2, B3, D3, E3, A4a, D4, E4, C5, D5, A6, C6b, D6, E6, A7b, D8, E8b, E8c, D9, and C10.

Complex Interactive Systems

Description

As noted in Chapter 1, as used in this report, a complex interactive system (also known as a system of systems) refers to an adaptive system consisting of a large, widespread collection or network of independent systems functioning together to achieve a common purpose. Complex interactive systems are distinguished from large, monolithic systems by the independent functioning of their components, which provides freedom for existing components to evolve and new components to emerge independent of a central configuration control authority. Complex interactive systems also tend to be distributed over a large geographic area and require effective communications and coordination protocols for the various components to interact efficiently (Maier, 2006).

To achieve the Strategic Objectives, the air transportation system must be understood as a complex interactive system, because its performance emerges from collective interactions among many independent systems and organizations, including aircraft of many different types, capabilities, and missions; pilots; air traffic controllers and air traffic flow managers; communication, navigation, and surveillance systems; airline operation control centers; manufacturers; labor organizations; and air carriers of many different sizes, capabilities, and operating philosophies. All of these "components" of the air transportation system loosely operate under a set of operating agreements, rules, regulations, and communications protocols established by international, national, and local government and nongovernmental organizations.

Background

As aeronautic systems become more complex, the following systems issues become more critical and more difficult to examine:

- When system performance is itself a complex, nondeterministic phenomenon emerging from the interaction of independent system components with stochastic behaviors, it may not be feasible to develop an analytical model of the entire system, making it difficult to describe, explain, and predict the system performance resulting from changes to any system component.
- Correspondingly, when a change to system behavior is desired, translation of this system-level representation into specific requirements for components can be difficult.
- Unlike centrally organized systems, which may be decomposed according to a hierarchy of control, decomposing the system model into design-manageable elements may be impossible when many different components interact in many different ways.
- The components' behaviors (especially human behaviors) will often be context dependent, especially when they are attempting to meet several competing objectives. Thus, a small change in one part of the system may change the operating context of several components, generating broader, unanticipated effects.
- The types of behaviors that can significantly impact system performance include not only the physical functioning of technologies but also the cognitive behaviors of humans in the systems; social and organizational dynamics; and economic dynamics.
- Complex interactive systems are typically collaborative—that is, they allow component systems to more or less voluntarily collaborate to fulfill agreed-upon central purposes. Agreements among the central players on service provision and rejection provide a primary enforcement mechanism to maintain standards.
- The linkages between components are typically created through communication and coordination protocols rather than mechanical linkages or command structures.

Looking specifically at the air transportation system, much of its structure has evolved over time, with each independent entity finding methods of operation that satisfy its own goals as much as possible within the overall constraints that are imposed upon them. Human–machine and machine–machine interfaces are often created after the development of the technologies and operating concepts, sometimes leading to problems when interface design is unduly difficult or expensive. Aircraft have been developed to meet market demands without full consideration of overall impact on the system of variant performance characteristics, which may reduce system capacity and efficiency. System models typically examine isolated effects or components within the system, and few models attempt to examine a large range of complex, interactive system effects, especially those involving nondeterministic behaviors. Additionally, current system models are not easily reconfigured or adaptable to real-time analysis.

Suggested approaches

Key to analyzing a complex interactive system such as the air transportation system is developing a suite of interacting models with comprehensive simulation and analysis capabilities. Such an interactive system of models should be capable of (1) assessing impacts locally within system components as well as globally across the system and (2) introducing new systems, operating procedures, and protocols for information transfer, communication, coordination, and collaboration. In addition, models suited to complex interactive systems are needed early in the conception and design of any technology intended to function within such a system, including systems intended to support human activity. This process will be facilitated by explicitly representing the anticipated contributions of the technology to the larger system.

The need for clear communication and coordination protocols within the system is another critical design consideration. System designs should also consider the need for collaborative decision making, the relative roles and authority of the components (including organizational structures and the role of technologies in mediating human interactions), and their information needs.

Benefits of synergy

The ability to analyze complex interactive systems is relevant to many R&T Areas, and methods of modeling and analyzing such systems can be broadly applicable. Redesigning the air transportation system will be difficult, but the ability to accurately and efficiently model it as a complex interactive system will help reduce program risk and allow coordinating design efforts across multiple agencies.

Relevant R&T Challenges

The following R&T Challenges are closely related to complex interactive systems: C1, E1, D2, E2, B3, C3, E3, E4, E6, A7b, E8a, E8b, E5, and D10.

KEY BARRIERS

The steering committee identified two key barriers to achieving the six Strategic Objectives that should guide civil aeronautics R&T: (1) certification and (2) change management, internal and external. If these barriers are not addressed, the Strategic Objectives will not be accomplished, even if individual R&T Challenges are successfully overcome. Although these barriers may not appear to be explicitly part of NASA's mission, if they are considered from the earliest stages of research, the civil aeronautics community will be more likely to use the results of NASA R&T in developing operational products and procedures. Furthermore, the barriers have technical aspects, which the R&T Challenges will address.

Certification

Certification is the demonstration of a design's compliance with regulations. For example, before it can be operated by U.S. airlines, a new aircraft must be shown to comply with U.S. federal aviation regulations. As systems become more complex and nondeterministic, methods to certify new technologies become more difficult to validate. Core research in methods and models for assessing the performance of large-scale systems, human-interactive systems, nondeterministic systems, and complex, software-intensive systems, including safety and reliability in all relevant operating conditions, is essential for NASA, because such research is currently beyond the capabilities of regulators such as the FAA. The ultimate utility of this research will be significantly enhanced through early and consistent coordination of technology maturation with the FAA and other organizations responsible for certification of operational systems. Furthermore, this research would be facilitated by collaboration with other organizations involved in advanced software development methods.

Certification can also be a major barrier to the ultimate implementation of new technologies and operating concepts. In some cases, such as low-cost avionics for general aviation, the cost of certification can be several times greater than the cost of developing and manufacturing the product itself. Furthermore, relying on empirical testing to demonstrate compliance with certification standards may not be feasible for large-scale systems (including complex, software-intensive systems and air traffic operating concepts) and human-in-the-loop behaviors, which are not the same in different operating contexts; in these cases, certification will be substantially aided by the use of design tools and design processes developed to mitigate concerns about design validity, safety, and reliability. Certification issues can be show-stoppers if not addressed early in the R&T process. Thus, NASA should address the following concerns in its aeronautics R&T program:

- Systematic documentation and publication of model and design assumptions from the earliest stage of R&T development, to aid in a technology's ultimate certification.
- Ongoing iterative validation of models and design tools—and their specifications—during their development, and verification of models and design tools relative to their specifications.
- Generation of databases and models from empirical data to provide a basis for validation and certification.
- Establishment of community-accepted metrics, criteria, and methods for validation and certification.

Change Management, Internal and External

The air transportation system includes large organizations with long-standing institutional cultures and business concerns that are impacted by—and sometimes resist—the introduction of new technologies. These organizations must be motivated to participate in new operating concepts and to accept the risk of change to improve performance. Changing an interactive system as complex as the air transportation system is difficult because it involves changing a large number of individual elements, including equipment of many different kinds, personnel training, institutional organization, and business models. Additionally, the end state of the air transportation system remains undefined, so R&T should create and maintain the flexibility to steer the system in any of several different directions. This requires interdisciplinary applications of large-scale systems engineering, organization design, economics, and financial analysis, an approach which in some ways is beyond the current state of knowledge. Even so, improved change management techniques are vital to a cost-effective, noncontentious, and safe transition to the air transportation system of the future.

Change management within the federal government is particularly important because of the major impact that federal agencies, regulations, and funding have on the operation of the air transportation system and the development of new aeronautical technologies. In addition, change management within the federal government is particularly difficult because of the complex internal organization of the federal government, with multiple independent agencies, competing national priorities, and political factors that are beyond the control of any one person or agency. One way to facilitate change in the midst of such complexity is to establish strong, focused leadership that establishes a public/private process for change that defines air transportation as a national priority, produces a widely endorsed long-term vision of the air transportation system, and coordinates action by interested organizations. The process should be carefully structured to accommodate the increasing complexity of the air transportation system, competing national and organizational priorities, and fiscal limitations. The process should produce validated R&T requirements, a clear understanding of government and industry roles, and a plan to implement new technologies, operational concepts, and system architectures (NRC, 2003). The establishment of the Next Generation Air Transportation System (NGATS) Joint Planning and Development Office (JPDO) is an example of federal efforts to change interagency relationships to improve change management in civil aviation.

The issues related to change management transcend NASA's role as a single agency. The federal government should continue to support the work of the JPDO while conducting a high-level review of organizational options for ensuring U.S. leadership in civil aeronautics.[1]

[1]A more detailed assessment of the management and organizational issues associated with NASA aeronautics R&T appears in another recent report, *Aeronautics Innovation: NASA's Challenges and Opportunities* (NRC, 2006).

REFERENCES

Maier, M. 2006. Architecting Principles for Systems-of-Systems. The Information Architects Cooperative (TiAC) White Paper Repository. Available online at <www.infoed.com/Open/PAPERS/systems.htm>.

National Research Council (NRC). 2003. Securing the Future of U.S. Air Transportation: A System in Peril. Washington, D.C.: The National Academies Press. Available online at <http://fermat.nap.edu/catalog/10815.html>.

NRC. 2006. Aeronautics Innovation: NASA's Challenges and Opportunities. Washington, D.C.: The National Academies Press. Available online at <http://fermat.nap.edu/catalog/11645.html>.

5

Findings and Recommendations

PRINCIPAL FINDINGS

R&T Challenges

The top 10 R&T Challenges for each R&T Area are listed in Table 5-1.[1] The quantitative scores for the Challenges are relative scores that are valid in an ordinal sense within each R&T Area. They represent the results of linked, but separate, comparative analyses within each R&T Area and therefore should not be used to make absolute comparisons of the relative priority of various Challenges from different R&T Areas.

The QFD rankings in Table 5-1 should be taken as a guide rather than a prescription. Many of the R&T Challenges are considerably dissimilar in scope and content. In some cases, progress will require success in overcoming multiple linked Challenges. Other Challenges stand on their own. All of the R&T Challenges are considered worthy of NASA attention. In addition, many of the high-priority R&T Challenges have uses in fields other than civil aeronautics and are applicable to the missions of other federal agencies—DoD, DHS, and FAA, among others. Cooperative research between NASA and other agencies could therefore produce substantial national benefits.

The steering committee believes that the highest-priority R&T Challenges in each R&T Area should be included in the "foundation for the future" that forms the core of NASA's aeronautics program. The steering committee made no specific budgetary recommendations, in accordance with the statement of task for this study.

Success will require stable funding and consistent research priorities and planning, with the intent to pursue identified challenges for a decade or longer, as long as satisfactory progress continues. For NASA to exert strong leader ship in key cutting-edge aeronautics R&T, it should focus on state-of-the-art research. Research plans should not be open-ended, however. An acceptable level of feasibility needs to be demonstrated (see suggested milestones in Appendixes A-E), and it is critical to have a process to stop or redirect efforts that fail to progress.

Common Themes

In Chapter 4 the steering committee identified threads of commonality among the R&T Thrusts and the R&T Challenges from different R&T Areas. These threads have been captured as five Common Themes:

- Physics-based analysis tools
- Multidisciplinary design tools
- Advanced configurations
- Intelligent and adaptive systems
- Complex interactive systems

Each Theme encompasses enabling approaches that will contribute to multiple R&T Challenges.

Exploiting the synergies identified in each Common Theme would enable NASA's aeronautics program to make the most efficient use of available resources.

Key Barriers

The steering committee identified two key barriers in Chapter 4: certification and change management. If these barriers are not addressed, the Strategic Objectives will not be accomplished, even if individual R&T Challenges are successfully overcome.

As systems become more complex, methods to ensure that new technologies can be readily applied to FAA-certified systems become more difficult to validate. NASA

[1] The top 11 R&T Challenges are listed for Area A because the NASA priority scores for Challenges A10 and A11 are very close, and there is a large gap between the scores for Challenges A11 and A12.

TABLE 5-1 Fifty-one Highest Priority Research and Technology Challenges for NASA Aeronautics, Prioritized by R&T Area

A Aerodynamics and Aeroacoustics	B Propulsion and Power	C Materials and Structures	D Dynamics, Navigation, and Control, and Avionics	E Intelligent and Autonomous Systems, Operations and Decision Making, Human Integrated Systems, Networking and Communications
A1 Integrated system performance through novel propulsion–airframe integration A2 Aerodynamic performance improvement through transition, boundary layer, and separation control A3 Novel aerodynamic configurations that enable high performance and/or flexible multi-mission aircraft A4a Aerodynamic designs and flow control schemes to reduce aircraft and rotor noise A4b Accuracy of prediction of aerodynamic performance of complex 3-D configurations, including improved boundary layer transition and turbulence models and associated design tools A6 Aerodynamics robust to atmospheric disturbances and adverse weather conditions, including icing A7a Aerodynamic configurations to leverage advantages of formation flying A7b Accuracy of wake vortex prediction, and vortex detection and mitigation techniques A9 Aerodynamic performance for V/STOL and ESTOL, including adequate control power A10 Techniques for reducing/mitigating sonic boom through novel aircraft shaping A11 Robust and efficient multidisciplinary design tools	B1a Quiet propulsion systems B1b Ultraclean gas turbine combustors to reduce gaseous and particulate emissions in all flight segments B3 Intelligent engines and mechanical power systems capable of self-diagnosis and reconfiguration between shop visits B4 Improved propulsion system fuel economy B5 Propulsion systems for short takeoff and vertical lift B6a Variable-cycle engines to expand the operating envelope B6b Integrated power and thermal management systems B8 Propulsion systems for supersonic flight B9 High-reliability, high-performance, and high-power-density aircraft electric power systems B10 Combined-cycle hypersonic propulsion systems with mode transition	C1 Integrated vehicle health management C2 Adaptive materials and morphing structures C3 Multidisciplinary analysis, design, and optimization C4 Next-generation polymers and composites C5 Noise prediction and suppression C6a Innovative high-temperature metals and environmental coatings C6b Innovative load suppression, and vibration and aeromechanical stability control C8 Structural innovations for high-speed rotorcraft C9 High-temperature ceramics and coatings C10 Multifunctional materials	D1 Advanced guidance systems D2 Distributed decision making, decision making under uncertainty, and flight path planning and prediction D3 Aerodynamics and vehicle dynamics via closed-loop flow control D4 Intelligent and adaptive flight control techniques D5 Fault-tolerant and integrated vehicle health management systems D6 Improved onboard weather systems and tools D7 Advanced communication, navigation, and surveillance technology D8 Human–machine integration D9 Synthetic and enhanced vision systems D10 Safe operation of unmanned air vehicles in the national airspace	E1 Methodologies, tools, and simulation and modeling capabilities to design and evaluate complex interactive systems E2 New concepts and methods of separating, spacing, and sequencing aircraft E3 Appropriate roles of humans and automated systems for separation assurance, including the feasibility and merits of highly automated separation assurance systems E4 Affordable new sensors, system technologies, and procedures to improve the prediction and measurement of wake turbulence E5 Interfaces that ensure effective information sharing and coordination among ground-based and airborne human and machine agents E6 Vulnerability analysis as an integral element in the architecture design and simulations of the air transportation system E7 Adaptive ATM techniques to minimize the impact of weather by taking better advantage of improved probabilistic forecasts E8a Transparent and collaborative decision support systems E8b Using operational and maintenance data to assess leading indicators of safety E8c Interfaces and procedures that support human operators in effective task and attention management

should anticipate the need to certify new technology before its introduction, and it should conduct research on methods to improve the confidence in and the timeliness of certification. Methods might include new approaches (e.g., "design for certification"). If the civil aeronautics R&T program does not address this certification barrier, manufacturers and users will not be able to effectively exploit new technology in operations.

As discussed in Chapter 4, changing a complex interactive system such as the air transportation system is becoming more difficult due to the growing complexity of interactions between the various elements and the growing number of internal and external constraints. To effectively exploit the benefits of R&T to meet the Strategic Objectives, new tools and techniques are required to anticipate and introduce change. Without research to better define and develop change management tools and techniques, the inability to introduce change in a timely manner will serve as a barrier to exploiting the benefits offered by meeting the R&T Challenges.

OTHER FINDINGS OF IMPORTANCE

Allocation of Resources and Workforce Issues

NASA's aeronautics program is likely to operate in an environment of constrained resources for the foreseeable future. Nonetheless, the committee believes that NASA must meet its commitment to the nation as the leader of cutting-edge aeronautics research. This requires NASA to carry out, at a minimum, the following missions:

1. Perform cutting-edge, high-value aeronautics research in support of the nation's future industrial and government aeronautics needs.
2. Maintain in-house technical expertise to advise other parts of the U.S. government, including the FAA, the Environmental Protection Agency, and DoD, on relevant aeronautics issues.
3. Maintain state-of-the-art research, testing, computational, and analytical capabilities in support of the U.S. civil aviation community, including industry, academia, and the general public.
4. Facilitate the exchange of information on civil aeronautics R&T among academia, industry, U.S. government agencies, and the international regulatory community.
5. Provide aeronautics expertise and capabilities in support of NASA's space program.

For NASA to complete these missions in a constrained fiscal environment, the committee believes that NASA must consider the criteria listed below when considering whether to perform the work in-house by NASA engineers and technical specialists or externally by industry and/or universities:

- Specialized technical expertise of in-house and external organizations.
- Specialized facilities and capabilities, such as wind tunnels, simulators, laboratories, and analytical methods, that are available in-house and at external organizations.
- The requirement for NASA to have the expertise and experience necessary to be an informed buyer of aeronautics R&T.
- The requirement for NASA to provide independent technical advice to other federal agencies on aeronautics issues.

As of January 2006, NASA seemed intent on allocating 93 percent of NASA's aeronautics research funding for in-house use.[2] While the committee has no specific recommendation on the in-house/external split, it does not believe that such a split would serve the best interests of NASA or the nation. NASA R&T would likely suffer from the absence of relevant, specialized technical expertise, facilities, and capabilities (the first two criteria, above) without procuring expertise and capabilities from academia and, to a lesser degree, industry. Also, NASA would likely be limited in its ability to provide technology that supports the nation's future industrial aeronautics needs (Mission 1, above) without greater inclusion of industry. Furthermore, an insular approach to R&T would not leverage the creativity and multiplicative ideas that a more inclusive approach would likely produce. Technology transfer among government, universities, and industry would be more effective if all three groups have significant roles in NASA's research programs.

NASA researchers in some cases possess world-class technical expertise, and this expertise should be maintained. Furthermore, some level of technical expertise in a wide range of subjects is required for NASA to meet its obligations for conducting cutting-edge research, advising the government, and facilitating outside collaboration (Missions 1, 2, and 4, above). However, NASA should consider the capabilities of other research organizations before deciding whether to outsource R&T related to cutting-edge research, state-of-the-art capabilities, and the space program (Missions 1, 3, and 5, above).

A more balanced allocation of aeronautics R&T funding would allow NASA to form stronger partnerships with academia and industry. Stable funding of academic research grants, graduate student fellowships, student internships, and

[2]NASA's Aeronautics Research Mission Directorate has established four levels of research. NASA plans to allocate 7 percent of the total aeronautics budget to university and small company research at Level 1 (foundational research) (Wlezien, 2006b). Research at Level 2 (develop discipline-specific technologies and tools) and Level 3 (develop integrated, multidisciplinary methods and technologies) will be performed in-house by NASA. Research at Level 4 (develop integrated solutions for airspace and airport systems) will include collaboration with industry and partnerships with other agencies, but participating companies will be expected to pay their own way (NASA, 2006; Wlezien, 2006a).

cooperative NASA-university research centers would expand the intellectual pool contributing to NASA R&T, improve the skills of the nation's future aeronautics workforce, recruit new talent into the NASA workforce, and foster R&T projects that could not be done by NASA working alone. NASA should strive to foster close collaborative research with university partners, continuing a tradition of supporting basic research and new research directions in the academic environment.

Stronger partnerships with industry would help (1) identify technologically important problems, the answers to which can benefit the nation, (2) advance important precompetitive R&T that would not otherwise be done, (3) leverage industry research funded by other agencies or industry itself, (4) ensure that the results of NASA aeronautics research take into account relevant standards and practices, and (5) facilitate the transfer of research results to industry so that they find valuable, real-world applications. Ideally, programs that involve university, industry, and NASA researchers would lead to long-term benefits to NASA and the nation. A more inclusive approach to NASA's research would also increase the return on the government's investment in aeronautics R&T.

Taking Advantage of Advances in Cross-Cutting Technology Funded by Others

Most of the federal government's civil aeronautics research is done by NASA, but operational products are developed by manufacturers and operated by industry or the FAA. The FAA and industry also have the lead when it comes to certification issues. Within the federal government, no agency is responsible for all of the federal government's aeronautics research, nor is one agency focused exclusively on aeronautics research. This organizational structure mandates close cooperation and coordination between NASA's civil aeronautics R&T program and other federal agencies that support related research.

DoD conducts research on and development of many technologies important to national defense. At a fundamental technology level, much of the work sponsored by DoD is synergistic with NASA's civil aeronautics R&T, especially in transition modeling, reacting-flow physics, multiphase flows, novel aerodynamic configurations, morphing aerodynamic surfaces, adaptive cycle engines, high-speed and high-performance propulsion systems, integrated power systems, network-centric operations, control systems, UAVs, rotorcraft, impact dynamics, high-temperature and multifunctional materials and structures, low-cost materials and manufacturing, sensors, and multidisciplinary optimization.

Interactions with and coordination of research conducted by the FAA should be pursued in areas such as ATM and the measurement and modeling of aircraft wakes and weather phenomena. The large-scale weather models developed by the National Oceanic and Atmospheric Administration (NOAA) and the National Center for Atmospheric Research (NCAR) should be coordinated with the local terminal weather models important for prediction of wake vortex trajectories. Where synergistic research is being conducted, structured interactions and collaborations should be pursued.

Software-intensive systems are being developed in many industries. Applications relevant to civil aviation include highly autonomous systems, advanced decision aids, morphing aircraft, and advanced guidance systems. NASA should support these applications by collaborating with other organizations that are supporting research to write and qualify complex, safety-critical software in a more timely and cost-effective manner.

Significant opportunities also exist for NASA to collaborate with international research organizations, especially at the level of foundational physics. Structured processes should be developed to monitor international activities and plan appropriate collaborations.

With so many organizations involved in the development of new civil aeronautics technologies, advances often occur piecemeal, in areas of particular interest to individual stakeholders. Singularity of vision is needed to ensure that R&T programs develop all of the pieces necessary for game-changing advancements across the board. Within NASA, this is complicated by the fact that NASA's vision, resources, and energy must be shared between aeronautics and the much larger space exploration and space science programs—and the aeronautics program does not have a clear vision akin to the space program's vision of the human exploration of the Moon and Mars. Although organizational issues are beyond the purview of this study, the steering committee noted in the course of its deliberations that the factors cited above represent a potential barrier to the pursuit and implementation of aeronautics R&T. Because the issues transcend NASA, the steering committee observes that in the national interest, organizational options should be reviewed by a senior group commissioned by or within the federal government.

How Far Should NASA Advance Research?

NASA's congressionally mandated charter directs it to "preserve the role of the United States as a leader in aeronautical science and technology and the application thereof." To achieve this goal, NASA should embrace a comprehensive roadmap of foundational research that develops discipline-specific and multidisciplinary capabilities, including system-level design. The roadmap should include (1) progressive empirical validation up to and including a limited number of flight demonstration vehicles (X-planes), (2) technology readiness metrics, such as NASA's technology readiness levels (TRLs) (see Table 5-2), and (3) research partnerships with industry, academia, and other federal agencies. X-planes have played and will continue to play a crucial role

TABLE 5-2 NASA Technology Readiness Levels 1 to 9 for Aeronautics Research

Key Player	Stage of System Development	TRL	Description
Industry	System test and operations (TRL 8-9)	9	Actual system flight proven on operational flight
		8	Actual system completed and flight qualified through test and demonstration
	System/subsystem development (TRL 6-8)	7	System prototype demonstrated in flight environment
Government	Technology demonstration (TRL 5-6)	6	System/subsystem model or prototype demonstrated/validated in a relevant environment
		5	Component and/or breadboard verification in a relevant environment
	Technology development (TRL 3-5)	4	Component and/or breadboard test in a laboratory environment
	Research to prove feasibility (TRL 2-3)	3	Analytical and experimental critical function or characteristic proof of concept
	Basic technology research (TRL 1-2)	2	Technology concept and/or application formulated (candidate selected)
		1	Basic principles observed and reported

SOURCE: NASA, 2000.

in the advancement of aeronautical research by validating the practicality and robustness of specific technological advances. It is important to note that they are not limited to high TRL research. While an X-plane may represent a system prototype (TRL 7), it may also be used to observe basic phenomena, prove concepts, or validate a component or subsystem (TRL 1-6). TRLs provide a consistent and objective measure of technology maturation and progress. Research partnerships with external organizations provide an important mechanism to maintain the crucial links between NASA and the organizations that will use the results of NASA's aeronautics R&T. These partnerships also help to ensure that the research priorities of the aeronautics community as a whole remain relevant.

NASA is also charged with identifying, encouraging, and fostering cutting-edge R&T that will address important national goals but cannot be justified by individual companies in terms of return on investment. NASA should have clear criteria and metrics for entering, continuing to support, and leaving a research area (because of lack of progress or because goals have been achieved). Emerging areas are characterized by a multitude of ideas and approaches. Setting clear criteria for success and a timescale for evaluation allows research to focus on the most fruitful areas without prematurely abandoning an idea that still holds promise.

As noted in Table 5-2, NASA has historically supported research through TRL 6 and then transferred research results to industry, with the expectation that industry would continue development of new technologies through TRL 9. The steering committee, however, believes that different transfer points are often appropriate, because industry's interest in developing new technologies varies based on urgency and expected payoff. For urgent, high-payoff applications, for example, it may be sufficient for NASA to mature technologies to TRL 5.

When NASA is developing technologies for transfer to operational federal agencies such as the FAA, the committee believes that research results should normally be transferred to industry first, to ensure product support, enhancement, integration with other systems, and certification. For government agencies that include an R&D mission, agency-to-agency transfer is appropriate, and such transfers may occur at reasonably low TRLs (e.g., TRL 3).

Ground and Flight Test Capabilities

Since the creation of the small wind tunnel used by the Wright brothers to investigate aerodynamics for the first powered aircraft, advances in aeronautics have been closely tied to ground test facilities, such as simulators, wind tunnels, combined-environment load facilities, propulsion test cells, and acoustic facilities. These facilities allow repeatable and accurate assessment of physical processes and play a vital role in validating physical and computational models. To conduct the cutting-edge research outlined above, NASA must maintain world-class facilities and diagnostic capabilities. It should invest in research associated with improved facilities and diagnostics in coordination with DoD and industry. Furthermore, NASA should establish facility access and pricing policies that enable and encourage industry and academia to use NASA facilities. If costs are too high, they drive away potential customers, causing the price for remaining customers to spiral higher still. Eventually, underutilized tunnels are mothballed, and NASA loses testing capabilities and the expertise of workers who move on to other jobs or industries. NASA should seek a business model that will generate the optimal combination of income and utilization.

Flight test capabilities are required for research that cannot be adequately simulated in ground facilities. This research includes atmospheric propagation of noise and sonic booms, reacting-flow hypersonic phenomena, and large-scale propulsion–airframe integration.

While this study did not include a detailed assessment of facilities, some key facilities—and their deficiencies—are noted in Appendixes A-E. Facility concerns have been addressed in detail by studies devoted solely to this topic

(Anton et al., 2004; Kegelman, 2006; NRC, 1994, 2004). These reports have identified critical issues associated with the capabilities, funding, and use of national aeronautical test facilities, many of which have yet to be resolved.

RECOMMENDATIONS

The steering committee offers eight recommendations:

1. NASA should use the 51 Challenges listed in Table 5-1 as the foundation for the future of NASA's civil aeronautics research program during the next decade.

2. The U.S. government should place a high priority on establishing a *stable* aeronautics R&T plan, with the expectation that the plan will receive sustained funding for a decade or more, as necessary, for activities that are demonstrating satisfactory progress.

3. NASA should use five Common Themes to make the most efficient use of civil aeronautics R&T resources:[3]

- Physics-based analysis tools
- Multidisciplinary design tools
- Advanced configurations
- Intelligent and adaptive systems
- Complex interactive systems

4. NASA should support fundamental research to create the foundations for practical certification standards for new technologies.

5. The U.S. government should align organizational responsibilities as well as develop and implement techniques to improve change management for federal agencies and to assure a safe and cost-effective transition to the air transportation system of the future.

6. NASA should ensure that its civil aeronautics R&T plan features the substantive involvement of universities and industry, including a more balanced allocation of funding between in-house and external organizations than currently exists.

7. NASA should consult with non-NASA researchers to identify the most effective facilities and tools applicable to key aeronautics R&T projects and should facilitate collaborative research to ensure that each project has access to the most appropriate research capabilities, including test facilities; computational models and facilities; and intellectual capital, available from NASA, the Federal Aviation Administration, the Department of Defense, and other interested research organizations in government, industry, and academia.

8. The U.S. government should conduct a high-level review of organizational options for ensuring U.S. leadership in civil aeronautics.

REFERENCES

Anton, P.S., E.C. Gritton, R. Mesic, and P. Steinberg. 2004. Wind Tunnel and Propulsion Test Facilities: An Assessment of NASA's Capabilities to Serve National Needs. Santa Monica, Calif.: RAND Corporation. Available online at <http://ntrs.nasa.gov/index.cgi?method=ordering&oaiID=oai:casi.ntrs.nasa.gov:20050199428>.

Kegelman, J. 2006. Wind Tunnel Enterprise. NASA Langley Research Center. Available online at <http://windtunnels.larc.nasa.gov/enterprise.htm>.

National Aeronautics and Space Administration (NASA). 2000. Ultra Efficient Engine Technology Program. Home and Home Visit Series. Glenn Research Center, April 4-5. Available online at <www.aero-space.nasa.gov/events/home&home/glenn/ueet/sld004.htm>.

NASA. 2006. ARMD [Aeronautics Research Mission Directorate] Requests for Information (RFI) Questions and Answers. Available online at <www.aero-space.nasa.gov/rfi_qa.htm>.

National Research Council (NRC). 1994. Aeronautical Facilities: Assessing the National Plan for Aeronautical Ground Test Facilities. Washington, D.C.: National Academy Press. Available online at <http://fermat.nap.edu/catalog/9088.html>.

NRC. 2004. Investments in Federal Facilities: Asset Management Strategies for the 21st Century. Washington, D.C.: The National Academies Press. Available online at <http://fermat.nap.edu/catalog/11012.html>.

Wlezien, R. 2006a. NASA's Fundamental Research Program. Presentation to 44th AIAA Aerospace Sciences Meeting and Exhibit, Reno, Nev., January 12. Available online at <www.aero-space.nasa.gov/reno_presentations.htm>.

Wlezien, R. 2006b. NASA's Fundamental Research Program. Remarks made during presentation to 44th AIAA Aerospace Sciences Meeting and Exhibit, Reno, Nev., January 12. Available online at <www.streammercials.com/aiaa/player.htm>.

[3]The Common Themes are defined in Chapter 4.

Appendixes

A

R&T Challenges for Aerodynamics and Aeroacoustics

A total of 19 R&T Challenges were prioritized in the aerodynamics and aeroacoustics Area. Table A-1 shows the results. The R&T Challenges are listed in order of NASA priority. National priority scores are also shown.[1] This appendix contains a description of each R&T Challenge, including milestones and an item-by-item justification for each score that appears in Table A-1.[2]

A1 Integrated system performance through novel propulsion–airframe integration

Flow interactions in the region of the propulsion–airframe interface during takeoff, climb, and cruise pose a complex design problem. Design compromises have a significant effect on the aircraft efficiency and on radiated noise. Research into improved techniques for propulsion–airframe integration would improve aircraft flexibility and performance, especially as aircraft speeds increase. To meet this objective, both computational fluid dynamics (CFD) and experimental tests are indispensable. Improvements in the accuracy of predictions of three-dimensional (3-D) steady and unsteady interactions between external and internal aerodynamics and aeroacoustics are required to enable design of future aeronautical systems. These interactions include the effects of steady and dynamic distortion on engine operations and the effects of hot, reacting exhaust flows on vehicle aerodynamics. They are particularly important in the design of vertical and short takeoff and landing (V/STOL),[3] extremely short takeoff and landing (ESTOL), supersonic, and hypersonic airplanes. On V/STOL airplanes, exhaust jets are placed near the trailing edge of the wing where the aerodynamic stiffness of the high-speed flow increases wing lift by what is called the Jet Flap Effect (Spence, 1956). At supersonic speeds, adverse interactions between shock waves and boundary layers can increase drag and cause engine unstart. For many proposed hypersonic aircraft, the aircraft forebody is the inlet compression surface and the aircraft afterbody is the nozzle expansion surface, so that the airframe is part of the propulsion system. This is particularly the case with waveriders (Kuchemann, 1978). Propulsion–airframe integration has a significant impact on aircraft radiated noise. Improvements in test techniques and instrumentation are needed to characterize complex 3-D flow fields and acoustic radiation patterns. Key milestones include

- Validate the predictive capablity for 3-D mean and dynamic distortion at the propulsion–airframe interface.
- Validate the predictive capability of the impact of reacting exhaust flows on external aerodynamics.
- Validate the predictive capability of acoustic radiation patterns from integrated propulsion–airframe configurations.

[1]The prioritization process is described in Chapter 2.

[2]The technical descriptions for the first 11 Challenges listed below contain slightly more detail than the technical descriptions for these Challenges as they appear in Chapter 3.

[3]VTOL airplanes can take off and land vertically. This includes tiltrotors, the AV-8 Harrier, and the JSF, for example. VTOL airplanes do not routinely take off or land vertically because of the range-payload penalty associated with the weight limitations of purely vertical operations. Rather, they use any available field length to develop some forward motion and wing lift during takeoff to increase the useful load (fuel plus payload). They tend to land vertically only at the end of the mission, when they are lighter, after burning fuel and/or dropping weapons.

STOL airplanes use high-lift systems to take off in less distance than conventional aircraft (typically a few thousand feet). Very few STOL aircraft can safely take off on runways shorter than 3,000 ft and none on runways less than 2,000 feet. (This class does not include ultralight aircraft, kit planes, etc. that can operate out of short fields due to their small size but do not have high-lift systems).

ESTOL airplanes would be able to safely take off on runways of 2,000 ft. They would have high-lift systems and thrust-to-weight ratios that are higher than conventional aircraft but not as high as VTOL aircraft. ESTOL aircraft have not yet been developed for commercial or military operations.

V/STOL refers to both VTOL and STOL airplanes that convert to fixed-wing flight after takeoff; it does not include helicopters.

TABLE A-1 Prioritization of R&T Challenges for Area A: Aerodynamics and Aeroacoustics

R&T Challenge	Capacity	Safety and Reliability	Efficiency and Performance	Energy and the Environment	Synergies with Security	Support to Space	National Priority	Supporting Infrastructure	Mission Alignment	Lack of Alternative Sponsors	Appropriate Level of Risk	Why NASA Composite Score	NASA Priority Score
	Strategic Objective							Why NASA?					
Weight	5		3		1			1/4 each					
A1 Integrated system performance through novel propulsion–airframe integration	9	3	9	9	9	9	**132**	3	9	3	9	**6.0**	**792**
A2 Aerodynamic performance improvement through transition, boundary layer, and separation control	9	3	9	9	3	3	**120**	3	9	3	9	**6.0**	**720**
A3 Novel aerodynamic configurations that enable high performance and/or flexible multimission aircraft	9	3	9	9	3	1	**118**	3	9	3	9	**6.0**	**708**
A4a Aerodynamic designs and flow control schemes to reduce aircraft and rotor noise	9	1	3	9	3	1	**90**	3	9	3	9	**6.0**	**540**
A4b Accuracy of prediction of aerodynamic performance of complex 3-D configurations, including improved boundary layer transition and turbulence models and associated design tools	3	3	9	3	3	3	**72**	9	9	3	9	**7.5**	**540**
A6 Aerodynamics robust to atmospheric disturbances and adverse weather conditions, including icing	9	9	3	1	9	1	**112**	3	9	3	3	**4.5**	**504**
A7a Aerodynamic configurations to leverage advantages of formation flying	3	1	9	9	3	1	**78**	3	9	9	3	**6.0**	**468**
A7b Accuracy of wake vortex prediction, and vortex detection and mitigation techniques	9	9	3	1	1	1	**104**	3	9	3	3	**4.5**	**468**
A9 Aerodynamic performance for V/STOL and ESTOL, including adequate control power	9	3	3	1	3	1	**76**	3	9	3	9	**6.0**	**456**
A10 Techniques for reducing/mitigating sonic boom through novel aircraft shaping	3	1	3	9	3	1	**60**	9	9	3	9	**7.5**	**450**
A11 Robust and efficient multidisciplinary design tools	3	3	9	9	3	3	**90**	3	9	3	3	**4.5**	**405**
A12 Accurate predictions of thermal balance and techniques for the reduction of heat transfer to hypersonic vehicles	1	1	3	1	9	9	**40**	9	9	3	9	**7.5**	**300**
A13 Low-speed takeoff and landing flight characteristics for access-to-space vehicles	1	3	1	1	3	9	**38**	3	9	9	9	**7.5**	**285**
A14 Efficient control authority of advanced configurations to permit robust operations at hypersonic speeds and for access-to-space vehicles	1	1	3	1	9	9	**40**	3	9	3	9	**6.0**	**240**
A15 Decelerator technology for planetary entry	1	1	1	1	3	9	**28**	3	9	9	9	**7.5**	**210**
A16 Low-Reynolds-number and unsteady aerodynamics for small UAVs	1	1	3	1	9	3	**34**	3	9	3	9	**6.0**	**204**
A17 Low-drag airship designs to enable long-duration stratospheric flight	1	3	1	3	9	1	**42**	3	3	3	9	**4.5**	**189**
A18 Prediction of communication capability through reentry trajectory and techniques to mitigate impact of communication blackouts	1	1	1	1	9	9	**34**	3	9	3	3	**4.5**	**153**
A19 Aircraft protective countermeasures based on a range of small deployed air vehicles	1	3	1	1	9	1	**36**	3	3	3	3	**3.0**	**108**

- Develop novel propulsion–airframe configurations for supersonic and hypersonic flight.

Relevance to Strategic Objectives

Capacity (9): Novel integration of the propulsion and airframe system will be required to meet the objectives of V/STOL and ESTOL airplanes and enable general operation on shorter runways, thereby enabling a significant increase in capacity.

Safety and Reliability (3): Development of techniques to monitor conditions and predict interactions between external and internal flows will allow automated responses to potentially dangerous flight conditions, which will enhance aircraft safety.

Efficiency and Performance (9): Improved integration of the propulsion system and airframe will facilitate development of aircraft with improved performance and efficiency.

Energy and Environment (9): Careful integration of engines with airframes offers the potential for significant noise reduction through selective shielding of the engine exhaust. In addition, lower overall aircraft drag will reduce fuel consumption and CO_2 emissions.

Synergies with National and Homeland Security (9): Predictive capabilities concerning engine–airframe integration will be applicable to all military aircraft.

Support to Space (9): Careful integration of the propulsion system with the airframe is required for air-breathing access-to-space vehicles.

Why NASA?

Supporting Infrastructure (3): NASA possesses both facilities and computational capabilities to investigate propulsion-airframe integration issues, but similar facilities and capabilities exist in DoD and, to a lesser extent, in industry.

Mission Alignment (9): Efforts aimed at development of an improved predictive capability that will lead to increased performance and efficiency of aircraft are very relevant to NASA's mission.

Lack of Alternative Sponsorship (3): Propulsion–airframe integration is being investigated by DoD and industry.

Appropriate Level of Risk (9): Improving the predictive capability of propulsion–airframe integration would likely enable a new class of highly integrated aircraft. The risk associated with the development of techniques to significantly improve performance, especially with respect to novel aircraft designs, is moderate.

A2 Aerodynamic performance improvement through transition, boundary layer, and separation control

Aircraft performance and efficiency strongly depend on the state of the boundary layer over different portions of the wing and fuselage. Viscous drag at subsonic, supersonic, or hypersonic speeds may be reduced by developing flow control techniques that actively detect and control the state of boundary layer flows with the goal of maintaining attached flow, controlling transition to turbulence, or reducing turbulent drag. For example, takeoff and landing distances strongly depend on the ability to maintain attached flow over the wing. On a typical commercial aircraft, approximately 25 percent of the drag is due to turbulent flow over the fuselage. If a drag reduction of just 1 percent were achieved in either of these areas, a large aircraft could reduce fuel consumption by up to 100,000 gallons per year while also reducing emissions.

Flow control has the potential for significant improvements in aircraft performance when coupled with revolutionary aircraft design configurations that are designed to optimally exploit the effects of flow control. New flow control techniques include steady and unsteady flow injection, vibrating elements such as piezoelectric and voice-coil actuators, and single-dielectric barrier discharge plasma actuators that are fully electric with no moving parts. All of these techniques have shown promise in laboratory experiments and some limited-scale flight tests. However, significant work is needed to refine these approaches and develop them further in order to transition them to full-scale designs.

Accurate models are needed for the actuator effects that can be incorporated into high-fidelity numerical flow simulations. The models would also be used to optimize the design of flow actuators to improve their flow authority and expand their usable flight regime. For flow control to achieve its full potential as an element in multidisciplinary design and optimization, the accuracy of numerical simulations and models needs to be validated, and they must be computationally efficient. Key milestones include

- Develop energy-efficient and flexible active flow control actuators.
- Develop improved models for the operation of flow actuators.
- Demonstrate techniques to incorporate these models into flow simulation schemes.
- Validate models and simulation schemes through comparison with experiments.

Relevance to Strategic Objectives

Capacity (9): Flow control has the potential to improve high-lift performance and therefore reduce takeoff and landing distances, which is a critical challenge for V/STOL and ESTOL airplanes and for accommodating larger conventional airplanes on existing runways. Flow control can also increase cruise efficiency, which reduces fuel usage even as capability increases.

Safety and Reliability (3): Some flow control concepts may improve safety and reliability by improving control of flow separation in unusual flight conditions. However, other concepts may introduce additional complexity that could have an adverse effect on reliability.

Efficiency and Performance (9): Flow control may greatly increase the efficient use of airport infrastructure through improved cruise efficiency. It also may be important for efficient supersonic flight.

Energy and Environment (9): Reduced cruise drag has a direct effect on fuel requirements and en route emissions. Reduced takeoff weight and improved low-speed performance reduces takeoff noise.

Synergies with National and Homeland Security (3): Flow control may improve the fuel efficiency of military aircraft and enhance the performance of aircraft with mission constraints related to separation control (e.g., short-field performance, aft loading ramps, and highly maneuverable aircraft). Transition control may be very important for military supersonic and hypersonic vehicles.

Support to Space (3): Transition management and separation control will impact air-breathing access-to-space vehicles and may be important to reentry heat transfer and to aerodynamics of vehicles in other planetary atmospheres.

Why NASA?

Supporting Infrastructure (3): NASA possesses relevant wind tunnels, infrastructure, and computational infrastructure. The tunnels permit high-Reynolds-number testing (e.g., the National Transonic Facility) and large-scale testing (e.g., National Full Scale Aerodynamics Complex operated by DoD at NASA Ames Research Center). NASA Dryden provides full flight testing capabilities. All of them are deemed critical to this Challenge. However, while NASA's infrastructure is very important, it is not unique in these areas.

Mission Alignment (9): This Challenge is broadly applicable to civil and military aeronautics.

Lack of Alternative Sponsorship (3): Because of the potentially high payoff, research relevant to this Challenge is conducted by many organizations outside NASA, including universities, industry, and DoD. Despite this, NASA could help coordinate these often disparate efforts and contribute directly to meeting this Challenge.

Appropriate Level of Risk (9): Despite the potential advantages of these technologies, they are not extensively used at present due mainly to a lack of understanding, tools, and validation—all areas to which NASA could contribute. This Challenge faces moderate risk.

A3 Novel aerodynamic configurations that enable high performance and/or flexible multimission aircraft

Most classes of aircraft configurations have remained constant for many years (e.g., the tube and wing of a subsonic transport, and the main rotor plus tail rotor of a helicopter). Novel aerodynamic configurations provide substantial opportunities to make long-term breakthroughs in aircraft capabilities. A number of innovative concepts have been proposed and pursued to differing levels. Examples include the blended wing body, canard rotor wing, oblique flying wing, and strut-braced wing. A sustained research program should be promoted to develop novel aircraft configurations, including further development of existing concepts where appropriate, with emphasis on achieving breakthroughs related to the high-priority Strategic Objectives. Specific examples of potential research include novel configurations with stepwise changes in performance, such as concepts with very high cruise efficiency to reduce fuel burn and emissions; STOL aircraft, V/STOL airplanes, and high-speed rotorcraft to achieve significant changes in capacity; and quiet supersonic aircraft to improve the efficiency of the air transportation system.

Other R&T Challenges would also contribute to enabling novel aerodynamic configurations. Advances in flight mechanics and propulsion–airframe integration (R&T Challenge A1) are required to make advanced concept airplanes viable and robust. Flow control (R&T Challenge A2) could significantly enhance the capability of novel configurations, since it could be assumed a priori in the design process rather than added as an improvement to an existing airplane. Research related to the Common Theme of physics-based analysis tools is needed to move beyond empirical design tools.[4] In addition, flight testing is a critical element of a successful research program in novel configurations. Key milestones include

- Develop a family of aircraft configurations with cruise efficiency twice as high as conventional aircraft.
- Demonstrate design approaches to develop novel configurations able to operate from small airfields.
- Validate the ability to predict the performance of novel airframe configurations using data from ground and flight tests.

Relevance to Strategic Objectives

Capacity (9): Novel configurations can enable stepwise changes in aircraft speed and payload, which are the primary independent variables of capacity. Additionally, V/STOL and ESTOL concepts can increase capacity because of their ability to use smaller airports.

Safety and Reliability (3): Advanced configurations can be designed to include safety and reliability requirements up front.

Efficiency and Performance (9): Existing aircraft configurations have been refined to the limits of today's technology. However, existing aircraft technologies can enable new

[4]"Physics-based" refers to the general use of scientific principles in the place of empirical data. It includes the use of principles from chemistry, biology, material science, etc.

classes of configurations that offer additional (and possibly significant) improvements.

Energy and Environment (9): Novel concepts offer significant potential for reducing fuel burn, noise, and emissions.

Synergies with National and Homeland Security (3): Some novel configurations may have military applications. For example, a multirole configuration has the potential to offer military and commercial capabilities from a common platform and a common crew.

Support to Space (1): This Challenge has no impact on this Objective.

Why NASA?

Supporting Infrastructure (3): NASA wind tunnels, simulators, and flight test facilities are important elements of the infrastructure required for the development of novel configurations, but industry and DoD also have relevant infrastructure.

Mission Alignment (9): Novel configurations offer significant potential for breakthroughs in aircraft performance. This Challenge is very relevant to NASA's mission.

Lack of Alternative Sponsorship (3): NASA has an important role to play in meeting this Challenge. However, some R&T will be pursued by industry. It is important that NASA pursue collaborative opportunities, where appropriate.

Appropriate Level of Risk (9): This Challenge faces high risk.

A4a Aerodynamic designs and flow control schemes to reduce aircraft and rotor noise

Reducing the aerodynamic noise from fixed- and rotary-wing aircraft at or near airports is a long-term issue that must be addressed to increase capacity at many airports. Design tools are needed at both the technology level and the aircraft system level, with particular attention to integrated solutions for aerodynamic and operational issues. This Challenge requires a balanced combination of physics modeling, tool development, and experiments.

For large fixed-wing transports, airframe noise on approach has become important, requiring efficient aerodynamic designs for flaps, ailerons, and landing gear that minimize noise radiation. Examples include smoothly varying section- and spanwise profiles to obtain high lift and low noise and high-drag/low-noise devices, which permit steep approaches, mitigating noise on the ground. Key research needs include a basic understanding of the fluid physics of cavity-like flows, unsteady flow–solid surface interactions, flow separation, development of physics-based source noise prediction methods, and development of improved computational aeroacoustic tools.

Many of today's airports now limit operations because of the noise emitted to the surrounding community. Future passenger growth at many airports will be limited if the noise levels emitted by the newer aircraft are not reduced further, thus adversely affecting capacity. Off-loading the main runway of regional jets by using ESTOL aircraft and rotorcraft, thus reducing congestion for larger passenger aircraft on the main runway, will dramatically increase capacity by allowing more takeoffs and landings at existing airports without increasing demand for runway usage (NRC, 2003; FAA, 2000). However, it will only be possible if these ESTOL aircraft and rotorcraft are quiet. To reduce takeoff and landing noise, such aircraft require the development of very high lift devices that are quiet and do not impose undue performance sacrifices Novel technologies are needed to decrease unsteady interactions between the propulsive lift devices and the lifting and control surfaces of the aircraft. These aircraft should also be designed to shield major sources of noise from the ground.

Minimizing impulsive noise generated by rotorcraft requires a better understanding of the aerodynamic state of the rotor, which is a strong function of the rotor blade structural properties. Both main and tail rotor noise are important, depending upon the configuration chosen. Methods of noise reduction include lowering the rotor rpm (which can degrade performance), reducing the major disturbances (shedding of tip vortices) to the following blades though rotor design and/ or vortex-blade position control, integrating advanced control schemes for active rotorcraft noise reduction, and reducing the rotor response to vortex-induced disturbances. Also required are advances is the ability to predict rotor dynamic stall, to predict wake vortex dynamics, and to design rotor blades that minimize the time derivative of the blade's aerodynamic loading response to sharp-edged disturbances. Key milestones include

- Improve techniques for prediction and control of the aeroacoustics associated with high-lift devices, protuberances, and cavities for fixed-wing aircraft.
- Develop techniques for the prediction and design of quiet drag devices for fixed-wing aircraft.
- Improve understanding and modeling of unsteady fluid–structure interactions and resulting noise radiation for rotorcraft and fixed-wing aircraft.
- Demonstrate novel rotor system design tools that can be used to reduce rotor noise with minimum performance sacrifices for rotorcraft.

Relevance to Strategic Objectives

Capacity (9): Reducing aerodynamic noise of both fixed- and rotary-wing aircraft increases capacity by allowing more flights in and out of noise-impacted airports and by facilitating expansion of flight operations at satellite airports.

Safety and Reliability (1): This Challenge has little or no impact on this Objective.

Efficiency and Performance (3): Reducing noise can in some cases also improve aircraft performance. In a more general sense, reducing noise can improve the performance of the overall system by removing noise constraints that are currently impeding system efficiency.

Energy and Environment (9): Reducing noise reduces the environmental impact of aviation.

Synergies with National and Homeland Security (3): Noise reduction is important to make military aircraft more difficult to detect and to make operations near noise-sensitive areas more acceptable to the public.

Support to Space (1): This Challenge has no impact on this Objective.

Why NASA?

Supporting Infrastructure (3): Although NASA does not own all the nation's best aeroacoustic facilities, it does have access to a unique aeroacoustic facility for rotor noise testing: the 40 × 80 × 120-foot acoustically treated tunnel whose size allows far-field acoustic measurements of medium- and large-scale rotorcraft. However, this tunnel is now operated by the U.S. Air Force.

Mission Alignment (9): This Challenge is very relevant to NASA's mission.

Lack of Alternative Sponsorship (3): Industry, the FAA, and the DoD also carry out and sponsor work related to this Challenge.

Appropriate Level of Risk (9): This Challenge faces high risk because most of the easily attained techniques have already been incorporated into aircraft design.

A4b Accuracy of prediction of aerodynamic performance of complex 3-D configurations, including improved boundary layer transition and turbulence models and associated design tools

The aerospace industry lacks computational analysis and design tools that can rapidly and accurately predict complex flow behavior driven by boundary layer transition, flow separation, novel configurations, off-design operation, and multidisciplinary interactions. To meet this need, physics-based design tools must be developed and systematically validated in representative environments. Ideally, they should have the following attributes:

- Adaptive, intelligent, self-generating grids that are easily implemented using simple computer-aided design surface instructions, minimal boundary condition definition, and desktop operation.
- Seamless applicability over the continuum of fluid flows (speed regimes, phase, periodicity) and reference frames.
- Ability to accurately predict transitional and separated flows, validated through experimentation.
- Ability to fully describe the state of the fluid at any point in the solution domain, with useful information on the surfaces.
- Inverse design capability.

The benefit of technologies developed by this Challenge would be enhanced by the parallel development of multidisciplinary design tools to address complex nonlinear interactions and of methods to handle parameter uncertainties in a computationally efficient way (see Challenge A11). Key milestones include

- Develop improved techniques for the prediction of boundary layer transition on 3-D configurations and validate them against ground and flight test data.
- Demonstrate computationally efficient techniques to couple aerodynamic and structural analysis tools.
- Develop structured techniques for predicting performance in the presence of parameter uncertainties.

Relevance to Strategic Objectives

Capacity (3): Meeting this challenge will enable aircraft designs with breakthrough performance, yielding higher capacity.

Safety and Reliability (3): A better understanding of these phenomena will help to identify and mitigate unexpected transitional and separated flows, leading to increased safety.

Efficiency and Performance (9): Improved analysis tools will enable designs with higher efficiency that go beyond current performance limitations caused by nonoptimized and compromised designs.

Energy and Environment (3): The analysis tools will enable designs with higher aerodynamic efficiency and hence reduced fuel burn.

Synergy with National and Homeland Defense (3): Improvements in aerodynamic prediction capabilities are expected to improve DoD aircraft development.

Support to Space (3): Boundary layer transitional behavior through multiple speed regimes remains one of the large uncertainties for air-breathing access-to-space vehicles. Improvements in predictive capabilities are likely to improve the potential performance of these vehicles by reducing design margins.

Why NASA?

Supporting Infrastructure (9): NASA possesses some of the most advanced computational fluids modeling and empirical validation tools in the world, across multiple speed regimes and environments.

Mission Alignment (9): NASA's historical CFD code development provides a foundation for much of industry's design codes today; this Challenge is very relevant to NASA's mission.

Lack of Alternative Sponsorship (3): Industry and DoD have a broad interest in this Challenge.

Appropriate Level of Risk (9): Transitional and separated flows remains a largely unsolved frontier in aeronautics. This Challenge faces significant risk.

A6 Aerodynamics robust to atmospheric disturbances and adverse weather conditions, including icing

Adverse weather conditions, including storms and icing conditions, significantly reduce the capacity and reliability of the air transportation system. Adverse weather also degrades system safety. This issue is of importance to both civil and military aviation. Research is needed to improve the ability to predict and monitor environmental conditions and develop aerodynamic designs and techniques that are robust to adverse conditions.

At present, wind-shear warning systems are built into commercial aircraft, icing hazards are handled by regulatory constraints on flight operations, and prediction techniques are largely empirical. Low-cost techniques to measure environmental conditions ahead of an aircraft should be developed. Examples of promising techniques include microwave, lidar, and laser-acoustic measurement techniques. Efforts to miniaturize and reduce the cost of the measurement equipment should be supported. Techniques to predict and mitigate the impact of adverse environmental conditions on the aircraft operation should be improved. Required improvements include the development of models to predict the impact of multiphase, nonequilibrium situations encountered under icing conditions; validation of icing prediction capabilities to enable a reduction in the high cost of aircraft and helicopter icing certification; and models for the complex-flow, time-dependent, 3-D interactions encountered during wind shear or ambient turbulence on the aircraft flowfield. Key milestones include

- Develop and validate 3-D icing prediction tools.
- Demonstrate systems with improved spatial and temporal measurements of upstream environmental conditions.
- Develop high bandwidth techniques to respond to and mitigate the impact of upstream environmental conditions.

Relevance to Strategic Objectives

Capacity (9): Improving operations in adverse weather conditions will increase capacity by allowing more on-time flights and fewer diversions to other airports.

Safety and Reliability (9): Identifying and mitigating adverse environmental conditions will reduce accident rates and increase the reliability of the air transportation system.

Efficiency and Performance (3): Through mitigating the adverse impacts of weather, air transportation resources can be optimally used with fewer operational constraints.

Energy and Environment (1): This Challenge has no impact on this Objective.

Synergies with National and Homeland Defense (9): DoD and DHS operations will be enhanced if adverse weather conditions become less of a constraint. In particular, DoD capabilities are substantially enhanced if U.S. military forces can operate effectively in weather conditions that degrade the effectiveness of enemy forces, and if enemy forces cannot use adverse weather as cover.

Support to Space (1): Most space operations have the luxury of waiting for favorable environmental conditions. This Challenge has little impact on this Objective.

Why NASA?

Supporting Infrastructure (3): NASA possesses icing wind tunnels and research aircraft that complement infrastructure in academia and industry.

Mission Alignment (9): This Challenge is very relevant to NASA's mission.

Lack of Alternative Sponsorship (3): Both the FAA and industry conduct work related to this Challenge.

Appropriate Level of Risk (3): Advances related to this Challenge can reasonably be expected within the next decade.

A7a Aerodynamic configurations to leverage advantages of formation flying

Formation flight is currently used by military airplanes for a variety of operational reasons, although rarely for drag reduction. Recent breakthroughs in accurate navigation and control make possible extended precision formation flight in cruise and permit exploiting favorable interference to reduce vortex drag. Although this phenomenon is well known, the magnitude of the potential savings is not widely appreciated. Three airplanes flying in formation and designed to best exploit these effects could reduce vortex drag by more than 50 percent in cruise, a greater reduction than that obtainable by extensive laminar flow control on the wing. This would mean roughly a 20 percent reduction in total drag under identical operating condition. However, with less induced drag, the optimum altitude increases, reducing viscous drag as well. The net result is almost a 30 percent reduction in total drag. Unlike the tight formations required for military applications, drag savings are possible even with longitudinal separations of several miles (Spalart, 1998), reducing safety concerns associated with formation flight. Initial NASA work on autonomous formation flight has identified some of the technology requirements for achieving these savings, but considerable research remains in both control methodology and aerodynamic design to take most advantage of the concept. Applications to cargo airplanes, rotorcraft, and even supersonic flight are possible but have not been studied extensively. Aerodynamic challenges include vortex location prediction, sensing and control, and wing design for efficient high-lift cruise. Key milestones include

- Develop improved methods to accurately predict wake vortex evolution.
- Demonstrate design tools for evaluation and optimization of multiple interacting airplanes.
- Validate models and tools for formation flying using ground and flight experiments to evaluate real atmospheric effects.

Relevance to Strategic Objectives

Capacity (3): Formation flying may have some effect on capacity through reduced takeoff weight, which will lead to reduced spacing requirements and fewer noise-related restrictions. Formation flying will also increase en route density, leading to increased capacity.

Safety and Reliability (1): Formation flying introduces additional safety concerns, some of which are ameliorated with large longitudinal spacing.

Efficiency and Performance (9): Reduced fuel consumption and the potential for more efficient cargo delivery systems will improve efficiency and performance.

Energy and Environment (9): Reduced cruise drag has a direct effect on fuel requirements and en route emissions. Reduced takeoff weight and improved low-speed performance reduces takeoff noise.

Synergies with National and Homeland Security (3): Formation flying may improve military airplane fuel efficiency in long-duration operations.

Support to Space (1): This Challenge has no impact on this Objective.

Why NASA?

Supporting Infrastructure (3): NASA has flight-testing and computational infrastructure important to this Challenge. Recent experience with formation flight testing is a key point.

Mission Alignment (9): This research is broadly applicable to civil and military aeronautics.

Lack of Alternative Sponsorship (9): Despite some interest from DoD, this R&T Challenge is sufficiently unconventional, new, and uncertain that sponsorship from other sources is unlikely.

Appropriate Level of Risk (3): There is a very high risk that it will not be possible to implement the technology in a practical way for civil aeronautics.

A7b Accuracy of wake vortex prediction, and vortex detection and mitigation techniques

Wingtip vortices produced by airplanes present a danger to following aircraft, so airplane designs and techniques that mitigate the strength of these vortices, techniques to locate and determine their strength, and techniques to predict their propagation and decay are important factors in minimizing aircraft separation and enhancing safety.[5] (Since aircraft lift is intimately tied to the production of circulation, these vortices cannot be completely eliminated.) Currently, aircraft separation standards are set by conservative estimates of the wake vortex trajectory (generally a sinking trajectory, but also affected by local weather conditions) and decay rate. Techniques to measure the characteristics of upstream wake vortices include lidar and laser-acoustic techniques, but these technologies are currently expensive (limiting their use to larger aircraft) and are less reliable than desired. Research into techniques to predict the formation, trajectory, and decay of vortices needs to be performed. Key milestones include

- Develop numerical techniques to predict accurately wingtip vortex trajectory, strength, and dissipation.
- Validate numerical methods with experiments and flight testing.
- Integrate local weather prediction techniques into larger-scale weather models.
- Demonstrate low-cost techniques for locating and measuring the strength of wake vortices for ground-based and aircraft-based applications.
- Investigate aircraft designs that mitigate the strength of wake vortices.

Relevance to Strategic Objectives

Capacity (9): Accurate prediction and measurement of wake vortex strength and trajectory can reduce airplane spacing requirements, increasing capacity.

Safety and Reliability (9): An improved understanding of wake vortex dynamics will improve safety of following aircraft. Techniques aimed at minimizing the strength or increasing the dissipation rate will present weaker vortices to following aircraft.

Efficiency and Performance (3): Technologies aimed at reducing aircraft spacing will improve the efficiency and performance of the air transportation system.

Energy and Environment (1): This Challenge has no impact on this Objective.

Synergy with National and Homeland Defense (1): This Challenge has minimal impact on this Objective.

Support to Space (1): This Challenge has no impact on this Objective.

Why NASA?

Supporting Infrastructure (3): NASA and the aircraft industry have conducted research into wake physics modeling

[5]The scope of this Challenge does not include and would not directly apply to helicopter blade wakes.

and measurement; both have important infrastructure to bring to this area.

Mission Alignment (9): This Challenge is very relevant to NASA's mission.

Lack of Alternative Sponsorship (3): Relevant research conducted by NASA is synergistic and closely aligned with similar work of the FAA.

Appropriate Level of Risk (3): Good progress can be expected in this area within the next decade.

A9 Aerodynamic performance for V/STOL and ESTOL, including adequate control power

Since 2001, the U.S. aviation industry has undergone a profound change. On many routes, regional jets have replaced propeller-driven aircraft, which used different runways and flew at lower altitudes than the large commercial transports. Regional jets are using the same runways and airways as the large transports. This increases congestion and delays at major airports, which degrades the performance of the entire air transportation system.

Powered lift (V/STOL and ESTOL) airplanes and rotorcraft may provide solutions to this problem. V/STOL jets with highly swept wings can operate from the short runways previously used by straight-wing propeller-driven transports. These aircraft, together with rotorcraft, may also be able to operate from taxiways and other paved areas at major airports or smaller regional airports. Any of these applications would relieve congestion on the main runways at major airports. In responding to natural disasters and carrying out military operations, low-cost VTOL tactical transports would be able to operate from short, austere landing fields near the focus of attention (e.g., the location of injured civilians or troops, battle areas, and landslides).

These aircraft will require advances in aerodynamics; propulsion; acoustics; stability and control; structures and materials; and guidance, navigation, and communications. Specific aerodynamic issues that require attention include development of a low-drag, high-lift system, simple boundary layer control systems to prevent wing leading-edge separation, systems to provide pitch trim and control power at low speeds, a reversing deflecting exhaust nozzle, and wing design and fuselage shaping to reduce cruise drag in the transonic regime.

An important task for research related to rotorcraft and fixed-wing VTOL aircraft is improving hovering and cruise efficiency. Reductions in downward forces in near-hovering flight dramatically improve the payload capability of tilt-rotor and powered-lift aircraft. Active control of large separation regions on these aircraft through blowing, zero-mass effectors and integrated mechanical devices are promising methods of reducing download. Active twist control of the rotor also allows the rotorcraft to be designed to better match the hover and cruise design conditions, thereby improving efficiency. Active control of separation regions and smart design guided by high-fidelity codes will decrease cruise drag and greatly improve the performance of V/STOL airplanes and rotorcraft. Validated codes require interdisciplinary research efforts as well as efforts to improve separation prediction and control. Key milestones include

- Develop low-drag, high-lift systems.
- Demonstrate systems to provide pitch trim and control power at low speeds.
- Develop new techniques for active twist control of rotors.
- Demonstrate low-cost, simple flow control techniques for prevention of leading-edge separation from V/STOL wings.
- Improve wing design and fuselage shaping to reduce transonic cruise drag.

Relevance to Strategic Objectives

Capacity (9): This Challenge could greatly increase capacity by shifting regional jets from the major runways to smaller runways and/or taxiways and by enabling the use of smaller regional airports.

Safety and Reliability (3): Developing adequate control power at high lift, low speed would increase safety.

Efficiency and Performance (3): Novel methods of improving the performance of V/STOL airplanes (e.g., download reduction and avoiding flow separation regions in cruise) and rotorcraft can improve their efficiency.

Energy and Environment (1): This Challenge has no impact on this Objective.

Synergies with National and Homeland Security (3): Mobility, especially over unprepared or short fields, is very important for quick-response situations. V/STOL and ESTOL airplanes would enhance these capabilities.

Support to Space (1): This Challenge has no impact on this Objective.

Why NASA?

Supporting Infrastructure (3): Large-scale testing is critical to this Challenge. The best large-scale ground testing facility is the 40 × 80 × 120 ft tunnel at NASA Ames. NASA has access to this facility, although it is now operated by the U.S. Air Force. NASA also has smaller scale facilities that can support this Challenge.

Mission Alignment (9): This Challenge capitalizes on NASA in-house expertise in powered lift and rotorcraft development and is very relevant to NASA's mission.

Lack of Alternative Sponsorship (3): Industry and DoD also carry out and sponsor work related to this Challenge.

Appropriate Level of Risk (9): This Challenge faces high risk.

A10 Techniques for reducing/mitigating sonic boom through novel aircraft shaping

Safe, efficient, cost-effective, environmentally acceptable supersonic flight over land remains elusive nearly 60 years after airplanes broke the sound barrier. The principal remaining problems are sonic boom mitigation, public acceptance, and sustained supersonic flight performance. Today, federal regulations prohibit civil supersonic flight over land. If this regulatory barrier can be overcome, it will probably stimulate investment that would overcome the other barriers and help usher in a new era of time-critical air travel. Building on the recent in-flight validation of NASA's shaped sonic boom persistence theory, a robust and comprehensive plan of research for technology maturation and tool development should be pursued to determine if practical supersonic airplanes can be developed whose sonic boom is acceptable to the public (Pawlowski et al., 2005). Such a plan should comprise the determination of what level of sonic boom is acceptable to the public; community exposure testing; aircraft shaping techniques that result in low-amplitude, acceptable acoustic signature with minimal performance impact; critical propulsion–airframe integration technologies commensurate with low-boom design; aircraft and acoustic scaling methodologies; sensitivities to off-design conditions under a variety of atmospheric conditions; rapid and inverse computational design tools that address multiple design constraints; systematic validation through ground and flight test; and metrics to assess progress and guide continuation according to the plan. This Challenge is closely tied to Challenge B8. Key milestones include

- Develop guidelines for allowable exposure of the public to sonic booms.
- Develop accurate techniques for the prediction of sonic boom propagation through the atmosphere under realistic environmental conditions.
- Demonstrate novel aircraft shapes that minimize sonic boom levels.

Relevance to Strategic Objectives

Capacity (3): Enabling supersonic flight over land will increase capacity by moving airplanes through the system more rapidly, although, at least initially, such a capability will affect only a small segment of the population.

Safety and Reliability (1): This Challenge has no impact on this Objective.

Efficiency and Performance (3): Success in this Challenge will usher in a new era of time-critical travel.

Energy and Environment (9): This Challenge will profoundly reduce the noise produced by supersonic airplanes.

Synergy with National and Homeland Defense (3): This Challenge will enable quiet supersonic airplanes, which will also benefit military missions.

Support to Space (1): This Challenge has no impact on this Objective.

Why NASA?

Supporting Infrastructure (9): NASA possesses unique empirical facilities, including a supersonic flight test corridor, and extensive code, test, and measurement resources, reflecting NASA's large historical investment in supersonic airplane programs such as the High Speed Research Program.

Mission Alignment (9): This Challenge is very relevant to NASA's mission.

Lack of Alternative Sponsorship (3): Broad interest has been shown by industry.

Appropriate Level of Risk (9): Achievability of efficient, cost-effective, environmentally acceptable supersonic cruise airplanes remains highly uncertain, and the risk associated with research aimed at sonic boom reduction is high.

A11 Robust and efficient multidisciplinary design tools

Multidisciplinary design tools are pervasive in aeronautics. A multidisciplinary, integrated system-level design approach to assessing potential costs, benefits, and risks would help advance aerodynamic technologies, shorten the design cycle time for conventional aircraft, and develop novel aircraft configurations. Tools that couple a small number of disciplines have reached a level of maturity and fidelity that should be exploited by design tools. For example, aeroelastic design tools are now within reach that couple CFD and finite-element analyses for full aircraft configurations. More recently, multidisciplinary design tools have begun to incorporate a broader range of disciplines, and techniques such as multidisciplinary design optimization (MDO) have been used to a limited extent in aircraft conceptual design.

One of the major limitations of past efforts to create MDO tools has been a low level of fidelity, driven by a lack of physics-based models that are sufficiently efficient for use at the system design level. In addition, MDO tools have often been developed and applied for very specific applications and flight conditions, so they lack flexibility. Key challenges associated with next-generation multidisciplinary design tools include tool fidelity, computational efficiency, and the ability to handle parameter uncertainties. The practical resolution of these challenges will require fundamental research efforts in physics-based models for use in design tools (see R&T Challenge A4b), new design methodologies that can seamlessly manage models of multiple fidelities for the various components of the system, methods to increase the computational efficiency of tools, methods to handle complex interactions with high accuracy, and automated techniques for handling and propagating parameter uncertainties throughout the design. Key milestones include

- Develop and validate physics-based models to predict performance for novel aircraft configurations.
- Assess a family of aircraft configurations with major improvement in cruise efficiency, including a quantitative description of the benefits and risks.
- Assess novel concepts for flexible multimission aircraft, including a description of potential benefits in performance and cost.
- Conceive design approaches to develop novel V/STOL and ESTOL configurations.
- Validate design codes to predict the performance of novel airframe configurations by comparing code predictions with ground and flight tests.

Relevance to Strategic Objectives

Capacity (3): The development of improved predictive capabilities and multidisciplinary tools will lead to flexible aircraft capable of increasing capacity.

Safety and Reliability (3): A capability for design under uncertainty will help improve safety and reliability.

Efficiency and Performance (9): Multidisciplinary design tools are a key enabling technology for achieving revolutionary aircraft designs with improved performance.

Energy and Environment (9): The novel designs that could be achieved with these tools could have a significant impact on energy and environmental issues.

Synergies with National and Homeland Defense (3): Multidisciplinary design tools are applicable to military aircraft.

Support to Space (3): Multidisciplinary design tools could also be used to improve space vehicle design.

Why NASA?

Supporting Infrastructure (3): NASA has a strong track record of R&T in system design tools and multidisciplinary design optimization. NASA has relevant computational infrastructure, although it is not unique.

Mission Alignment (9): This Challenge is very relevant to NASA's mission.

Lack of Alternative Sponsorship (3): Both industry and other government agencies are pursuing work in this Challenge, although NASA has a unique opportunity to provide a bridge between academic research and industrial needs.

Appropriate Level of Risk (3): This Challenge faces low risk.

A12 Accurate predictions of thermal balance and techniques for the reduction of heat transfer to hypersonic vehicles

Air-breathing access-to-space and reentry vehicles must operate in a stressing aerothermodynamic environment that requires high-performance, robust thermal protection systems (TPSs). The cost and feasibility of air-breathing launch vehicles are extremely sensitive to mass. Therefore, more accurate techniques are needed to (1) predict aerothermal loads (and thus decrease the design margins associated with the TPS) and (2) minimize both local and integrated heat transfer to the vehicle, which can significantly increase system performance. Specific needs include improved predictions of how the following factors affect heat transfer in the hypersonic environment: ablation, boundary layer transition, highly cooled walls (which affects boundary layer turbulence), wall chemistry (including catalytic effects), and radiating shock layers.

Relevant techniques include novel aerodynamic shaping, active flow control, transpiration cooling, and emissivity control. Research is also required to determine the utility of plasma aerodynamic and magnetohydrodynamic flow manipulation for heat transfer reduction. Key milestones include

- Improve models for predicting the effects of ablation on heat transfer.
- Develop a high-fidelity model for radiating shock layers.
- Develop turbulence models validated against experimental data for highly cooled walls.

Relevance to Strategic Objectives

Capacity (1): This Challenge has little impact on this Objective.

Safety and Reliability (1): This Challenge has little impact on this Objective.

Efficiency and Performance (3): Development of accurate heat transfer prediction techniques and mitigation of local high-heat-transfer regions will improve the performance of supersonic and hypersonic vehicles.

Energy and Environment (1): This Challenge is principally aimed at hypersonic vehicle applications, so this Challenge has no impact on this Objective.

Synergies with National and Homeland Defense (9): Development of robust supersonic and hypersonic systems can help address DoD missions in areas such as missile defense, time-critical strike, prompt global strike, and access to space.

Support to Space (9): Heat transfer is an important issue associated with air-breathing access-to-space and reentry systems.

Why NASA?

Supporting Infrastructure (9): NASA possesses unique capabilities related to hypersonic aerothermodynamics, including NASA Langley's Mach 10 aerothermodynamic wind tunnel and 8-ft high-temperature tunnel.

Mission Alignment (9): Hypersonic aerothermodynamics is a fundamental enabling R&T Challenge that is very relevant to NASA's mission.

Lack of Alternative Sponsorship (3): Both NASA and DoD conduct research to improve techniques to predict and mitigate the adverse impacts of heat transfer.

Appropriate Level of Risk (9): This Challenge faces moderate risk.

A13 Low-speed takeoff and landing flight characteristics for access-to-space vehicles

Air-breathing access-to-space vehicles capable of horizontal takeoff and landing hold significant promise in providing low-cost access to space. Mission and operational flexibility is greatly enhanced by the ability to operate from sites similar to those utilized for conventional aircraft. However, factors such as high wing sweep, sharp leading edges, and use of a propulsion system designed for hypersonic flight significantly increase runway length and require greatly modified flight corridors relative to conventional aircraft.

Research in this area should include, but not be limited to, development and evaluation of high-lift systems, active and passive flow control, consideration of two-stage-to-orbit configurations, optimum fuselage and wing shaping, morphing structures, and configurations with enhanced propulsion–airframe integration for improved low-speed flight characteristics. Strategies to improve takeoff and landing performance without significantly impacting payload capacity, range, fuel, and structural weight and cost are vital because the overall efficiency and cost of space launch vehicles is very sensitive to weight. More accurate tools are needed to predict the effects of flow control, vehicle shaping, and propulsion–airframe integration techniques on boundary layer behavior and flow separation. In addition, integrated system analysis tools must be developed and validated to predict how modifications to improve takeoff and landing performance will affect vehicle performance throughout the flight profile. The aerodynamic tools and novel strategies for improving low-speed performance must be validated by relevant ground and flight tests. Key milestones include

- Validate predictive capability for integrated vehicle aerodynamics in the presence of the runway.
- Develop strategies to maintain efficient low-speed aerodynamic performance for hypersonic vehicle designs.

Relevance to Strategic Objectives

Capacity (1): This Challenge will have no impact on this Objective in the time frame considered.

Safety and Reliability (3): Some of the flow control and separation mitigation techniques that address this Challenge may be applicable to subsonic commercial aircraft low-altitude flight and provide improvement in safety and reliability during the takeoff and landing phases of flight.

Efficiency and Performance (1): This Challenge has little to no impact on this Objective.

Energy and Environment (1): This Challenge has little to no impact on this Objective.

Synergies with National and Homeland Security (3): Some of the flow control and separation mitigation techniques that address this Challenge may be applicable to current and future high-speed military aircraft.

Support to Space (9): This Challenge could significantly improve the operational flexibility of access-to-space missions.

Why NASA?

Supporting Infrastructure (3): NASA possesses wind tunnels that are well suited for this Challenge. NASA expertise and computational facilities are appropriate but not unique.

Mission Alignment (9): This Challenge is very relevant to NASA's space exploration mission.

Lack of Alternative Sponsorship (9): Horizontal takeoff and landing vehicles will not be developed without federal investment, and the fundamental physics studies on boundary layers and separation are appropriate long-term research efforts for NASA. No significant, sustained work is done in this area by other government entities or industry.

Appropriate Level of Risk (9): This Challenge faces high risk.

A14 Efficient control authority of advanced configurations to permit robust operations at hypersonic speeds and for access-to-space vehicles

Hypersonic vehicles have the potential to provide affordable access to space, safe and predictable entry from space, flight in other planetary atmospheres, prompt global reach, and missile defense. Aerodynamic configurations optimized for hypersonic cruise or acceleration present significant design challenges. They can prove inefficient at off-design conditions, their performance is affected by interactions with the aerodynamic flow through the engine flow path, and they frequently require control at high altitudes.

Control of vehicles operating in this regime will require a better physics-based understanding of the flow field characteristics. The characterization and prediction of the effect of boundary layer transition on aerodynamic configurations remains a significant challenge. More accurate knowledge of the state of the boundary layer, transition location, areas of separation, and certain viscous interactions, including shock-wave–boundary-layer interaction, will facilitate development of configurations with adequate control authority. The implementation of flow control concepts through the use of novel actuation (e.g., plasma/magnetohydrodynamic concepts) may prove useful. In addition, characterizing the salient physics of improved measurement techniques and the use of high-quality test data to validate computational techniques will be required to adequately design and develop configurations with control authority at hypersonic speeds. Key milestones include

- Develop techniques to accurately predict flow control authority in shock-dominated flows and in transitional flows.
- Demonstrate novel flow control techniques applicable to hypersonic vehicles.
- Develop novel ground and flight test instrumentation techniques for validation of analytical and computational models.

Relevance to Strategic Objectives

Capacity (1): This Challenge has no impact on this Objective.

Safety and Reliability (1): This Challenge has no impact on this Objective.

Efficiency and Performance (3): Research in this Challenge will extend the understanding of complex fluid physics, which will increase understanding at lower speeds as well.

Energy and Environment (1): This Challenge has no impact on this Objective.

Synergies with National and Homeland Security (9): Most high-speed applications will benefit from the improved control authority techniques investigated by this Challenge.

Support to Space (9): This Challenge provides a significant contribution to space exploration and access-to-space missions.

Why NASA?

Supporting Infrastructure (3): NASA possesses some relevant capabilities in the areas of hypersonic aerodynamics, but several important assets exist at DoD as well.

Mission Alignment (9): Hypersonic aerodynamics is a fundamental enabling technology that is very relevant to NASA's mission.

Lack of Alternative Sponsorship (3): Both NASA and DoD conduct research associated with improvements in stability and control of hypersonic vehicles to better understand and control vehicle flight dynamics.

Appropriate Level of Risk (9): This Challenge faces high risk.

A15 Decelerator technology for planetary entry

Effective and reliable decelerator technologies for Earth reentry and planetary entry are needed to support NASA's space exploration mission. Such technologies must yield acceptable deceleration loads (which are most stringent for crewed vehicles). In addition, some missions require the generation of lift for improved cross-range and control of the entry trajectory. The key task for this Challenge is to provide the required aerodynamic loads in a robust and reliable system while yielding efficient, low-mass thermal protection.

Relevant research includes characterization of planetary atmospheric conditions and chemistry; development and evaluation of optimum vehicle shapes and novel configurations; use of parachutes, parafoils, ballutes; active and passive flow control; and accurate prediction of aerodynamic and thermal loads and trajectories during aerocapture and aero-assisted orbital transfer operations. Improved accuracy is needed in tools designed to predict thermal loads in the planetary atmosphere under consideration and to predict unsteady aeroelastic effects on structures that are highly flexible or whose shape may vary (e.g., due to ablation). In addition, integrated system analysis tools are also needed. Validation of the aerodynamic tools and novel strategies for improving performance and reliability of decelerator technologies requires ground and flight testing. Key milestones include

- Conceive novel approaches for deceleration in planetary atmospheres.
- Improve the computational efficiency of time-dependent aerothermal prediction tools.
- Develop integrated system analysis tools for planetary entry system design.

Relevance to Strategic Objectives

Capacity (1): This Challenge has no impact on this Objective.

Safety and Reliability (1): This Challenge has no impact on this Objective.

Efficiency and Performance (1): This Challenge has no impact on this Objective.

Energy and Environment (1): This Challenge has no impact on this Objective.

Synergies with National and Homeland Security (3): Ballistic and hypersonic cruise missiles will derive some benefit from the thermal protection elements of this Challenge.

Support to Space (9): This Challenge significantly contributes to space exploration and access-to-space missions.

Why NASA?

Supporting Infrastructure (3): NASA possesses relevant and unique high-enthalpy tunnels and reacting flow expertise. Other tunnels, materials testing, and computational facilities needed for this topic also exist elsewhere (e.g., the DoD facilities at the Arnold Engineering and Development Center and the Calspan–University of Buffalo Research Center).

Mission Alignment (9): The Challenge is very relevant to NASA's space exploration mission.

Lack of Alternative Sponsorship (9): Only NASA has the mission of space exploration, so no work is done in this Challenge by other government entities or by industry without NASA's involvement.

Appropriate Level of Risk (9): The fundamental physics studies on heat transfer and prediction require substantial long-term research. NASA, however, has unique thermal protection and planetary operations experience. This Challenge faces moderate risk.

A16 Low-Reynolds-number and unsteady aerodynamics for small UAVs

This Challenge deals with the special aerodynamic issues associated with small UAVs (wing spans on the order of 6 inches) that are capable of high maneuverability and flight in confined spaces. These vehicles are of interest to DoD and DHS for missions that involve autonomous reconnaissance in urban areas, including inside buildings. The vehicles are also relevant to flight in the martian atmosphere. Prevalent concepts include flapping wings that mimic birds or insects and rotating, lifting rotors. The dominant aerodynamics for these vehicles involves highly unsteady, dynamic-stall, vortex-driven flows. The lifting capability of state-of-the-art flapping-wing vehicles is too low. This Challenge seeks to enhance dynamic lift through different concepts that might involve reflexive wing structures, flow-energy extraction, and active flow control. This will require physics-based flow models and time-resolved experiments for highly unsteady flows to maximize lift, maneuverability, and flight control of these vehicles. Key milestones include

- Generate and validate time-resolved experimental data for highly unsteady low-Reynolds-number flows.
- Develop feasible approaches for flow-energy extraction and reflexive wing structures for small UAVs.

Relevance to Strategic Objectives

Capacity (1): This Challenge has no impact on this Objective.

Safety and Reliability (1): This Challenge has no impact on this Objective.

Efficiency and Performance (3): Some of the unsteady aerodynamics associated with this Challenge also applies to rotorcraft.

Energy and Environment (1): This Challenge has no impact on this Objective.

Synergies with National and Homeland Security (9): This Challenge addresses highly maneuverable autonomous flight in confined urban environments, which is relevant to the missions of DoD and DHS.

Support to Space (3): This Challenge is relevant to space missions involving aircraft in the martian atmosphere.

Why NASA?

Supporting Infrastructure (3): NASA has facilities in which to perform this research, although they are not unique.

Mission Alignment (9): This Challenge is very relevant to NASA's mission.

Lack of Alternative Sponsorship (3): DoD also supports relevant research, although flight in planetary atmospheres is only of interest to NASA.

Appropriate Level of Risk (9): This Challenge faces moderate risk; reasonable progress can be expected during the next decade of research.

A17 Low-drag airship designs to enable long-duration stratospheric flight

Airships capable of operating in the stratosphere for extended periods of time are being investigated for communications relay and surveillance applications. They offer the ability to provide wide area coverage from a persistent platform, while enabling economical retrieval of payloads for repair or replacement. These vehicles build on technologies available for existing airships that operate at lower altitudes, but require advances in lightweight hull fabrics, efficient energy generation and storage, low-drag aerodynamic configurations, and efficient propulsion systems capable of operation in a low-Reynolds-number environment.

Minimizing the drag of these vehicles is important, since a significant portion of the required onboard energy is expended by the propulsion system for station-keeping against winds. Aerodynamic issues of interest include boundary layer transition prediction for flexible thin-wall hulls, prediction of viscous drag in the regions of turbulent flow, boundary layer separation control, and unsteady aerodynamics associated with gossamer structures. Key milestones include

- Develop techniques for prediction of boundary layer transition in unsteady flows.
- Develop coupled aerodynamic and structural analysis tools for predicting the aerodynamics associated with deformable vehicles.

Relevance to Strategic Objectives

Capacity (1): This Challenge has no impact on this Objective.

Safety and Reliability (3): Stratospheric airships may provide persistent surveillance of low-flying aircraft or weather patterns, but the expected impact on this Objective is limited.

Efficiency and Performance (1): This Challenge has no impact on this Objective.

Energy and Environment (3): This Challenge could enhance environmental sensing.

Synergies with National and Homeland Security (9): The development of efficient and affordable stratospheric airships will provide a new class of vehicle for providing persistent surveillance and communication relay. Both DoD and DHS are investigating high-altitude airships for these applications.

Support to Space (1): This Challenge has no impact on this Objective.

Why NASA?

Supporting Infrastructure (3): NASA has both computational and experimental tools relevant to the development of airship technology. DoD and industry are capable of contributing infrastructure as well.

Mission Alignment (3): Development of fundamental technologies for gossamer vehicles is very relevant to NASA's mission.

Lack of Alternative Sponsorship (3): Both DoD and DHS are currently investing in the relevant technologies.

Appropriate Level of Risk (9): This Challenge faces high risk, but significant progress in understanding the underlying limitations of the technology could be made within the next 10 years.

A18 Prediction of communication capability through reentry trajectory and techniques to mitigate impact of communication blackouts

Vehicles returning from space through the atmosphere will encounter regions where communication is greatly reduced and at times nonexistent due to interaction with the atmosphere. At hypersonic velocities, aerothermal stresses create a charged flow field around the vehicle, eventually eliminating communication through this highly charged shear layer. Maintaining continuous communication with these vehicles is important for accurate control, targeting, and continuous health monitoring. As hypersonic concepts move to flight experimentation, robust communication with test vehicles will be increasingly important for the test range to have adequate control, destruct authority, and capability to download sufficient data throughout the flight regime.

Research is needed to understand and characterize the shear and boundary layers, and the relationship between signal transmission and the flow physics. Several promising concepts are under investigation for minimizing the effects of this charged flow field, creating innovative designs and providing information through test and analysis.

Currently, communication blackout is tolerated and systems have been designed around this problem by accepting a ballistic trajectory until communication is restored. Several technologies have been demonstrated in limited capacity on the ground and in flight but not one has completely alleviated the problem. Key milestones include

- Improve computational tools for predicting electron number density around 3-D shapes in hypervelocity flows.
- Develop sharp-leading-edge technology to minimize electron production in stagnation regions.

Relevance to Strategic Objectives

Capacity (1): This Challenge has no impact on this Objective.

Safety and Reliability (1): This Challenge has no impact on this Objective.

Efficiency and Performance (1): This Challenge has no impact on this Objective.

Energy and Environment (1): This Challenge has no impact on this Objective.

Synergies with National and Homeland Security (9): This Challenge is relevant to DoD strategic and strike systems.

Support to Space (9): This Challenge is relevant to space exploration and specifically reentry from space.

Why NASA?

Supporting Infrastructure (3): NASA possesses relevant capabilities in hypersonics, but relevant assets exist in DoD as well.

Mission Alignment (9): This Challenge is very relevant to NASA's mission.

Lack of Alternative Sponsorship (3): Both NASA and DoD conduct research associated with improvements in reentry physics.

Appropriate Level of Risk (3): This Challenge faces very high risk.

A19 Aircraft protective countermeasures based on a range of small deployed air vehicles

This Challenge deals with the special aerodynamic issues associated with small subsonic or supersonic flight vehicles that might be deployed from commercial aircraft as a countermeasure to an attack from ground- or aircraft-launched missiles. The vehicles need to be highly maneuverable, autonomous, able to station-keep long enough to allow the passenger aircraft to escape the airspace, and inexpensive enough to be adopted by a wide range of civil aircraft. Relevant technologies include flow control approaches for flight control in subsonic (and, perhaps, supersonic) regimes. Possible vehicle configurations include small missile bodies and miniature delta-wing aircraft. Achieving maximum flight control will be critical and may involve advanced flow controls that manipulate coherent vortices and shock waves to produce large asymmetric surface pressure loading and resultant force vectoring. Key milestones include

- Develop validated techniques for deployment of small, lightweight vehicles from larger aircraft.

- Design novel aerodynamic configurations for small missile defense vehicles.

Relevance to Strategic Objectives

Capacity (1): This Challenge has no impact on this Objective.

Safety and Reliability (3): This Challenge would enhance safety and reliability in the face of a terrorist attack.

Efficiency and Performance (1): This Challenge has no impact on this Objective.

Energy and Environment (1): This Challenge has no impact on this Objective.

Synergies with National and Homeland Security (9): This Challenge enables highly maneuverable, autonomous flight vehicles that have national and homeland defense applications.

Support to Space (1): This Challenge has no impact on this Objective.

Why NASA?

Supporting Infrastructure (3): NASA has facilities that can perform this research, although they are not unique.

Mission Alignment (3): This Challenge is relevant to NASA's mission.

Lack of Alternative Sponsorship (3): DoD already supports relevant R&T, although it is focused on protecting military aircraft.

Appropriate Level of Risk (3): This Challenge faces low risk.

REFERENCES

Federal Aviation Administration (FAA). 2000. Federal Aviation Administration Office of System Capacity, Aviation Capacity Enhancement Plan. Washington, D.C.: FAA. Available online at <www.faa.gov/ats/asc/00ACE.html>.

Kuchemann, D. 1978. The Aerodynamic Design of Aircraft. Oxford, England: Pergamon Press.

National Research Council (NRC). 2003. Securing the Future of U.S. Air Transportation: A System in Peril. Washington, D.C.: The National Academies Press. Available online at <http://fermat.nap.edu/catalog/10815.html>.

Pawlowski, J., D. Graham, C. Boccadoro, P. Coen, and D. Maglieri. 2005. Origins and Overview of the Shaped Sonic Boom Program. AIAA-2005-0005, 43rd AIAA Aerospace Sciences Meeting and Exhibit, Reno, Nev., January 10-13.

Spalart, P.R. 1998. Airplane trailing vortices, Annual Review of Fluid Mechanics 30: 107-138.

Spence, D.A. 1956. The lift of a thin, jet-flapped wing. Proceedings of the Royal Aeronautical Society A238: 46-48.

B

R&T Challenges for Propulsion and Power

A total of 16 R&T Challenges were prioritized in the propulsion and power Area. Table B-1 shows the results. The R&T Challenges are listed in order of NASA priority. National priority scores are also shown.[1] This appendix contains a description of each R&T Challenge, including milestones and an item-by-item justification for each score that appear in Table B-1.[2]

B1a Quiet propulsion systems

The adverse environmental by-products of aviation—primarily noise and emissions—are major constraints on the growth of aviation. Public concerns over the environmental impact of aircraft and airport operations, along with increasingly strict legal and regulatory requirements, can severely constrain the ability of civil aviation to meet national and global needs for mobility, increased market access, and sustained economic growth. Aircraft noise concerns include takeoff and landing noise; taxi and engine run-up noise; flyovers at cruise altitude over very quiet areas; and sonic booms associated with supersonic flight.

Figure B-1 shows how the impact of aviation noise on people living around airports has declined in the United States. It contrasts the growth of air travel with the reduction in the number of people exposed to 65-decibel (dB) day-night average sound level (DNL), which is what the federal government has defined as the "significant noise level." In 1975, approximately 7 million people were exposed to significant aircraft noise. Since 1975, the number of persons exposed to significant noise levels has greatly declined even as air travel has grown dramatically. One of the most effective federal policies implemented to reduce aviation noise was the transition of commercial aircraft to quieter models. The availability of low-noise technologies, such as high-bypass-ratio engines, contributed significantly to this transition.

Assuming the industry's continued recovery, and given the goal of doubling capacity over the next 10 to 35 years, future abatement efforts may need to achieve noise levels, as recognized by authorities both in the United States (NASA, 2003) and Europe (ACARE, 2001). The environmental impact of aircraft noise is projected to remain roughly constant in the United States for the next several years and then increase as air travel growth outpaces expected technological and operational advancements (Waitz et al., 2004). Furthermore, the public currently reports considerable annoyance even when DNLs are below 65 dB. Regulatory actions to limit or reduce noise exposure will likely lead to even more stringent limits.

Meeting future noise targets will be extremely challenging and will require continued fundamental research in noise generation and transmission phenomena and advanced propulsion technologies. Since the revolutionary introduction of the turbofan, engine source noise reductions have been more evolutionary, with incremental advances such as high-bypass-ratio engines and better acoustic liner technology. The development of validated noise prediction tools by NASA will greatly aid the development of quieter engines. NASA should emphasize physics-based noise source models that can distinguish core noise from other engine noise sources to identify source mechanisms. Research is needed to reduce the noise of engine systems, including fan noise, jet noise, and core noise. Research should also encompass systems analysis; advanced concepts, such as adaptable chevrons; the community impact of aircraft noise; and improved metrics to quantify and mitigate these impacts.

Noise and emissions are not independent phenomena in aircraft engines. There are physical interrelationships between noise and emissions and among various types of emissions, so that when one is decreased, another may be increased.

[1] The prioritization process is described in Chapter 2.

[2] The technical descriptions for the first 10 Challenges listed below contain substantially more detail than the technical descriptions for these Challenges as they appear in Chapter 3.

TABLE B-1 Prioritization of R&T Challenges for Area B: Propulsion and Power

R&T Challenge	Strategic Objective: Capacity	Strategic Objective: Safety and Reliability	Strategic Objective: Efficiency and Performance	Strategic Objective: Energy and the Environment	Strategic Objective: Synergies with Security	Strategic Objective: Support to Space	National Priority	Why NASA?: Supporting Infrastructure	Why NASA?: Mission Alignment	Why NASA?: Lack of Alternative Sponsors	Why NASA?: Appropriate Level of Risk	Why NASA Composite Score	NASA Priority Score
Weight	5		3		1			1/4 each					
B1a Quiet propulsion systems	9	1	3	9	3	1	**90**	3	9	3	9	**6.0**	**540**
B1b Ultraclean gas turbine combustors to reduce gaseous and particulate emissions in all flight segments	9	1	3	9	3	1	**90**	3	9	3	9	**6.0**	**540**
B3 Intelligent engines and mechanical power systems capable of self-diagnosis and reconfiguration between shop visits	3	9	3	3	3	1	**82**	3	9	3	9	**6.0**	**492**
B4 Improved propulsion system fuel economy	3	1	9	9	3	1	**78**	3	9	3	9	**6.0**	**468**
B5 Propulsion systems for short takeoff and vertical lift	9	1	3	3	3	1	**72**	3	9	3	9	**6.0**	**432**
B6a Variable-cycle engines to expand the operating envelope	3	1	9	3	3	9	**68**	3	9	3	9	**6.0**	**408**
B6b Integrated power and thermal management systems	3	1	9	3	3	9	**68**	3	9	3	9	**6.0**	**408**
B8 Propulsion systems for supersonic flight	3	1	3	1	9	9	**50**	9	9	3	9	**7.5**	**375**
B9 High-reliability, high-performance, and high-power-density aircraft electric power systems	1	3	9	3	3	3	**62**	1	9	3	9	**5.5**	**341**
B10 Combined-cycle hypersonic propulsion systems with mode transition	1	1	3	1	9	9	**40**	9	9	3	9	**7.5**	**300**
B11 Alternative fuels and additives for propulsion that could broaden fuel sources and/or lessen environmental impact	3	1	3	9	3	1	**60**	3	3	3	9	**4.5**	**270**
B12 Hypersonic hydrocarbon-fueled scramjet	1	1	3	1	9	9	**40**	9	3	3	9	**6.0**	**240**
B13 Improved propulsion system tolerance to weather, inlet distortion, wake ingestion, bird strike, and foreign object damage	3	9	3	1	3	1	**76**	3	3	3	3	**3.0**	**228**
B14 Propulsion approaches employing specific planetary atmospheres in thrust-producing chemical reactions	1	1	1	1	1	9	**26**	3	9	9	9	**7.5**	**195**
B15 Environmentally benign propulsion systems, structural components, and chemicals	1	1	1	9	3	1	**44**	3	3	3	3	**3.0**	**132**
B16 Reduced engine manufacturing and maintenance costs	3	3	3	3	3	1	**52**	3	1	1	3	**2.0**	**104**

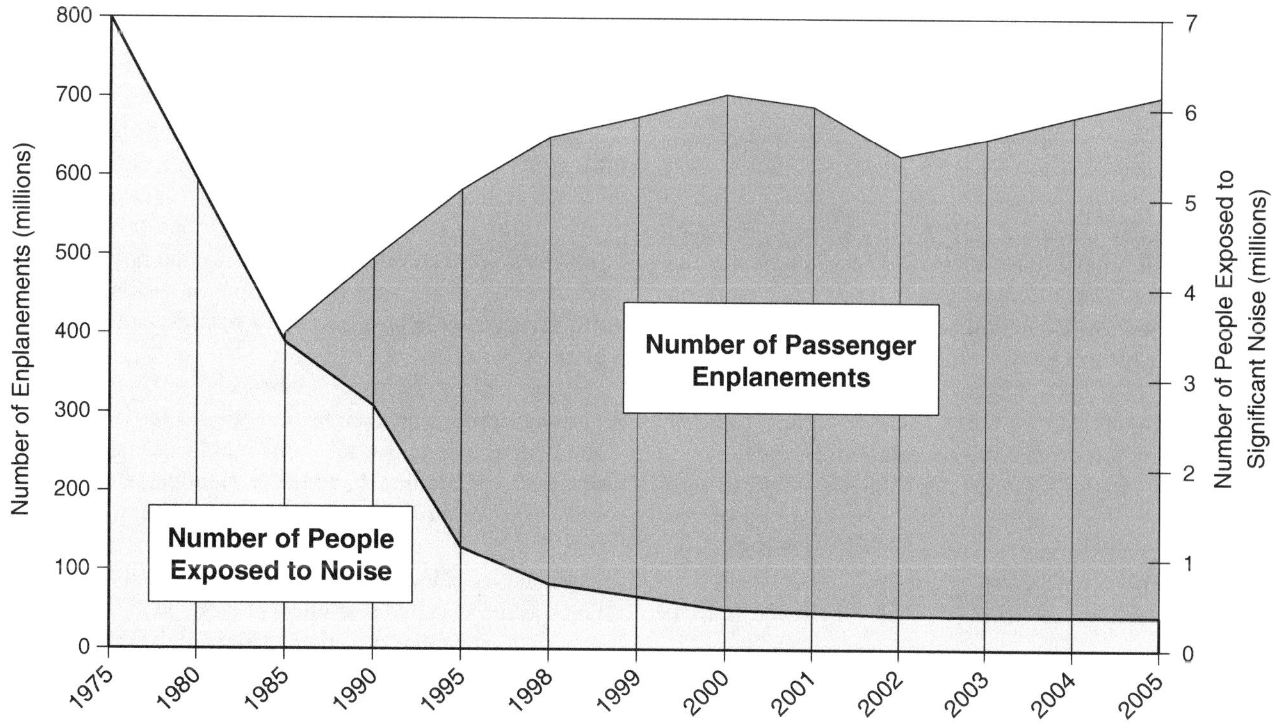

FIGURE B-1 Actual and predicted exposure to significant noise (65-dB day-night average sound level) and enplanement trends for the United States, 1975-2005. SOURCE: C. Burleson, FAA, "Aviation environmental challenges," Presentation to Panel B on December 13, 2005.

Adequately understanding and mitigating the environmental impact of aviation requires an integrated approach to noise and emissions research that considers these tradeoffs.

High-risk, long-term research is required to meet future demands. Close collaboration between government and industry is required to mature and transition promising technologies. NASA plays a critical role in supporting fundamental source noise abatement research at universities, which can lead to both revolutionary technology advances and a workforce that can answer new technical questions. Key milestones include

- Develop validated physics based models to predict engine noise and conduct trade-off studies.
- Improve understanding and prediction capabilities, and develop propulsion cycles compatible with noise and emissions reduction.
- Develop advanced low-noise fan designs, liner concepts, and active control technologies.
- Develop concepts to reduce installed noise (e.g., adaptable chevrons).
- Develop and demonstrate propulsion designs that show the feasibility of technologies to reduce noise by 10 dB (in 15 years) from Boeing 777/GE 90 levels.

Relevance to Strategic Objectives

Capacity (9): In the absence of breakthroughs, increasingly strict noise requirements will constrain aviation system capacity.

Safety and Reliability (1). This Challenge will not help to achieve this objective, though equipment designed to reduce noise must be compatible with safety and reliability requirements.

Efficiency and Performance (3): Some noise reduction approaches (e.g., higher bypass ratio) increase efficiency while others (e.g., acoustic liners) increase weight, which may reduce efficiency. In addition, engines often run at nonoptimal conditions to reduce noise. Innovative noise solutions may permit new, optimized operating approaches.

Energy and the Environment (9): Aircraft noise directly impacts the environment.

Synergies with National and Homeland Security (3): Noise constraints impact DoD operations in civil airspace. This Challenge will alleviate this constraint.

Support to Space (1): This Challenge has no impact on this Objective.

Why NASA?

Supporting Infrastructure (3): NASA is well poised to conduct engine source noise abatement research. It has excellent facilities, a large staff of qualified personnel working in this area, and a track record of contributing to advancements. However, strong capabilities also exist at universities and in industry.

Mission Alignment (9): This Challenge is very relevant to NASA's mission to improve aircraft performance.

Lack of Alternative Sponsors (3): NASA is well-qualified to support this Challenge, but industry has a strong incentive to conduct noise reduction research, even if NASA does not.

Appropriate Level of Risk (9): This Challenge is high risk.

B1b Ultraclean gas turbine combustors to reduce gaseous and particulate emissions in all flight segments

Emissions from aircraft constrain the growth of aviation due to their environmental impacts and potential human health consequences. While aviation sources remain a very small percentage of transport emissions, local worries about the environmental impact of these emissions can impede airport improvements to increase capacity. About 25 percent of U.S. commercial airports are in areas that are in nonattainment or maintenance for national ambient air quality standards—including 43 of the top 50 airports. Airports located in air quality nonattainment or maintenance areas increasingly find that air emissions add to the complexity, length, and uncertainty of the environmental review and approval of expansion projects (Akin et al., 2003). Furthermore, it is increasingly difficult for airport development projects to conform to Clean Air Act requirements, and air quality regulators in some states are working to directly or indirectly control growing aircraft emissions.

Key pollutants of concern include oxides of nitrogen and sulfur (NO_x and SO_x), carbon monoxide (CO), unburned hydrocarbons (UHCs), hazardous air pollutants, and particulate matter (PM). In addition, emissions of carbon dioxide (CO_2) and water vapor (H_2O) in the upper troposphere and stratosphere are of concern because of their potential impact on Earth's climate (IPCC, 1999). Both CO_2 and H_2O are inherent combustion products of hydrocarbon fuels, and their emissions can only be reduced through improvements in overall cycle efficiency (see R&T Challenge B4) or a change in fuels.

Emissions of SO_x and, possibly, PM can be reduced through fuel processing (e.g., desulfurization), fuel additives, or both. However, an improved understanding of PM formation and destruction mechanisms is required to reduce PM emissions. This a difficult problem given the inherent chemical complexity of aviation jet fuels and the lack of well-validated measurement techniques for PM.

Emissions of NO_x, CO, UHC, and PM from the combustor can be reduced through the development of ultraclean combustion approaches, a critical step to mitigate the environmental impacts of aviation. Understanding (1) the tradeoffs between different emissions and noise and (2) the health and welfare impacts of various emissions and noise at different levels is also necessary to make informed design choices.

Low NO_x emissions can be achieved with both lean-burning combustor designs and those that run rich in the front end (but lean overall)—the main point being low combustion temperatures. In addition, catalytic combustion systems have ultralow emissions levels, but durability and cost considerations make them unlikely candidates for aviation applications, at least for several decades. Rich-burn concepts (such as the so called rich-burn/quick-quench/lean-burn concept) use sequential rich, then lean combustion and, to some extent, are realized in most commercial engines using a rich primary zone followed by dilution. Key technical issues with this concept involve PM emissions and quench zone mixing (Lefebvre, 1999). Lean combustion concepts attempt to create a lean premixed fuel-air mixture, either upstream of the combustion chamber with lean, premixed, prevaporized (LPP) designs, or in the combustion chamber with multipoint, lean direct injection (LDI) approaches. Lean premixed approaches have received substantial market penetration in land-based gas turbine applications over the last two decades. While the majority of these devices use natural gas, similar LPP concepts have been used for liquid fuels by vaporizing the fuel. Such systems have demonstrated ultralow levels of NO_x, CO, UHC, and PM. The key issues associated with LPP combustors are unsteady combustion phenomena, including combustion instability, flame blow-off, flashback, and autoignition, which are major operability concerns; autoignition is a key concern in high-pressure-ratio engines. These unsteady combustion issues are prominent concerns in commercial land-based applications and have degraded engine reliability and availability relative to more polluting alternatives (i.e., non-premixed flame combustors). LDI approaches, which have been extensively explored at NASA, avoid flashback and autoignition, but at the price of increased complexity. A variant of these concepts is to heavily dilute the fuel-air mixture with combustion products prior to combustion, sometimes referred to by the misnomer flameless combustion.

These combustion approaches share a number of common issues that should form the basis of future NASA research. These include mixing, PM formation and inhibition, and unsteady combustion phenomena. Fast, effective fuel-air and combustion product–reactant–quench air mixing is a key enabling technology for all of the above-mentioned combustion technologies. Unsteady combustion phenomena are quite complex and, while heuristic explanations have allowed for an understanding of the basic physics, more in-depth understanding of the underlying dynamic processes is required to develop true predictive capabilities. For example, the conditions under which combustion instabilities occur and the amplitudes of instabilities cannot currently be predicted. In addition, some phenomena, such as blowoff or flashback, are well understood in fundamental laboratory burners but not at all in practical swirling devices, where new mechanisms and physics occur.

Developing effective mitigation for PM is predicated on understanding the formation of particles and their composition, growth, and transport mechanisms. Effective measurement techniques are also needed to assess aviation's contribution to PM concentrations and potential interrelationships between PM and other aviation emissions, as well as noise. Metrics for human health and atmospheric impacts should also be established and correlated with particulate emissions from aviation. Finally, mitigation strategies to address all aviation emissions, taking into account interdependencies, need to be defined and developed. Key milestones include

- Understand PM formation mechanisms and kinetics and develop fuel additives to disrupt formation.
- Understand air toxicity measurement techniques and the impact of PM on human health and welfare.
- Improve understanding and prediction capabilities and develop optimized approaches for mixing in multiphase flows.
- Develop large eddy simulations (LES) with optimized subgrid models that contain key physics needed to capture chemical reactions, mixing, and unsteady combustor phenomena.
- Develop physics-based, reduced-order combustor models, including emissions, combustion instability, blowoff, and flashback, for inclusion in intelligent engine control systems.
- Develop validated chemical mechanisms that describe fuel kinetics.
- Develop and demonstrate combustor designs that show the feasibility of technologies to reduce NO_x emissions by 85 percent while also reducing PM, relative to 1996 International Civil Aviation Organization (ICAO) limits for future large and regional subsonic engines (with pressure ratios of 55:1 and 30:1, respectively).

Relevance to Strategic Objectives

Capacity (9): This Challenge will help create breakthroughs that are necessary to prevent increasingly strict emissions requirements from constraining the capacity of the air transportation system.

Safety and Reliability (1): Reliability issues have been a prominent issue for commercialized low-emissions combustors for ground-based applications. Low-emission combustion approaches for aircraft engines are unlikely to enhance safety and need to be well engineered so safety and reliability are not compromised.

Efficiency and Performance (3): Efficiency improvements reduce the fuel burn and pollutants emitted for a given mission, all other things remaining equal.

Energy and the Environment (9): Combustor emissions directly impact the environment.

Synergies with National and Homeland Security (3): Improved understanding of dynamic combustion processes and mixing will contribute to DoD goals for main engine and augmentor combustors.

Support to Space (1): This Challenge has no impact on this Objective.

Why NASA?

Supporting Infrastructure (3): NASA has excellent facilities and a large staff working in this area, but strong capabilities also exist at other university, DoD, and DOE laboratories.

Mission Alignment (9): This Challenge is very relevant to NASA's mission to improve aircraft performance.

Lack of Alternative Sponsors (3): NASA is well qualified to support this Challenge, but DOE and DoD are also supporting similar programs for power, energy, and military applications.

Appropriate Level of Risk (9): This Challenge is quite challenging.

B3 Intelligent engines and mechanical power systems capable of self-diagnosis and reconfiguration between shop visits

In the future, advances in sensing, control, and information technology will lead to engines that are more sophisticated and more intelligent. Research thrusts should investigate how more intelligent systems can (1) improve engine health diagnostics and remedial actions in flight, (2) optimize the mission, and (3) use flight data to improve maintenance on the ground.

For current engines, the focus will be very much on diagnostics. Better physics-based modeling will be essential. Development of better CFD tools, better life-prediction tools, and better performance, steady-state, and dynamic checks will be the keys to success. Reducing in-flight shutdowns by a factor of 3 and unscheduled engine removals and delays and cancellations by a factor of 5 should reduce maintenance costs by 50 percent. Requirements include (1) smaller sensors, with better response and higher operating temperatures and (2) better materials with narrower properties tolerances. This should increase the lives of disks and airfoils by 50 percent.

Intelligent engine development will include active combustor control, which will permit operation with leaner burners, leading to lower NO_x emissions. Active stall control will enable compressors at higher pressure ratios, increasing propulsion efficiency. For engines with current architectures, intelligent engines will put more emphasis on variable systems. The goal will be to have an engine morph itself between takeoff and cruise, for example, to accommodate the individual point requirements. On new engines with new architectures, the ultimate intelligent engines will be the variable-cycle engines, which are discussed in Challenge B6a below.

Another technology to develop will be closed-loop clearance control. A pressing need in this regard is the development of a three-point probe system to monitor turbine clearances online. Software will be developed to control turbine clearances by modulating the cooling air on the casing. In-service deterioration will be reduced by accommodating the clearance loss due to rubs in the airfoils and the casing and by minimizing large clearances due to transients. This should significantly reduce operating temperature margins. Reducing the required engine margins by 50°F would increase on-wing life by about 3 years for most engines. Such a system would improve turbine efficiencies in flight and reduce fuel burn as much as 2 percent. Other relevant technologies include variable exhausts, active cooling control, improved aircraft–engine integration, better electric power generation, and better noise and emissions controls. Key milestones include

- Develop better computational simulation tools to understand operability limits.
- Develop better life prediction tools.
- Develop improved steady-state and dynamic performance checks.
- Develop improved health diagnostics systems.
- Develop new health prediction systems.
- Develop improved clearance control systems.
- Develop active compressor stall control.
- Develop active combustion control.

Relevance to Strategic Objectives

Capacity (3): Intelligent engines can improve capacity by preventing in-flight shutdowns and reducing delays and cancellations caused by unscheduled maintenance.

Safety and Reliability (9): Intelligent engines will provide new diagnostics systems and life-prediction capabilities that will greatly improve aviation safety and reliability.

Efficiency and Performance (3): Intelligent engines will help increase efficiency by reducing aircraft downtime, and they may provide a small reduction in fuel burn through engine optimization.

Energy and the Environment (3): Intelligent engine technology may reduce engine noise or control combustion for lowered pollutant formation.

Synergies with National and Homeland Security (3): This Challenge is relevant to military aircraft.

Support to Space (1): Space propulsion systems are generally highly instrumented so contributions from intelligent engine technology are likely to be minor, particularly for expendable vehicles.

Why NASA?

Supporting Infrastructure (3): NASA has very good sensor development and modeling capabilities.

Mission Alignment (9): This Challenge is very relevant to NASA's mission to increase aircraft performance and operability.

Lack of Alternative Sponsors (3): This Challenge involves far-reaching technology that industry expects NASA to develop. DoD and DOE also perform relevant research.

Appropriate Level of Risk (9): This is challenging research that requires breakthroughs to succeed.

B4 Improved propulsion system fuel economy

The fuel economy of gas turbine propulsion systems is a function of engine efficiency, propulsion-induced drag, and propulsion weight. Overall engine efficiency is the product of the efficiency of creating hot, high-pressure gases (thermal or cycle efficiency), the efficiency of transferring energy from the hot high-pressure gases to a more desirable form (transfer efficiency), and the efficiency of creating thrust from the engine fan and core flows (propulsion efficiency). The thermal efficiency for a gas turbine (Brayton cycle) is primarily a function of the overall engine pressure ratio. That is, as long as the turbine can tolerate the inlet temperature corresponding to a given pressure ratio, the overall pressure ratio sets the efficiency of the cycle. Figure B-2 illustrates very clearly that state-of-the-art gas turbines have not reached the theoretical limits of thermal efficiency. The technologies identified in the figure have the potential to improve the thermal efficiency of gas turbines, to significantly increase fuel economy, and to decrease the environmental impact of the air transportation system.

The pressure ratio for state-of-the-art gas turbines is approximately 46:1 (for large engines) and 18 to 1 (for small engines). To reach fuel economy goals, the overall pressure ratios for large engines must be increased to between 60:1 and 65:1, and small engines must be increased to between 30:1 and 40:1. The technology that limits the overall pressure ratio is compressor disk material stress at operating temperature. Maximum disk temperature must be increased from 1350°F to 1500°F, and turbine blade materials, coatings, and cooling configurations must withstand 3600°F. Thus, advances in materials technology are key enablers of enhanced fuel economy.

Transfer efficiency is determined by the component efficiencies of the fan and low-pressure turbine and the losses of

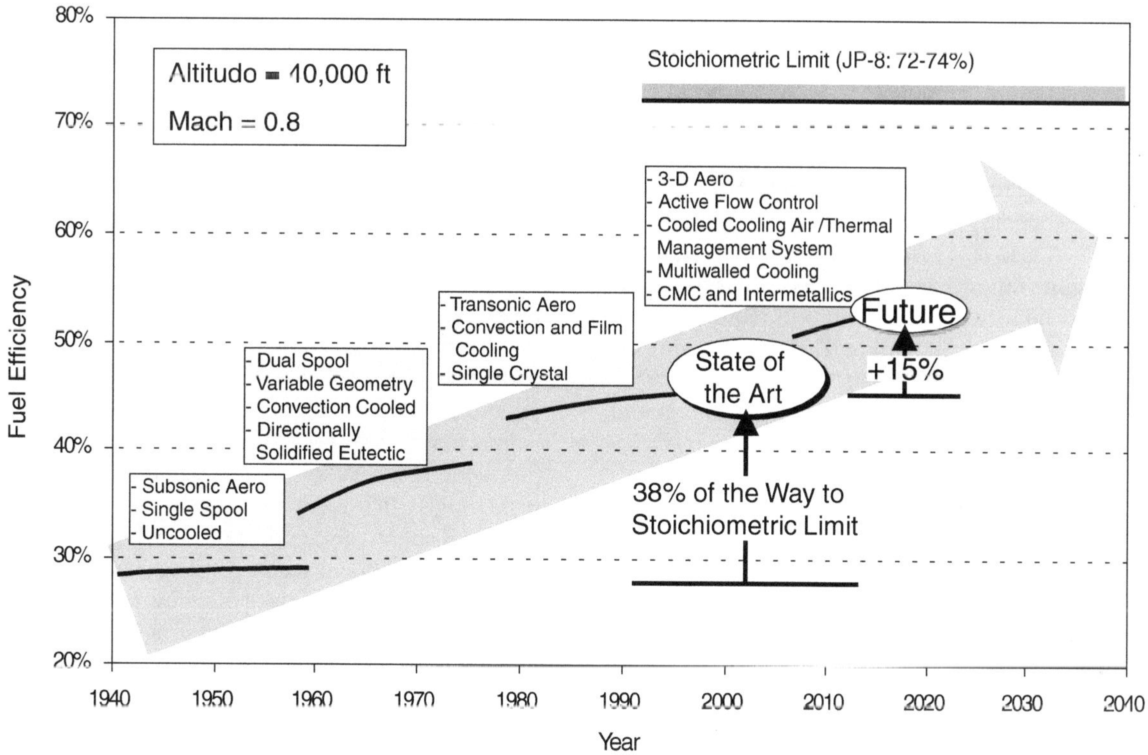

FIGURE B-2 Considerable gas turbine fuel efficiency improvements are still possible. SOURCE: J. Stricker, Air Force Research Laboratory, Private communication to panel member D. Crow, February 2006.

the shaft bearings. High-efficiency, low-pressure turbines need high rotor speeds, but highly efficient fans require low rotor speeds. Therefore, engines with high transfer efficiency must have reduction gearboxes or other technologies that permit different rotor speeds for the fan and low-pressure turbine.

Propulsion efficiency is a function of the difference between the velocity of engine exhaust and the forward velocity of the aircraft. Increasing the mass flow of air through the system at slower speed improves propulsion efficiency and decreases noise. However, doing so increases the diameter of the engine, which increases friction and flow blockage. Since larger engines will also be heavier, the use of composites or other lightweight materials for construction of the large structural pieces of the turbofan will also be necessary.

As shown in Figure B-2, improving thermal efficiency by 15 percent requires advances in several technologies: 3-D aerodynamics, active flow control, cooled cooling air and a thermal management system, multiwalled cooling, and ceramic matrix composites (CMCs) and intermetallics. Also, unconventional engine architectural arrangements, such as unducted fan engines, have demonstrated high performance potential and should be considered.

Over the long term, advances in all three efficiencies (thermal, transfer, and propulsion) should be able to improve fuel economy by 30 percent relative to the GE 90 for large commercial engines and 30 percent relative to T700/CT7 for small engines. Key milestones include

- Demonstrate laboratory-scale materials for 1500°F compressor disks.
- Demonstrate materials for full-scale, 1500°F compressor disks.
- Perform 1,000-hour test of a 50-horsepower per pound speed reduction gearbox.
- Test a reduced-weight, high-bypass-ratio engine and nacelle-to-wing configuration in a wind tunnel.
- Demonstrate an acceptably low-cost, advanced high-pressure turbine cooling system.

Relevance to Strategic Objectives

Capacity (3): Improving the fuel economy of civil aircraft will reduce operating costs and increase capacity by permitting airlines to increase flight schedules and fleet sizes profitably.

Safety and Reliability (1): This Challenge has no impact on this Objective.

Efficiency and Performance (9): The fuel economy of the propulsion system and the drag of the aircraft determine the fuel burned for air travel. Based on FAA projections (FAA, 2006), a 20 percent increase in fuel economy would decrease fuel consumption by U.S. civil aviation by 4 to 6 billion gallons a year between 2006 and 2016. As gas prices approach $2 per gallon, this amounts to $8 billion to $12 billion dollars.

Energy and the Environment (9): Increasing fuel economy will significantly decrease aircraft emissions. Also, because increasing fuel economy requires engines with higher bypass ratios, it will also decrease noise.

Synergies with National and Homeland Security (3): Increasing fuel economy in civil aircraft will require engines that operate at higher pressure ratios. The high-pressure engine core is applicable to military aircraft engines.

Support to Space (1): This Challenge has no impact on this Objective.

Why NASA?

Supporting Infrastructure (3): NASA programs such as the Energy Efficient Engine and the High Speed Civil Transport have led to technologies that have greatly improved civil aviation. NASA has the complete set of analytical and experimental tools to undertake this Challenge, including excellent staff and facilities for the development of gearboxes. However, DoD and industry also have many of the necessary tools.

Mission Alignment (9): This Challenge is very relevant to NASA's mission.

Lack of Alternative Sponsors (3): No organization other than NASA is supporting high-bypass-ratio research to improve fuel economy. However, DoD is supporting research to increase overall engine pressure ratio.

Appropriate Level of Risk (9): Relevant R&T related to materials and rotating machinery aerodynamic issues faces significant risk.

B5 Propulsion systems for short takeoff and vertical lift

The use of short (STOL), extremely short (ESTOL), or vertical (VTOL) takeoff and landing airplanes (collectively called V/STOL)[3] and increased use of helicopters could greatly increase the capacity of the air transportation system by allowing more takeoffs and landings at existing airports without increasing demand for runway usage (NRC, 2003). V/STOL airplanes include tilt-wing aircraft, tilt-rotor aircraft, vertical-lift fan aircraft, and blown-wing aircraft.[4] Currently, the fuel economy of V/STOL propulsion systems is not on par with that of fixed-wing commercial airplanes. Propulsion systems for all new aircraft must also demonstrate extremely high levels of reliability. Propulsion systems for V/STOL aircraft are in an early state of development or do not exist for civil airplanes. In addition, engine-out strategies need to be developed and verified for certification.

This Challenge should support development of V/STOL and helicopter propulsion systems with fuel economy comparable to future small commercial aircraft—namely, 20 percent better than the CT7 family of engines that is currently in production for small conventional aircraft. Many of the same technologies that apply to large and small engines for conventional aircraft also apply to V/STOL propulsions systems. However, additional technologies such as high-efficiency, angled gearboxes; high-efficiency reduction gearboxes, large-bleed systems; thrust vectoring systems; noise reduction both inside and outside the aircraft; and fan-tip-driven turbines will be required to put V/STOL airplanes into affordable, large-scale commercial service with minimal environmental impact.

There are three major technology efforts to be undertaken in support of V/STOL aircraft for civil aviation. The most important is to demonstrate an engine in the 3,000-shaft-horsepower (hp) range that meets the fuel economy goals. The important characteristics of this demonstration engine are to achieve overall pressure ratios of 25:1 or 30:1 and turbine inlet temperatures of 2800°F. This will require some combination of the following technologies: (1) new compressor disk materials, (2) greatly improved turbine cooling configurations, (3) new turbine blade alloys and coatings, (4) component aerodynamics designed with the latest computational models, and (5) highly effective, low-pressure-drop dirt separation devices. Such an engine would benefit helicopters as well.

Secondly, the powertrain system of most V/STOL airplanes (as well as helicopters) will consist of shafting with speed reduction gearboxes, angled gearboxes, and perhaps clutch systems. Reliable clutch operation would enable many new types of V/STOL aircraft. NASA should develop the design tools and demonstrate candidate gearboxes and clutch systems.

Thirdly, engine-assisted wing lift, such as the blown wing, offers the simplest, most energy-efficient short takeoff. Wing aerodynamics need to be developed and the bleed or suction

[3]VTOL airplanes can take off and land vertically. They include tilt-rotors, the AV-8 Harrier, and the JSF. VTOL airplanes do not routinely take off or land vertically because of the range-payload penalty associated with the weight limitations of purely vertical operations. Rather, they use any available field length to develop some forward motion and wing lift during takeoff to increase the useful load (fuel plus payload). They tend to land vertically only at the end of the mission, when they are lighter, after burning fuel and/or dropping weapons.

STOL airplanes use high-lift systems to take off in less distance than conventional aircraft (typically a few thousand feet). Very few STOL aircraft can safely take off on runways shorter than 3,000 ft and none on runways less than 2,000 feet. (This class does not include ultralight aircraft, kit planes, etc. that can operate out of short fields due to their small size but do not have high-lift system.)

ESTOL airplanes would be able to safely take off on runways of 2,000 ft. They would have high-lift systems and thrust-to-weight ratios that are higher than conventional aircraft but not as high as VTOL aircraft. ESTOL aircraft have not yet been developed for commercial or military operations.

V/STOL refers to both VTOL and STOL airplanes that convert to fixed-wing flight after takeoff; it does not include helicopters.

[4]Blown-wing V/STOL aircraft use engine exhaust directed to specific locations on the wing to increase lift during takeoff and landing.

locations and quantities required need to be demonstrated for blown-wing V/STOL airplanes.

The tools, techniques, and devices demonstrated in the paragraphs above would enable new families of V/STOL aircraft to enter the civil aviation market. The capabilities of these aircraft would greatly increase the capacity of the civil air transportation system and decrease door-to-door travel time for the flying public. Key milestones include

- Demonstrate pressure ratios between 25:1 and 30:1 and turbine inlet temperatures of 2800°F for 3,000-shaft-hp-class engine components.
- Develop and validate the design tools required for candidate gearboxes and clutch systems.
- Demonstrate highly reliable gearboxes, which have transfer efficiencies of about 99.8 percent and power: weight ratios of about 50 hp per pound.
- Demonstrate clutch system technologies with 10,000-cycle life and a probability of failure of 1×10^{-6} over the life of the system.

Relevance to Strategic Objectives

Capacity (9): V/STOL civil airplanes would allow more takeoffs and landings at existing airports and enhance the role of small or regional airports within the air transportation system.

Safety and Reliability (1): New V/STOL aircraft would be certified for commercial use only if they meet the existing high standards for safety and reliability. For vertical lift approaches, engine out strategies need to be demonstrated and validated. Congestion relief could increase safety.

Efficiency and Performance (3): Efficient V/STOL civil airplanes will decrease door-to-door travel time.

Energy and the Environment (3): Properly designed and engineered V/STOL airplanes could help restrict noise to within the boundaries of the airport, but they might increase noise within airport boundaries.

Synergies with National and Homeland Security (3): DoD and DHS have used and will continue to use many V/STOL airplanes.

Support to Space (1): This Challenge has no impact on this Objective.

Why NASA?

Supporting Infrastructure (3): NASA has many analytical and experimental tools to develop propulsion systems for V/STOL airplanes. DoD also has some relevant tools. In the past, NASA and the Army shared funding and leadership in basic rotorcraft research. However, NASA eliminated all rotorcraft funding in FY 2006.[5]

Mission Alignment (9): This Challenge is very relevant to NASA's mission.

Lack of Alternative Sponsors (3): DoD will develop some—but not all—of the technologies needed by civil V/STOL airplanes.

Appropriate Level of Risk (9): Propulsion technology required to develop affordable and environmentally benign civil V/STOL airplanes does not yet exist.

B6a Variable-cycle engines to expand the operating envelope

Variable-cycle engines have two or three flow paths through the engine, variable vanes, and variable exhaust nozzles, all of which allow them to vary engine bypass ratios and pressure ratios. Variable-cycle engines can improve the performance of both military and civil aircraft in many flight regimes by changing the bypass ratio and pressure ratio as a function of speed, altitude, and mission requirements. For the long-range JSF, this should permit a twofold increase in rapid response radius, an eightfold increase in loiter capability, and a 30 percent reduction in gross weight. For a JSF follow-on aircraft, a 25 percent increase in lift and a 10-25 percent increase in range, depending on the mission, appear possible.

Variable-cycle engines have the potential to increase subsonic engine fuel economy. They also appear attractive for a supersonic commercial aircraft that has to accommodate stringent takeoff noise requirements and still achieve reasonable performance at supersonic speeds. For access to space, variable-cycle engines could provide a large reduction in payload costs as well as marked safety improvements.

This Challenge will lower noise at takeoff while maintaining good fuel consumption at cruise, and it will enable optimized engine configurations during climb and descent. Engines will be able to run cooler, which will reduce maintenance costs.

This Challenge requires the development of numerous technologies: integrated thermal management approaches; reliable prime air-to-fuel heat exchangers; low-pressure-drop air-to-air heat exchangers; improved JP-8 heat sink capability; CMC technologies and associated life-prediction tools for operation above 2400°F; complex shape fabrication; high-speed bearings; improved turbine cooling; better engine health predictions; probabilistic life analysis, in-flight data analysis; low-emission, high-temperature combustors; variable-geometry fan systems; and improved airframe–engine integration. This Challenge would benefit from the development of smart engines (Challenge B3). Key milestones include

- Develop variable exhaust nozzle technology to optimize fuel burn.
- Develop improved thermal management systems.
- Develop CMC technologies for hot section components.
- Develop highly loaded, high-speed bearings.

---

[5]R. Flater, American Helicopter Society, Letter to Curt Weldon, Chairman, Tactical Air and Land Forces Subcommittee, U.S. House of Representatives Committee on Armed Services, February 18, 2005. Contained in an appendix to NIA (2005).

- Develop probabilistic analysis for more accurate designs and life prediction.
- Develop improved turbine cooling technology.
- Develop high-temperature combustors to accommodate increased operating pressure ratios.
- Develop improved aircraft–engine integration tools.

Relevance to Strategic Objectives

Capacity (3): The reduced maintanence needed by these engines would decrease delays and increase availability.

Safety and Reliability (1): This Challenge has no impact on this Objective.

Efficiency and Performance (9): This Challenge will allow the aircraft to attain constant on-design performance by adapting to flight conditions.

Energy and the Environment (3): Variable-cycle engines may be able to tailor conditions to reduce noise or emissions, although they will primarily be used to improve performance and efficiency.

Synergies with National and Homeland Security (3): There will be some impact here as they will become our future weapon systems.

Support to Space (9): Variable-cycle engines are very relevant to potential two-stage-to-orbit systems.

Why NASA?

Supporting Infrastructure (3): NASA has relevant facilities in place, but they are not unique.

Mission Alignment (9): This Challenge is very relevant to NASA's mission, especially for supersonic systems and space.

Lack of Alternative Sponsors (3): Currently the only other agency that would sponsor relevant technologies would be DoD. Industry will not fund relevant R&T until it is more advanced.

Appropriate Level of Risk (9): This Challenge faces high risk.

B6b Integrated power and thermal management systems

The goal of this Challenge is to integrate and optimize, at the aircraft system level, the traditionally severable airframe power and thermal management system. An integrated systems approach optimizes aircraft cost, weight, and performance rather than optimizing individual components. This approach also enables integrated prognostics and health monitoring, thereby improving safety and reliability. The current state of the art involves architecture of federated systems, with separate component machinery for auxiliary and emergency power; environmental control; engine start; accessory drive units; waste heat rejection; and so on.

"Integration" refers to the physical, functional, and requirements integration of key propulsion and power system components, with those components combined into fewer multifunctional units all tied together in a more-electric architecture (see Challenge B9). Key components and functions include engine starting; electrical power generation, power conditioning, and routing; air cycle environmental control; avionics, fuel, and oil cooling; ventilation; flight control actuation; and overall vehicle and propulsion system thermal management, especially waste heat recovery and/or rejection. For example, engine start, auxiliary power, and environmental control systems may be combined into an airframe-mounted integrated power package that is physically coupled to the engine through power extraction and waste heat recovery. In this integrated approach, flight control systems are likely to be driven by electric or electrohydrostatic actuation, and thermal management is addressed in a seamless, system-level fashion. At the propulsion system level, electric power must be generated and integrated with airframe needs in the most efficient manner. This may be by a generator mounted on the shaft of the low-pressure turbine or, eventually, by fuel-cell-driven generators distributed within the airframe.

This Challenge includes airframe thermal management and waste heat recovery for higher speed applications. At hypersonic speeds, the thermal energy generated by the high-enthalpy flow over the airframe must be dissipated. The primary method for thermal management involves heating the fuel prior to combustion using structural cooling or heat exchange with working fluids used to cool the structure.

Today's modeling tools are derived from legacy approaches in which numerous component suppliers individually design, develop, and validate their product based on component-level requirements and specifications. New modeling and simulation infrastructures are necessary to allow these tools to be used in a system-level design framework, accommodating multiple platforms across multiple sites. A robust modeling framework is necessary to justify the system-level benefit of a given integrated component, which may need to weigh or cost more than a traditional component or have different or enhanced functionality.

Integrated systems also defy traditional business models in which hardware and software design, development, and validation responsibilities are clearly defined. In the integrated approach, some hardware manufactured or procured by the airframe manufacturer will be engine mounted, and some engine hardware may be mounted on the airframe. Although the engineering product is physically and functionally integrated, contractual responsibilities must still be divided between business units, and it is unclear how to do this. Key milestones include

- Identify and mature new business models for the design, development, validation, and support of hardware and software components of integrated systems.
- Develop an objected-oriented modeling infrastructure that allows networking resources to operate across different hardware platforms and geographic sites.

- Develop new engine-airframe systems integration architectures for both subsonic and higher speed flight.
- Develop physics-based subsystem component models that can analyze transient operations.
- Develop and mature concepts for the integration of fuel cell technology as secondary power sources.
- Develop advanced electric or electromechanical actuators that have rapid response, high power-to-weight, and low heat rejection.
- Develop subsystem components that can survive in more stressful thermal environments, require less cooling, and reject less waste heat, including thermally efficient fuel pumps and high-temperature electronics for power management and distribution systems.
- Develop lightweight, high-energy-density batteries.
- Develop advanced heat exchanger technologies.

Relevance to Strategic Objectives

Capacity (3): This Challenge will reduce aircraft weight and cost, increasing the number of passengers a given aircraft can carry.

Safety and Reliability (1): This Challenge will have little impact on system safety but should result in modest gains in system reliability.

Efficiency and Performance (9): This Challenge will significantly reduce fuel consumption, particularly with waste heat recovery rather than rejection.

Energy and the Environment (3): This Challenge will reduce emissions by increasing efficiency and reducing fuel consumption.

Synergies with National and Homeland Security (3): This Challenge is applicable to military aircraft.

Support to Space (9): General principles associated with this Challenge apply to high-Mach-number space launch vehicles.

Why NASA?

Supporting Infrastructure (3): NASA has led the development of new turbomachinery computing infrastructure, including the Numerical Propulsion System Simulation tool, now used for many U.S. aircraft engine development and integration efforts. NASA's code framework could be expanded to fully encompass the modeling requirements of integrated systems.

Mission Alignment (9): This Challenge is very relevant to NASA's mission. Novel configurations offer significant potential for breakthroughs in aircraft performance.

Lack of Alternative Sponsors (3): NASA has an important contribution to make to this Challenge. DoD and industry will also support relevant R&T. Collaboration is suggested whenever possible.

Appropriate Level of Risk (9): This Challenge faces high risk.

B8 Propulsion systems for supersonic flight

Commercially viable supersonic propulsion remains an elusive goal. To be successful, a commercial supersonic aircraft must simultaneously meet environmental standards related to local air quality, noise, sonic boom, and high-altitude emissions; performance requirements in terms of T/W, specific fuel consumption, etc.; and FAA certification requirements (NRC, 2001). With the last flight of the Concorde in 2003, the world entered a hiatus from commercial flight at a Mach number greater than 1. At least two American companies are trying to build and fly a supersonic business jet with a capacity of about 12 people by 2012 (<www.aerioncorp.com>; <www.saiqsst.com>). No efforts to build a commercial supersonic transport (with a capacity, for example, of more than 100 people) are under way.

Faster travel is a natural progression of any transportation system. In the case of affordable supersonic flight, shorter travel times, especially on long transoceanic and transcontinental routes, are highly desirable. A profitable supersonic transport would open a new avenue of growth for the U.S. aerospace industry. Two previous NRC studies (NRC, 1997, 2001) specifically focused on commercial supersonic flight, its complexities, and a possible roadmap forward.

Today, federal regulations (14 CFR 91 ¶817) ban civil supersonic flight over the continental United States. Furthermore, since 1994, the FAA has had a supersonic noise policy stating that any future supersonic airplane must have no greater noise impact on a community than a subsonic airplane. After January 1, 2006, that means the aircraft design must also meet Stage 4 noise standards. Defining and achieving acceptable sonic boom levels and reducing community noise to Stage 4 levels, with sufficient margin to account for additional noise restrictions that may be imposed in the future, are critical to making supersonic commercial flight viable. Particularly for supersonic flight, propulsion systems development needs to be integrated with the design of the rest of the aircraft in a multidisciplinary effort to find an optimal trade-off between performance, efficiency, noise, emissions, and thermal management. Engine–airframe integration becomes more critical as the flight speed increases. This Challenge requires validated physics-based numerical simulation codes for component-level analysis and the improvement of multidisciplinary, system-level design tools for vehicle analysis. Technology development should proceed in close coupling with psychoacoustic research to establish acceptable noise levels, especially for sonic booms in inhabited areas, and with climate impact research to establish appropriate emissions levels. As the cruise Mach number increases, a more integrated approach to thermal management is needed to efficiently reject or use the increased amounts of both aerodynamic heat and waste heat generated by the propulsion and power systems. Gas turbine research topics of interest include

- Variable-cycle engines optimized for both subsonic and supersonic flight with low specific fuel consumption, high T/W, and low noise.
- Lightweight, low-noise, efficient inlets and nozzles that also reduce wave drag and help in efficient sonic boom shaping.
- Integrated airframe and propulsion controls to actively reduce vibration mode interactions between the engine and the plane (NIA, 2005).
- Noise and emissions data to validate models for sonic boom signature and its effect on humans (psychoacoustics), to assess the interaction of combustion products with ozone, and to help establish or confirm noise and emissions regulations.
- Electric actuation systems to eliminate the need for high-temperature hydraulic actuation systems.
- Active flow control to improve engine efficiency, reduce noise, and enable different airframe–propulsion integration concepts.
- Combustion process physics: modeling and experimental validation of injection, mixing, ignition, finite-rate kinetics, turbulence–chemistry interactions, and combustion instability to improve efficiency and life.
- Advanced materials and coatings (including high-temperature alloys for compressor and turbine disks) that meet requirements for operating temperature, service life, strength, and propulsion system noise.
- Alternative engine cycles for supersonic flight might replace or enhance traditional gas turbines.

Many of these technologies are included in other R&T Challenges; much of the research proposed for subsonic engines will build a foundation for supersonic flight. In addition, knowledge gained through NASA's High Speed Research Program and DARPA's Quiet Supersonic Platform Program should be leveraged in the search for a new generation of commercial supersonic aircraft.

The technology issues for commercial supersonic transports become more difficult as cruise speed increases. The technology issues for commercial supersonic transports with cruise speeds below approximately Mach 2 are more tractable than those for higher cruise speeds. Key milestones include

- Establish needed boundary conditions, initial conditions, and other inputs and outputs for each module of multidisciplinary, system-level design tools.
- Develop technology that will enable supersonic aircraft to meet Stage 4 noise standards.
- Validate boundary layer control techniques for inlet performance and drag reduction.
- Demonstrate a supersonic variable-cycle engine with specific fuel consumption of 1.1 or lower and a T/W of at least 6 (NIA, 2005).
- Demonstrate high-performance, low-drag, noncircular inlet designs (NIA, 2005).
- Obtain flight test data on noise, emissions, human annoyance caused by sonic boom, and system interactions across the flight regime.

Supersonic aircraft represent the next step toward hypersonic aircraft. Most of the technologies matured for supersonic aircraft can become the starting point for hypersonic aircraft. To cite a few examples, variable-cycle engines proposed for optimized supersonic flight could be a starting point for combined cycles for access to space. Some high-temperature composites or alloys developed for supersonic flight will also carry over to hypersonic flight. Finally, a more-electric engine will likely transition to hypersonic applications.

In summary, the development of propulsion systems for supersonic transports may require NASA or some other federal agency to support the multidisciplinary, multiyear effort described by this Challenge.

Relevance to Strategic Objectives

Capacity (3): Increasing the speed with which people and goods are moved from one place to another directly increases capacity.

Safety and Reliability (1): This Challenge is not relevant to this Objective.

Efficiency and Performance (3): Supersonic flight improves performance but reduces efficiency, because higher speed increases fuel consumption.

Energy and the Environment (1): Supersonic flight has potentially negative environmental impacts, which need to be mitigated for supersonic flight to become viable.

Synergies with National and Homeland Security (9): This Challenge is very relevant to supersonic military aircraft.

Support to Space (9): This Challenge will be applicable to combined cycles, including air-breathing supersonic flight, for access to space. Many of the technologies developed for supersonic flight can also be transitioned to hypersonic flight.

Why NASA?

Supporting Infrastructure (9): NASA has a unique collection of facilities tailored for supersonic flight research, such as the Langley Unitary Plan Wind Tunnel, the Supersonic Low Disturbance Tunnel, and the 20-inch Supersonic Wind Tunnel. NASA also has staff that know how to operate such facilities and have done extensive research in this area (e.g., the High Speed Research program).

Mission Alignment (9): This research is very relevant to NASA's mission to transform our nation's air transportation system and to support future air and space vehicles.

Lack of Alternative Sponsors (3): DoD already supports supersonic R&T for military applications, and industry could sponsor work in this area, especially for development of supersonic business jets.

Appropriate Level of Risk (9): Commercial supersonic flight is a long-range, high-risk Challenge, but it is achievable.

B9 High-reliability, high-performance, and high-power-density aircraft electric power systems

Future aircraft power systems must be able to meet the demands of what is being called the "more-electric aircraft" (MEA). Future aircraft will progressively replace more and more mechanical and hydraulic systems with electrical systems, and electrical loads imposed by conventional systems will also continue to grow, to improve performance, convenience, and reliability. The higher power requirements of conventional loads is being driven by advances in avionics as well as by passenger entertainment and productivity needs. For example, the electric power demand on Boeing's 787 is nearly 1 MW, which is double that of the Boeing 777 and many times that of the first U.S.-built commercial jet, the Boeing 707 (Ames, 2005). The growth of new MEA loads is being driven by advances in the capabilities of electric actuators and controls, and it is being enabled by the development of more flexible and reliable aircraft generators. This Challenge can be met by improving key components and system-level technologies.

Below is a representative list of the potential benefits of future advanced aircraft power systems and a sampling of the technology developments that will enable them.

- *Power efficiency.* Reduction of heat dissipated by the power system to minimize the on-board thermal management problems.
- *Energy efficiency.* Reduction of fuel consumed for electric power generation by up to 20 percent.
- *Power density.* Reduction of power system weight and volume per unit of power generated, processed, and delivered to the load.
- *Energy density.* Fivefold reduction of weight and volume of (1) energy storage components, such as batteries and ultracapacitors, and (2) static electric power plants, such as fuel cells.
- *Flexibility.* Ability to upgrade or evolve the power system as the component technologies or the system mission changes.
- *Reliability.* Ability of the power system to perform without malfunctions and to recover from or adapt to full or partial faults and failures (short circuits, open circuits, control and component failures, aging, and so on).
- *Stability.* Ability of the system to maintain performance integrity in the presence of deleterious dynamics caused by sudden change in the states of the loads or in the power management and distribution (PMAD) system.
- *Advanced system engineering and development methodologies.* Advanced analytical and computer modeling of multiconverter aircraft power systems and controls.
- *Advanced component development.* Wireless control systems, compact, high-efficiency electric motors and generators, advanced sensorless electric machine controls for improved performance, and advanced PMAD systems, including model-referenced control of power systems.
- *High-power-density electric generators.* Integrated engine–generator architectures, such as a high-power-density electric generator on the low-pressure turbine shaft.

Present aircraft power systems are similar to other vehicle power systems, with a generator coupled to the engine through drives and gears. Generators on transport aircraft are generally connected to a 400-Hz power bus, which feeds the loads through manual or electronic switches, with some automatic PMAD functionality. Conventional aircraft power systems will become too large, heavy, inefficient, and inflexible if they are scaled up to supply the power demands of future MEA. Therefore, fundamentally better architectures and technologies for aircraft power systems must be developed to improve the weight and volume density of power systems by factors of 10 and 2, respectively (Emadi et al., 2003). Key milestones include

- Demonstrate tenfold increase in power density for suitable electric generators and motors.
- Demonstrate fivefold increase in energy and power density of suitable batteries and hybrid storage systems (e.g., the battery–ultracapacitor).
- Demonstrate an order of magnitude lighter optimized power system architectures (including, for example, a DC power bus, remotely controlled loads, and a wireless system control).
- Demonstrate intelligent PMAD using advanced system models and wireless sensors or sensorless control technologies for graceful degradation and failsafe operation.
- Demonstrate advanced analysis and simulation tools for multiconverter power systems, which can predict new modes of system dynamics and instability.

Relevance to Strategic Objectives

Capacity (1): This Challenge is not relevant to this Objective.

Safety and Reliability (3): Aircraft safety, to a moderate extent, and aircraft reliability, to a greater extent, will be improved by aircraft power systems with improved stability and fault tolerance.

Efficiency and Performance (9): Aircraft efficiency and performance will be improved by this Challenge due to improved aircraft design, improved PMAD, improved electric power generation, and intelligent system behavior.

Energy and the Environment (3): More efficient electrical systems will reduce fuel burn and emissions.

Synergies with National and Homeland Security (3): This Challenge is relevant to manned and unmanned military aircraft and will improve the performance of reconnaissance and surveillance aircraft for homeland security missions.

Support to Space (3): This Challenge is relevant to space launch vehicles.

Why NASA?

Supporting Infrastructure (1): NASA does not have any significant supporting infrastructure for the development of MEA technologies.

Mission Alignment (9): This Challenge is very relevant to NASA's mission, due to its key role in the advancement of aircraft technology.

Lack of Alternative Sponsors (3): Industry is supporting some R&T related to this Challenge. However, some of the basic knowledge and tools, such as multiconverter power system analysis and simulation models, are unlikely to be developed without NASA's support.

Appropriate Level of Risk (9): This Challenge faces significant risk.

B10 Combined-cycle hypersonic propulsion systems with mode transition

The United States has made significant progress in hypersonic flight technology over the past 40+ years; however, a renewed effort is needed if it is to continue to progress toward making hypersonic flight viable and to maintain world leadership in this challenging and critical technology. This is especially relevant in today's environment, where Japan (Kakuda Space Center) and Germany (DLR) have some of the best high-enthalpy test facilities in the world. Australia (University of Queensland), like the United States, has also flight-tested a hydrogen-fueled scramjet. The two pacing technologies for hypersonic flight are the propulsion system, the topic of this R&T Challenge, and high-temperature materials, which is one of the R&T Thrusts in this Area, propulsion and power.

The primary NASA hypersonics mission is for access to space in support of the space initiative and in placing and maintaining scientific payloads in low Earth orbit. A two-stage-to-orbit (TSTO) vehicle using a hydrogen-fueled, air-breathing first stage and a hydrogen-fueled rocket second stage, could double the payload fraction to low Earth orbit relative to a two-stage hydrogen-fueled rocket (P. Buckley, AFRL, "Payload mass fraction vs. staging velocity for TSTO vehicles to 51.7° orbit," Presentation to the DoD Technology Area Review and Assessment on March 29, 2004). This greatly reduces the cost of putting a payload into orbit. In addition, air-breathing hypersonic vehicles offer the potential for airplanelike operations, with increased safety and efficiency, robust operation, and mission flexibility relative to rockets. A secondary mission for NASA hypersonics is to provide synergy with the DoD programs in the development of missiles for time-critical mobile targets; global strike and rapid resupply aircraft; and routine, on-demand space launch for placing, maintaining, and protecting key satellites in orbit.

NASA's X-43A hypersonic vehicles demonstrated thrust greater than drag at Mach 7 and Mach 10. These were the first in-flight tests of scramjets on a flight vehicle. The X-43A propulsion system was designed to operate at a fixed Mach number (rather than accelerate the vehicle to higher speeds), had enough fuel for just a few seconds of powered flight, and used a heat-sink structure (instead of a fuel-cooled structure, which is needed for a cruise vehicle). Designing a vehicle to accelerate from takeoff to hypersonic speeds, while managing the thermal loads on the aircraft, is still a considerable problem.

One combined-cycle hypersonic propulsion system under study for access to space is a turbine-based combined-cycle (TBCC) system. This system uses a turbine to accelerate the vehicle from takeoff to Mach 3+. At about Mach 3.5, the system transitions to operate as a ramjet and then operates in a dual mode (mixed subsonic and supersonic) from about Mach 4.5 to 5.5. At about Mach 5.5, the system transitions again to operate as a scramjet for TSTO or for single stage to orbit (SSTO). A lot of research has been conducted on steady-state engine operation in the three modes, but transients associated with mode transitions are very difficult to study experimentally or to model numerically. Since all hypersonic vehicles will experience mode transitions on acceleration and deceleration between Mach 3 and Mach 6, it is critical that these transients be well understood.

In order to design complex, combined-cycle hypersonic propulsion systems, experimentally validated, physics-based tools must be developed and refined, because steady, full-enthalpy, clean air conditions cannot be reproduced in hypersonic ground test facilities. Experiments must be conducted on unit problems (e.g., jet injection into a supersonic stream) that contain the relevant flow physics but are amenable to simulation. Facility upgrades, such as for long duration, high-temperature testing of engine materials and structures, should be completed to conduct the unit experiments under near-realistic flight conditions. Advanced instrumentation must be developed and used to obtain detailed databases in unit problem experiments for complete validation of computational tools that can then be used for the vehicle design. Multiple-point validations are needed to verify that the tools produce results that can be extrapolated to conditions not available on the ground. Ultimately, flight testing must be conducted in order to obtain results under realistic operating conditions. Low-cost flight experiments on suborbital rockets should be exploited in lieu of experiments on expensive flight vehicles.

Figure B-3 delineates some of the fundamental problems that must be addressed during development and validation of computational tools for engine components:

- *Inlet*. Shock wave–boundary layer interactions, prediction of turbulence amplification by shocks (currently overpredicted by Reynolds-averaged Navier-Stokes (RANS) methods), 3-D spillage, starting mechanism, and off-design performance.
- *Isolator*. Prediction of shock train generated by combustion backpressure or engine contraction ratio, and unsteady flow due to cowl door or lip movement.
- *Isolator/combustor*. Dual-mode operation governed by complex interactions of boundary layer separation, shock-boundary layer interaction, shock–shock interactions, fuel-air mixing and combustion efficiency, and chemical kinetics.
- *Combustor*. Modeling of injection, mixing, ignition, and flameholding by LES and other techniques, subgrid scale modeling of turbulent combustion, RANS/LES transition methodology, excessive dissipation in LES modeling of mixing and combustion, probability density function transport modeling of species mixing, turbulence–chemistry interactions, and finite-rate chemistry.
- *Nozzle*. Thermal and chemical nonequilibrium, and boundary layer relaminarization.

Once these component-level models have been developed and validated, they can be integrated to create a full vehicle, tip-to-tail computational tool.

Milestones should mark the completion of validated component-level models using appropriate facilities, unit experiments, and advanced diagnostics. Key milestones include

- Develop advanced diagnostics capable of measuring time-averaged and time resolved flow parameters and their correlations.
- Demonstrate ramjet-scramjet (dual-mode) transition and isolator performance for a simplified geometry with alternately clean and vitiated air.
- Conduct transient experiments to simulate cowl door movement for turbine-ramjet mode transition and cowl lip movement to control inlet contraction.
- Demonstrate injection, mixing, and combustion using simple fuel injectors and with alternately clean and vitiated air.
- Conduct inlet studies with variable angles of attack and sideslip angles.
- Investigate new engine configurations using inward-turning inlets, elliptical cross-sections, etc.

Relevance to Strategic Objectives

Capacity (1): This Challenge has no impact on this Objective.

Safety and Reliability (1): This Challenge has no impact on this Objective.

Efficiency and Performance (3): This Challenge will improve the understanding of complex fluid and structural physics at hypersonic speeds, which will be useful for understanding lower speeds as well.

Energy and the Environment (1): This Challenge has no impact on this Objective.

Synergies with National and Homeland Security (9): Hypersonic propulsion systems being developed by the DoD will benefit significantly from a NASA research program in combined-cycle propulsion systems with mode transition.

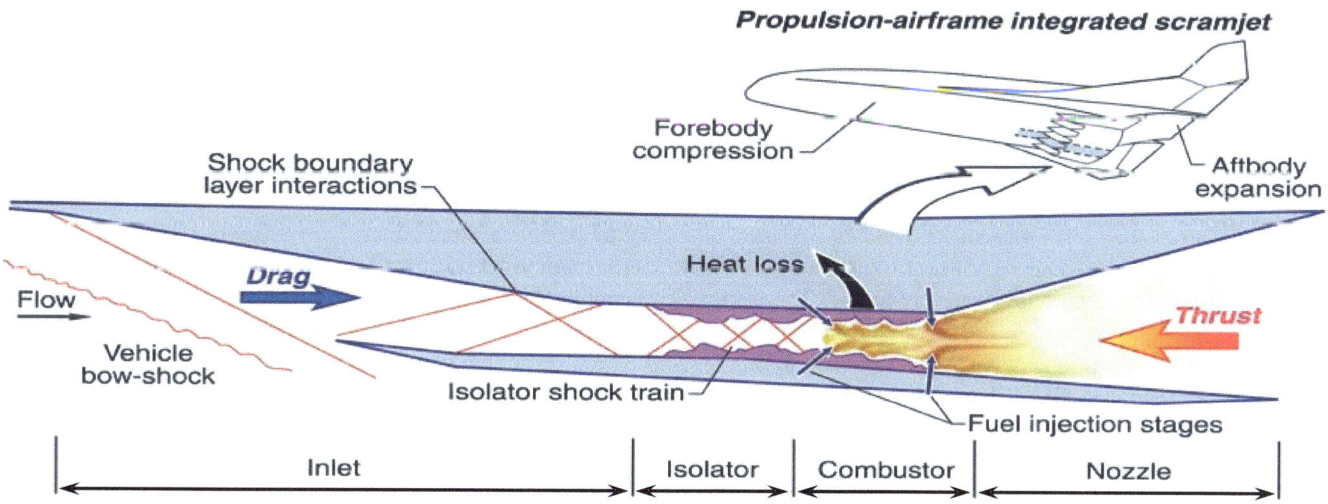

FIGURE B-3 Technology issues in supersonic combustion ramjets. SOURCE: NASA, 2006.

Support to Space (9): This Challenge is very relevant to NASA's space exploration and access-to-space missions.

Why NASA?

Supporting Infrastructure (9): NASA has relevant facilities (e.g., Langley's 8-Foot High-Temperature Tunnel, Arc-Heated Scramjet Test Facility, Combustion-Heated Scramjet Test Facility, Direct Connect Supersonic Combustion Test Facility, 15-Inch Mach 6 High-Temperature Tunnel, 20-Inch Mach 6 CF4 Tunnel, 20-Inch Mach 6 Tunnel, 31-Inch Mach 10 Tunnel) and expertise that are uniquely capable of supporting the development of combined-cycle hypersonic propulsion systems.

Mission Alignment (9): This Challenge is very relevant to several items in NASA's charter.

Lack of Alternative Sponsors (3): Both NASA and the DoD support the development of hypersonic propulsion systems.

Appropriate Level of Risk (9): This Challenge faces high risk, too much for industry to take the lead, but it has a good chance of success if the program is supported adequately over the next decade.

B11 Alternative fuels and additives for propulsion that could broaden fuel sources and/or lessen environmental impact

Current aircraft are designed to operate on kerosene, which, like other transportation fuels in the United States, is currently derived from petroleum. The U.S. transportation sector increases U.S. dependence on foreign oil. The environmental impact of current fuels, which emit, among other things, NO_x, SO_x, particulates, and greenhouse gases (CO_2 and H_2O), is coming under increasing scrutiny. These issues, coupled with the long-term inability of gains in energy efficiency to fully offset increasing demand, provide impetus for alternative fuels for transportation in general and aviation in particular. Alternative fuels for transportation include liquid fuel derived from domestic shale oil and coal (e.g., kerosene produced from gasified coal via Fisher-Tropsch chemistry), biomass-derived fuels, natural gas, hydrogen, methanol, and ethanol. That last two have significantly lower energy densities (heating values of 22.6 and 29.7 MJ/kg, respectively, relative to about 43 MJ/kg for gasoline) and are therefore less likely candidates for aviation fuel. The energy density of hydrogen by weight is high (120 MJ/kg), but the energy densities by volume of hydrogen and natural gas are very low compared to liquid fuels, which creates storage problems that are difficult to solve when it comes to aviation. Key performance metrics for any alternative fuel include cost, availability, sustainability, energy density, pollutant emissions, greenhouse gas emissions, and safety.

Use of alternative fuels such as synthetic kerosene, methane, or hydrogen could reduce aircraft emissions and mitigate U.S. dependence on imported crude oil, thereby increasing sustainability. For example, hydrogen would enable engines with zero emissions of CO, SO_x, particulates, and CO_2, but large emissions of water vapor, which could exacerbate environmental issues associated with contrails (NRC, 2002). Synthetic kerosene derived from domestic resources could be cleaner and emit lesser amounts of particulates and SO_x. The environmental benefits of an alternative fuel would, of course, need to be quantified. Alternative aviation fuels would provide significantly less benefit than alternative fuels for ground-based transportation, because aviation accounts for only 2.6 percent of U.S. greenhouse gas emissions compared to the 28 percent share accounted for by the total transportation sector (EPA, 2006).

DOE, DoD, and industry support most alternative fuels research. Within DOE, significant efforts have been under way to develop alternative fuels for the automotive sector, including compressed natural gas, methanol, ethanol, biomass-derived fuels, hydrogen (the hydrogen fuel initiative was reviewed in a recent NRC report (NRC/NAE, 2004)). The U.S. Air Force conducted extensive research in the late 1970s and early 1980s to develop alternative aviation fuels from shale oil, coal, and tar sands, and tested them extensively in military engines. The DoD also has a new initiative in clean fuels (Barna et al., 2005).

The key technical questions associated with alternative fuels for civilian aviation are cost, availability, the ground transportation and storage infrastructure, onboard storage, combustion, quantification of environmental impacts, and certification. All of the potential alternative fuels are currently more expensive than conventional jet fuel (Saynor et al., 2003), so large-scale, cost-effective production of alternative fuels (particularly synthetic kerosene and hydrogen) still requires significant research. Recent reports address in detail the technical barriers for hydrogen production (NRC/NAE, 2004; DOE, 2003).

No commercial aircraft or engines have been designed to operate using alternative fuels, so airframe and auxiliary systems as well as new engines and fuel injection systems for handling these fuels remain to be developed. Onboard storage systems for natural gas and hydrogen would require significant modifications to existing aircraft or new airframe designs (NRC, 2002). Hydrogen-based fueling concepts have been advanced and investigated since the 1950s (Saynor et al., 2003; Faass, 2001). Aircraft combustors capable of handling gaseous fuels such as methane and hydrogen or prevaporized liquid fuels will need to address a variety of dynamic combustor operability phenomena, such as blow-off, flashback, and combustion instabilities—problems that are still poorly understood in ground-based turbines that already use gaseous fuels. Fundamental combustion research, with particular focus on pollutant formation and unsteady combustor phenomenon, would be required to address these issues. Fuel specifications would also need to be defined to assure quality and consistency worldwide.

Research is needed to quantify the costs and environmental benefits of alternative fuels. For example, production and delivery of hydrogen as currently practiced could increase overall greenhouse gas emissions (NRC/NAE, 2004). Alternative fuel development for civilian aviation will, with reason, lag that for ground-based transportation due to the significantly greater problems associated with aviation applications and the larger research efforts necessary to solve them. NASA should monitor the progress of the DOE and DoD programs and take advantage of synergies, as appropriate. Key milestones include

- Develop mechanisms to monitor and interact with ongoing efforts in DOE, DoD, and elsewhere to develop alternative fuels with possible application to civil aviation.
- Develop specifications for alternative civil aviation fuels.
- Develop understanding of and predictive capabilities for correlating the molecular composition of fuels with their bulk properties (e.g., density, lubricity, stability, and emissions).
- Understand the various means, including additives, of enhancing the performance (e.g., lubricity, stability, emissions, performance) of alternative fuels.
- Understand chemical mechanisms and develop validated models that describe combustion for alternative fuels.
- Develop advanced testing methods and standards for alternative fuels.

Relevance to Strategic Objectives

Capacity (3): A long-term supply of sustainable fuels with reduced emissions would help eliminate constraints on growth in capacity.

Safety and Reliability (1): This Challenge has little or no relevance to this objective.

Efficiency and Performance (3): Development of alternative fuels is motivated primarily by emissions and sustainability concerns. However, alternative fuels would also affect the efficiency and performance of aircraft engines and, indirectly, the air transportation system as a whole.

Energy and the Environment (9): The long-term impact of alternative fuels with reduced emissions could greatly reduce the environmental effects of aviation.

Synergy to National and Homeland Security (3): Alternative fuels developed for civil aviation could probably also be used by military aircraft, and this Challenge would support ongoing work by the DoD on alternative fuels.

Support to Space (1): This Challenge has little relevance to this Objective because the amount of fuel used to support the space program is quite small compared to that for ground and air transportation.

Why NASA?

Supporting Infrastructure (3): NASA has the technical expertise to address engine combustor issues and, in collaboration with industry, aircraft design issues associated with alternative fuels.

Mission Alignment (3): Some aspects of this Challenge, such as the investigation of combustion issues, are very relevant to NASA's mission, but other aspects, such as fuel production and delivery, are not.

Lack of Alternative Sponsors (3): DOE, DoD, and industry are supporting research relevant to this Challenge, although NASA support is necessary to address all issues concerning the use of alternative fuels for civil aviation.

Appropriate Level of Risk (9): Use of alternative fuels is a long-term problem that faces moderate risk.

B12 Hypersonic hydrocarbon-fueled scramjet

DoD has had operational hypersonic systems for the past 40+ years in the form of intercontinental ballistic missiles, launch vehicles, and reentry vehicles. The Air Force's *Vision 2020: Global Vigilance, Reach and Power* (USAF, 2000) stated that the service should strive for "controlling and exploiting the full aerospace continuum." NASA programs to develop hypersonic propulsion systems should be coordinated with similar DoD efforts.

In the near term, hydrocarbon-fueled scramjets can be used to power rapid response aircraft, missiles, and expendable space lift vehicles. In the medium term, combined-cycle engines can propel rapid global response and reconnaissance aircraft. These engines, such as the TBCC, operate in several engine modes (see R&T Challenge B10). In the far term, combined-cycle engines can be used for access-to-space vehicles.

Air-breathing hypersonic propulsion systems for space access could enable aircraftlike operations, increasing mission flexibility and payload fraction relative to rocket-based propulsion systems. For NASA and DoD access to space, a TSTO vehicle could use a hydrocarbon fuel in the first stage and a hydrogen fuel for the second stage. The fuel of choice for most DoD applications will be hydrocarbons because of their high energy density, good heat capacity, and storability.

Significant ground testing has been done by the DoD on hydrocarbon-fueled scramjets. In addition, the Air Force scramjet-engine-demonstrator is being developed to demonstrate scramjet operation in flight, with a scramjet takeover at Mach 5.5 and cruise at Mach 6.5 to 7.0. NASA has already demonstrated in-flight operation of a scramjet at Mach 7 and Mach 10 in the X-43A program.

The development of hydrocarbon-fueled hypersonic propulsion systems will require many of the same basic technologies as those that are hydrogen-fueled. See B10 for a detailed listing of the issues involved. Key milestones include

- Develop advanced instrumentation capable of measuring time-averaged and time-resolved flow parameters to validate design tools.
- Complete unit experiments based on generic inlets, isolators, combustors, and nozzles to provide benchmark data sets for model validation.
- Conduct experiments on mode transition for validation of unsteady models.
- Assist DoD flight demonstration programs that are currently in progress.

Relevance to Strategic Objectives

Capacity (1): This Challenge has no impact on this Objective.

Safety and Reliability (1): This Challenge has no impact on this Objective.

Efficiency and Performance (3): This Challenge will improve the understanding of complex fluid and structural physics at hypersonic speeds, which will be useful for understanding lower speeds as well.

Energy and the Environment (1): Environmental impact is likely to be small given the small number of hypersonic vehicles likely to be produced.

Synergies with National and Homeland Security (9): Hypersonic propulsion systems being developed by the DoD will benefit significantly from this Challenge.

Support to Space (9): This Challenge is very relevant to NASA and DoD development programs for access-to-space vehicles.

Why NASA?

Supporting Infrastructure (9): NASA has relevant facilities (e.g., Langley's 8-Foot High-Temperature Tunnel, Arc-Heated Scramjet Test Facility, Combustion-Heated Scramjet Test Facility, Direct Connect Supersonic Combustion Test Facility, 15-Inch Mach 6 High-Temperature Tunnel, 20-Inch Mach 6 CF4 Tunnel, 20-Inch Mach 6 Tunnel, 31-Inch Mach 10 Tunnel) and expertise that are uniquely capable of supporting the development of DoD hypersonic propulsion systems.

Mission Alignment (3): This R&T Challenge has some relevance to the NASA mission, but would mainly be in support of DoD research.

Lack of Alternative Sponsors (3): Both NASA and the DoD support the development of hypersonic propulsion systems.

Appropriate Level of Risk (9): This Challenge faces high risk, too high for industry to take the lead, but it has a good chance of success if the program is supported adequately over the next decade.

B13 Improved propulsion system tolerance to weather, inlet distortion, wake ingestion, bird strike, and foreign object damage

Over the last 20 years, considerable progress has been made in engine and propulsion system design to address issues related to tolerance to weather, inlet distortion, wake ingestion, bird strike, and foreign object damage (FOD). Requirements are becoming more and more stringent, however, and more work remains to be done. This is particularly the situation with the advent of larger inlets on commercial engines, which makes engines more susceptible to bird ingestion and FOD.

To accommodate adverse weather (e.g., rain, ice, hail, and crosswinds), better analytical models are needed to predict the impact of the elements on fans, compressors, and combustor stability. These models should be physics-based and validated with experimental data. Detailed weather data as a function of altitude are needed for altitudes up to 20,000 ft; data on water concentration and droplet size are especially important. Analyses of the impact of ingested rain, ice, and so on as they traverse the propulsion system need to be improved. Better models will lead to more robust engine designs and improved operational procedures.

Better designs are also needed to toughen turbomachinery against bird ingestion and FOD without a significant loss in performance. This will require better materials and better design techniques. Development of improved aircraft and engine controls should also be considered to maximize the ability of aircraft to withstand to these events. Key milestones include

- Improve analytical tools to model more accurately the effects of rain, ice, and hail ingestion on engine behavior.
- Improve fan, compressor, and combustor stability to anomalous events.
- Collect detailed data about weather at altitude, particularly water concentration and droplet size, up to 20,000 ft.
- Improve impact resistance of turbomachinery.
- Improve engine and aircraft controls to adjust for impact events and erosion.

Relevance to Strategic Objectives

Capacity (3): Reduced sensitivity to weather and operational anomalies would increase capacity during adverse weather.

Safety and Reliability (9): This Challenge will increase safety and reliability by reducing the probability and severity of malfunction when the engine encounters anomalies.

Efficiency and Performance (3): This Challenge will minimize performance degradation through increased robustness.

Energy and the Environment (1): Improved robustness should maintain current levels of noise and emissions.

Synergies with National and Homeland Security (3): This Challenge will also benefit DoD aircraft.

Support to Space (1): This Challenge has very little relevance to this objective because space launch and return-to-Earth can generally avoid poor weather by altering schedules.

Why NASA?

Supporting Infrastructure (3): NASA has a few relevant facilities, such as icing tunnels.

Mission Alignment (3): This Challenge is relevant to NASA's mission of improving the safety and capacity of the air transportation system.

Lack of Alternative Sponsors (3): Industry and DoD are much more active than NASA in supporting research relevant to this Challenge. It has enough payoff and impact for industry to pursue.

Appropriate Level of Risk (3): This Challenge faces low risk.

B14 Propulsion approaches employing specific planetary atmospheres in thrust-producing chemical reactions

The three types of power sources for producing vehicle thrust in planetary atmospheres are electric, chemical, and nuclear. Electrical sources include solar cells, fuel cells, and batteries. Chemical sources involve combustion of a fuel in the presence of an oxidizer (bipropellant) or a catalyst (monopropellant) in an internal combustion engine, a piston expander, a gas turbine, or a rocket. Only chemical propulsion can produce the thrust levels needed for practical flight in planetary atmospheres.

The inner terrestrial planets—Mercury, Venus, Earth, and Mars—are too small to have prevented the light gases, hydrogen and helium, from being blown away by the solar wind; in fact, Mercury has only a trace atmosphere. Venus and Mars have about 96 percent CO_2 atmospheres, with Venus's atmosphere being very acidic. Titan's atmosphere is about 95 percent nitrogen but contains many hydrocarbons. The outer (Jovian) planets, Jupiter, Saturn, Uranus, and Nepture, are large enough to have retained gases from a nearby nebula; therefore, their atmospheres are all about 80-97 percent hydrogen, with several percent helium and methane. Mars and Venus, therefore, have oxidizing atmospheres, whereas Titan and the Jovian planets contain hydrogen or hydrocarbon, which can be used as fuels in a planetary flight vehicle. The method of propulsion from chemical reaction is, therefore, quite different for each planet.

Propulsion systems for planetary flight vehicles on Mars have received a lot of study and increasingly so with NASA's Space Exploration Initiative. Combustion of metals in CO_2 is the most promising source of energy for these vehicles. Using the CO_2 in the atmosphere of Mars averts the need to transport an oxidizer, greatly simplifying the spacecraft system and increasing its payload fraction. Magnesium is currently recognized as the best candidate fuel owing to its high adiabatic flame temperature, high specific impulse at high oxidizer:fuel ratios, high heat per unit mass, and low ignition temperature. The effective specific impulse of magnesium combustion in carbon dioxide is 1,190 seconds for an oxidizer:fuel ratio of 6. Magnesium also combusts at pressures suitable for internal combustion or turbine engines on Mars. It can be liquefied for use in a bipropellant rocket motor. It may be possible that, in the future, magnesium will be produced directly on Mars since the Viking Landers found that martian soil is 5 percent magnesium. The combustion of magnesium in CO_2 could become the main source of energy for human exploration on Mars.

Fundamental studies have been conducted on the burning of magnesium particles in CO_2. Simplified particle combustion models include two reactions zones: an outer zone, where magnesium reacts at a transport-limited rate to form condensed magnesium oxide plus carbon monoxide, and an inner zone, at the particle surface, where carbon monoxide reacts with liquid magnesium to form solid carbon and solid magnesium oxide, which remain with the particle. Kinetic mechanisms have been investigated and burning rates calculated and measured.

Considerable research needs to be done on engine cycles that operate with small magnesium and magnesium oxide particles. A gas turbine would have to operate with these erosive particles and would need a large inlet and exhaust nozzle owing to the rarefied atmosphere on Mars (about 1 percent of that on Earth). Internal combustion engines would have to be supercharged owing to the rarefied atmosphere. Binders for the magnesium and methods for supplying the gaseous or liquefied carbon dioxide to a rocket is another topic for research. Key milestones include

- Conduct detailed investigations, numerical and experimental, of the fundamental combustion characteristics of fuels or oxidizers that could be used in conjunction with the specific materials that make up the planetary atmospheres.
- Perform research on engine cycles that can operate with the indigenous atmospheric materials.
- Demonstrate propulsion systems found to be most promising for planetary flight vehicles in atmospheres of interest.

Relevance to Strategic Objectives

Capacity (1): This Challenge has no impact on this Objective.

Safety and Reliability (1): This Challenge has no impact on this Objective.

Efficiency and Performance (1): This Challenge has no impact on this Objective.

Energy and the Environment (1): This Challenge is not relevant to this objective.

Synergies with National and Homeland Security (1): This Challenge is not relevant to this Objective.

Support to Space (9): This Challenge provides a significant contribution to NASA's space exploration initiative to Mars and beyond.

Why NASA?

Supporting Infrastructure (3): NASA has relevant facilities and expertise, but they are not unique as there are also many ongoing university programs.

Mission Alignment (9): This Challenge area has major relevance and impact on several items in NASA's charter for space exploration as well as aeronautics.

Lack of Alternative Sponsors (9): This is a challenge where NASA would have to take the lead.

Appropriate Level of Risk (9): This Challenge faces high risk.

B15 Environmentally benign propulsion systems, structural components, and chemicals

This R&T Challenge is often referred to as the "green" (sometimes "evergreen") engine. The aim is to minimize the environment impact of gas turbine engines from cradle to grave (i.e., spanning the complete spectrum from manufacturing processes to operations to product end of life and disposal). The green concept generally includes minimizing noise and emissions, but these have already been identified separately as high-priority R&T Challenges and are not explicitly included here. The aim of this Challenge is to use, to the greatest degree possible, structural and maintenance materials that are environmentally benign and that can be recycled or safely disposed of at the end of life; environment-friendly manufacturing methods (e.g., the use of aqueous solvents, cutting fluids, and degreasers); and engine lubricants and working fluids that are environmentally safe. In some cases, human factors issues, such as ergonomically sound manufacturing and maintenance procedures, are included in comprehensive green engine programs.

It is difficult to quantify the environmental benefits of green engineering approaches, but their economic benefits may be easier to quantify. These accrue from less waste and disposal, more recycling of rare or precious materials, and a healthier and safer environment for manufacturing and maintenance personnel.

Significant progress has been made in the last decade in both engine materials and manufacturing processes. New materials include lead-free antigallants; lead-free, nonsilver dry film lubricants; chrome-free coatings; alternatives to cadmium plating and chromium anodizing; and nonchromate primers and coatings. In the manufacturing arena, solvents with low volatile organic content; closed-loop alkalai cleaning; and closed-loop acid pickling, milling, and stripping have been introduced. Key milestones include

- Assess current manufacturing processes and engine bills of materials to identify elements or compounds whose elimination would be environmentally beneficial.
- Quantify environmental benefits likely with elimination of targcted compounds.
- Assess feasibility and estimate the cost-benefit ratio of eliminating targeted compounds.
- Replace targeted compounds with validated substitutes.

Relevance to Strategic Objectives

Capacity (1): Although there may be some economic benefit to the environmental stewardship envisioned in this Challenge, it is unlikely to be sufficiently large to impact the cost of air travel in any significant way.

Safety and Reliability (1): This Challenge is not relevant to the safety of aircraft, although the safety of ground support personnel may be increased slightly (e.g., through the simplification of some maintenance procedures).

Efficiency and Performance (1): This Challenge is not relevant to this Objective.

Energy and the Environment (9): Modest energy savings in manufacturing may accrue from the recycling of precious or rare materials. This Challenge will produce major environmental benefits by reducing the use and disposal of environmentally harmful materials during every phase of product life, from design to end-of-life disposal.

Synergies with National and Homeland Security (3): The adverse environmental impact of DoD gas turbine engine design, manufacturing, maintenance, and disposal would be reduced.

Support to Space (1): This Challenge is likely to have very little relevance to this Objective because of the small number of access-to-space vehicles and the low volume of operations.

Why NASA?

Supporting Infrastructure (3): NASA has a strong materials capability that may be able to contribute to eliminating certain hazardous elements from gas turbine materials. It does not, however, have a comparably strong manufacturing support capability.

Mission Alignment (3): This Challenge is relevant to NASA's mission.

Lack of Alternative Sponsors (3): The Environmental Protection Agency, the Occupational Safety and Health Administration, and industry are likely to continue supporting environmental stewardship issues. Industry, in particular, is motivated by potential economic benefits. However, aeronautics may not be a priority for these groups.

Appropriate Level of Risk (3): This Challenge faces very high risk. In the structural materials area, some chemical constituents are very likely to be absolutely necessary for high performance and cannot be eliminated, so the return on investment may be problematic.

B16 Reduced engine manufacturing and maintenance costs

This Challenge will develop the ability to design engines with simpler, fewer parts that are easily serviceable while maintaining good performance and efficiency. To reduce manufacturing costs, better simulation and modeling tools will be developed to permit the design of engines with higher aerodynamic loadings in compressors and turbines. This will lead to fewer turbomachinery stages and fewer airfoils within each stage. Better materials, less expensive materials, and materials with less variability in properties will reduce the amount of expensive materials that must be used. Better machining and manufacturing techniques will reduce machining time and costs and reduce the amount of scrap, which can be expensive, particularly when exotic materials are involved. Maintenance costs will generally be lowered if engines contain fewer parts (e.g., fewer turbine blades that need to be replaced periodically). In addition, better predictions of the on-wing life of the parts will permit them to be used longer and more efficiently without an increase in malfunctions. More intelligent systems, described in R&T Challenge B3, will permit maintenance personnel to better assess the exposure of engine parts and, therefore, to better predict life. In addition, improved health diagnostics systems will permit maintenance staff to troubleshoot problems quickly and accurately, which will reduce overhaul and repair time. This produces a dual benefit: reducing the costs of overhauls and minimizing engine downtime. Key milestones include

- Improve turbomachinery design tools for reduced stage and part count.
- Develop better materials with narrower tolerances to reduce unnecessarily large design margins.
- Develop low-cost materials with the same high-performance properties.
- Develop better manufacturing techniques to reduce scrap.
- Develop better on-wing life predictions to minimize premature retirement of parts.
- Develop improved health diagnostics to allow performing maintenance as required rather than as scheduled (predictive maintenance).
- Develop intelligent engines that can adjust operation to minimize degradation of parts.

Relevance to Strategic Objectives

Capacity (3): Lower acquisition and maintenance costs will affect affordability of operations and should allow for profitable increases in capacity.

Safety and Reliability (3): Simpler designs are easier to maintain and are more reliable.

Efficiency and Performance (3): Improved, predictive maintenance should minimize performance degradation.

Energy and the Environment (3): Improved, predictive maintenance should minimize performance degradation and maintain engines near design noise and emissions levels.

Synergies with National and Homeland Security (3): DoD faces similar issues and would benefit from research on civil aircraft.

Support to Space (1): Some materials and techniques, particularly improved health diagnostics, might find applications on space vehicles, but because of differences in materials and flight regimes, benefits would be indirect.

Why NASA?

Supporting Infrastructure (3): NASA has strength in the development of the analytical tools, materials, sensors, and controls.

Mission Alignment (1): Achieving simpler, high-performance designs is very relevant to NASA's mission, but manufacturing technologies are not particularly relevant.

Lack of Alternative Sponsors (1): Industry and DoD are already key players in such initiatives.

Appropriate Level of Risk (3): The risk of this work can range from somewhat incremental to very challenging.

REFERENCES

Advisory Council for Aeronautical Research in Europe (ACARE). 2001. European Aeronautic—A Vision for 2020. Available online at <www.acare4europe.com/docs/Vision%202020.pdf>.

Akin, Gump, Strauss, Hauer, and Feld, LLP. 2003. Airport Air Quality Guide—A Guide to Air Quality Issues Under the Clean Air Act.

Ames, B. 2005. Power electronics drive next-generation vehicles. Military and Aerospace Electronics, July 1. Available online at <http://mae.pennnet.com/Articles/Article_Display.cfm?ARTICLE_ID=232549&p=32&pc=ENL>.

Barna, T., T. Sheridan, W. Harrison, P. Serino, and C. Bauer. 2005. OSD Clean Fuel Initiative and the Air Force: National Mining Association Briefing. Available online at <www.nma.org/attach/DOD briefing.pdf>.

Department of Energy (DOE). 2003. Hydrogen, Fuel Cells and Infrastructure: Multi-Year Research, Development and Demonstration Plan. DOE/GO-102003-1741. Washington, D.C.: U.S. Department of Energy, Energy Efficiency and Renewable Energy. Available online at <www.eere.energy.gov/hydrogenandfuelcells/mypp/>.

Emadi, A., M. Ehsani, and J. Miller. 2003. Vehicular Electric Power Systems: Land, Sea, Air, and Space Vehicles, New York, N.Y.: Marcel Dekker.

Environmental Protection Agency (EPA). 2006. Inventory of U.S. Greenhouse Gas Emissions and Sinks: 1990-2004. Available online at <http://

yosemite.epa.gov/oar/globalwarming.nsf/content/ResourceCenterPublicationsGHGEmissions.html>.

Faass, R. 2001. Cryoplane. Available online at <www.haw-hamburg.de/ers/Scholzz/dglr/hh/text_2001_12_06_cryoplane.pdf>.

Federal Aviation Administration (FAA). 2006. FAA Aerospace Forecasts FY 2006-2017. Available online at <www.faa.gov/data_statistics/aviation/aerospace_forecasts/2006-2017/>.

Intergovernmental Panel on Climate Change (IPCC). 1999. Special Report: Aviation and the Global Atmosphere. Geneva: World Meteorological Organization.

Lefebvre, A. 1999. Gas Turbine Combustion. Philadelphia, Pa.: Taylor and Francis.

National Aeronautics and Space Administration (NASA). 2003. Aerospace Technology Enterprise Strategy, NP-2003-01-298-HQ. Washington, D.C.: NASA Headquarters. Available online at <www.aero-space.nasa.gov>.

NASA. 2006. NASA Hyper-X program demonstrates scramjet technologies. NASA Facts. Available online at <www.nasa.gov/centers/langley/pdf/142891main_X43A_2006.pdf>.

National Institute of Aerospace (NIA). 2005. Responding to the Call: Aviation Plan for American Leadership. Available online at <www.nianet.org/nianews/AviationPlan.php>.

National Research Council (NRC). 1997. Assessing NASA's High Speed Research Program. Washington, D.C.: National Academy Press. Available online at <www.nap.edu/catalog/5848.html>.

NRC. 2001. Commercial Supersonic Technology: The Way Ahead. Washington, D.C.: National Academy Press. Available online at <http://books.nap.edu/catalog/10283.html>.

NRC. 2002. For Greener Skies: Reducing Environmental Impacts of Aviation. Washington, D.C.: National Academy Press. Available online at <http://books.nap.edu/catalog/10353.html>.

NRC. 2003. Securing the Future of U.S. Air Transportation: A System in Peril. Washington, D.C.: The National Academies Press. Available online at <http://fermat.nap.edu/catalog/10815.html>.

NRC/National Academy of Engineering (NAE). 2004. The Hydrogen Economy: Opportunities, Costs, Barriers, and R&D Needs. Washington, D.C.: The National Academies Press. Available online at <http://books.nap.edu/catalog/10922.html>.

Saynor, R., A. Bauen, and M. Leach. 2003. The Potential for Renewable Energy Sources in Aviation. Imperial College Centre for Energy Policy and Technology. Available online at <www.iccept.ic.ac.uk/pdfs/PRESAV%20final%20report%2003Sep03.pdf>.

U.S. Air Force. 2000. America's Air Force: Vision 2020: Global Vigilance, Reach and Power.

Waitz, I., J. Townsend, J. Cutcher-Gershenfeld, E. Greitzer, and J. Kerrebrock. 2004. Aviation and the Environment; A National Vision Statement, Framework for Goals and Recommended Actions. Report of the Partnership for AiR Transportation Noise and Emissions Reduction (PARTNER) to the United States Congress, December. Available online at <http://partner.aero>.

C

R&T Challenges for Materials and Structures

A total of 20 R&T Challenges were prioritized in the materials and structures Area. Table C-1 shows the results. The R&T Challenges are listed in order of NASA priority. National priority scores are also shown.[1] This appendix contains a description of each R&T Challenge, including milestones and an item-by-item justification for each score that appears in Table C-1.[2]

C1 Integrated vehicle health management

Integrated vehicle health management (IVHM) refers to the ability to monitor, assess, and predict the structural and material health[3] of primary and secondary structures for individual missions and lifetime durability using networks of onboard sensors. A fully integrated approach to IVHM relies on a multidisciplinary set of analysis, testing, and inspection tools, including miniaturized sensors and distributed electronics; sophisticated signal processing; data acquisition, integration, and database maintenance; artificial intelligence; damage science; and the mechanics of structures and their failure.

In addition to the obvious benefit of increasing safety and reliability, IVHM holds the promise of reducing vehicle cost, weight, and maintenance downtime, as well as speeding the introduction of new material systems and structural concepts. Real-time onboard sensor systems that monitor the actual state of materials and structural components enable more efficient use of material, including novel concepts. Moreover, with a national fleet of aging aircraft and infrastructure in an industry with low profit margin, IVHM is increasingly important due to its ability to increase safety and reliability. IVHM promises low-cost, real-time sensing and inspection methods to detect damage before catastrophic failure occurs.

IVHM currently means putting a variety of sensor systems onboard an aircraft, along with the artificial intelligence that automatically interprets the various sensor output streams. These data are used to provide input to prognostic systems that then draw conclusions about structural integrity issues. Two main features distinguish the next generation of IVHM from traditional nondestructive evaluation (NDE): (1) Sensor packages will be very small and exceedingly lightweight and (2) the reliance on humans to interpret the sensor output and assess the impact on structural integrity will be reduced or eliminated. Sensors and software are available, e.g., fiber-optic (Wood et al., 2000; Stewart et al., 2003; Carman and Sendeckyi, 1995) and piezoelectric (Lin and Chang, 2002; Giurgiutiu and Zagrai, 2002). The next major hurdle is integrating IVHM systems in flight structures. Laboratory tests have demonstrated that several classes of IVHM systems are available. Downselects to designs appropriate for aircraft structures are needed.

Three classes of IVHM systems warrant attention over the next 10 years. The first class includes fiber-optic sensor systems that can use multiplexed fibers attached to or embedded in the structure, each with numerous multiphysics sensing sites interrogated in turn by a single electro-optic module. The second class includes locally self-powered, wireless microelectromechanical sensors of various types tiny enough that very large numbers of sensors become practical. Each sensor mote performs a point measurement, so many are used to effectively cover large areas. The third class includes discrete active or passive remotely powered sensor modules (e.g., by means of guided-wave ultrasonic or acoustic emission) that may be large compared to sensor motes but can interpret multimode vibrations or multiphysics parameters (temperature, stress, humidity, etc.) that propagate over relatively long distances within the key structural

[1]The prioritization process is described in Chapter 2.

[2]The technical descriptions for the first 10 Challenges listed below contain substantially more detail than the technical descriptions for these Challenges as they appear in Chapter 3.

[3]"Health" in this context implies either an absence of measurable material flaws or an ability to coordinate the growth rate of flaws with the safe life remaining for the element in question.

TABLE C-1 Prioritization of R&T Challenges for Area C: Materials and Structures

	Strategic Objective							Why NASA?					
	Capacity	Safety and Reliability	Efficiency and Performance	Energy and the Environment	Synergies with Security	Support to Space	National Priority	Supporting Infrastructure	Mission Alignment	Lack of Alternative Sponsors	Appropriate Level of Risk	Why NASA Composite Score	NASA Priority Score
R&T Challenge **Weight**	5		3		1			$^1/_4$ each					
C1 Integrated vehicle health management	9	9	3	1	9	3	**114**	9	9	1	9	**7.0**	**798**
C2 Adaptive materials and morphing structures	9	3	9	3	9	3	**108**	9	9	1	9	**7.0**	**756**
C3 Multidisciplinary analysis, design, and optimization	9	3	9	1	3	3	**96**	9	9	3	9	**7.5**	**720**
C4 Next-generation polymers and composites	9	3	9	1	9	3	**102**	9	9	1	9	**7.0**	**714**
C5 Noise prediction and suppression	9	1	3	9	3	1	**90**	9	9	3	9	**7.5**	**677**
C6a Innovative high-temperature metals and environmental coatings	3	9	3	1	9	3	**84**	9	9	3	9	**7.5**	**630**
C6b Innovative load suppression, and vibration and aeromechanical stability control	3	9	3	1	9	3	**84**	9	9	3	9	**7.5**	**630**
C8 Structural innovations for high-speed rotorcraft	9	1	3	1	9	1	**72**	9	9	3	9	**7.5**	**540**
C9 High-temperature ceramics and coatings	3	1	9	3	3	9	**68**	9	9	3	9	**7.5**	**510**
C10 Multifunctional materials	3	3	9	3	9	9	**84**	3	9	3	9	**6.0**	**504**
C11 Novel coatings	3	9	3	3	1	1	**80**	3	9	3	9	**6.0**	**480**
C12 Innovations in structural joining	3	3	9	1	3	3	**66**	3	9	3	9	**6.0**	**396**
C13 Advanced airframe alloys	9	1	9	1	3	1	**84**	1	3	1	9	**3.5**	**294**
C14 Next-generation nondestructive evaluation	3	9	1	1	3	1	**70**	3	9	1	3	**4.0**	**280**
C15 Aircraft hardening	1	9	1	1	9	1	**66**	3	3	1	9	**4.0**	**264**
C16 Multiphysics and multiscale modeling and simulation	3	3	3	3	3	1	**52**	3	3	3	3	**3.0**	**156**
C17 Ultralight structures	3	1	3	1	3	3	**38**	3	9	1	3	**4.0**	**152**
C18 Advanced functional polymers	1	3	1	1	3	1	**30**	9	3	3	3	**4.5**	**135**
C19 Advanced engine nacelle structures	3	1	3	1	1	1	**34**	1	9	1	3	**3.5**	**119**
C20 Repairability of structures	3	3	3	1	1	1	**44**	3	3	1	3	**2.5**	**110**

elements. Over the near term, IVHM sensors are more likely to be discrete items integrated with the structure rather than just another function of the structural materials themselves. Over the next decade, however, there is much work to be done in making things small enough, smart enough, and with enough multifunctionality to be acceptable to the airframe designers and owners.

Successful application of IVHM also relies on continued research and refinement in fundamental structural mechanics and the mechanics of damage and failure for accurate interpretation of IVHM sensor data and to support autonomous decision making for damage recovery and mitigation.[4]

NASA's research program should target key applications where IVHM is most likely to make a difference rather than develop monitoring systems for unforeseen problems in existing structures. Detecting and characterizing multidamage states, composite debonding, corrosion, long-term fatigue, and impact damage are all important areas that can be mitigated with IVHM. Assessing the integrity of structural repair patches and finding latent faults in aging wiring are also key issues that need to be addressed over the next decade. IVHM approaches for aging aircraft are typically in response to a problem that has been identified. Regardless of the particular problem, however, the goals are still the same: putting a variety of sensor systems onboard aircraft along with the artificial intelligence that automatically interprets the various sensor output streams to provide input to prognostic systems that then assess structural integrity and makes life predictions. Key milestones include

- Develop lightweight sensor networks that characterize the state of materials and structures over large areas.
- Develop very-low-power or self-powered wireless sensors capable of operation in harsh environments.
- Develop artificial intelligence to automatically assess structural integrity from sensor responses and implement damage mitigation protocols.
- Develop components and sensors that are cost competitive and available from multiple vendors.
- Flight test full-scale IVHM systems to detect multisite damage.

Relevance to Strategic Objectives

Capacity (9): IVHM increases operating flexibility by permitting a wider range of operating conditions and environments due to increased confidence in the actual state of the structural elements. Monitoring system performance in real time allows one to do new things with confidence, such as enabling larger aircraft sizes and new aircraft concepts (for example, V/STOL aircraft, variable-cycle engines, and new structural configurations). These new designs may operate in currently unexplored regimes of operation and greatly increase the flexibility of the air transportation system. In addition, health monitoring could reduce maintenance time and costs. Aircraft could report the predicted lifetimes of their own parts and report the need for replacement parts. IVHM could quickly diagnose root problems, minimizing flight delays (Powrie and Fisher, 1999; Simon, 2000).

Safety and Reliability (9): Early detection of impending failures in aircraft materials, structures, and wiring is critical for avoiding fatalities as a part of the aging aircraft program. IVHM also reduces time lost to scheduled maintenance and reduces the likelihood of unscheduled downtime.

Efficiency and Performance (3): Once confidence in IVHM systems has been established, they could allow aircraft to operate closer to performance margins and with greater structural efficiency, which would reduce operating costs. IVHM will change future aircraft designs by reducing the margin of safety required in design, thereby reducing weight and increasing performance. It will also allow better design of engine fan blades, which now must be overdesigned for fear of fatigue failure. Finally, IVHM could increase the efficiency of aircraft maintenance, reducing operating costs and downtime.

Energy and the Environment (1): This Challenge will increase structural efficiency, which could save energy and thus reduce environmental effects, but overall, the impact is indirect.

Synergies with National and Homeland Security (9): DoD and others are already supporting research relevant to this Challenge but NASA has an opportunity to contribute when it comes to civil aeronautics and with a focus on the more sophisticated sorts of sensor systems that require a high-level understanding of the measurement physics involved. Reducing downtime has a significant impact on mission availability. Because IVHM systems for civil aviation will have to comply with stringent national and international certification requirements, advances will likely appear first in military aircraft.

Support to Space (3): IVHM could allow spacecraft to operate safely closer to performance margins and to be confidently designed with greater structural efficiency. However, differences in operating conditions mean that aeronautical IVHM systems would only be partially applicable to space vehicles.

Why NASA?

Supporting Infrastructure (9): NASA has many experts in technical fields related to IVHM. Facilities of particular relevance include the fiber-optic draw tower at Langley's Nondestructive Evaluation Sciences Branch and the NASA Dryden flight research facility, which provides a platform to test various IVHM approaches in flight.

Mission Alignment (9): IVHM is a natural part of NASA's aeronautics mission and also dovetails nicely with

[4] See R&T Challenges D4 and D5.

NASA's multifunctional materials and multiphysics analysis research, although over the near term IVHM sensors are more likely to be discrete items integrated with the structure rather than just another function of the structural materials themselves. Over the next decade, there is much work to be done in making things small enough, smart enough, and with enough multifunctionality to be acceptable to the airframe designers and owners.

Lack of Alternative Sponsors (1): DoD and others are supporting IVHM research, including applications for a wide variety of nonaerospace industries. Significant industrial commitment related to this Challenge is also evident.

Appropriate Level of Risk (9): By targeting key forward-looking applications where IVHM is most likely to make a difference over the medium term, rather than focusing on solving previously unforeseen problems in existing structures by coming up with monitoring systems for them, this Challenge faces high risk.

C2 Adaptive materials and morphing structures

Use of adaptive materials and morphing structures to change the aircraft shape (outer mold lines) and functions on demand represents a revolutionary approach for enabling optimal performance over a range of flight missions. Efficient, multipoint adaptability allows optimal performance for a variety of diverse, often contradictory, mission objectives (Lin and Crawley, 1995). Historically, morphing devices have included retractable landing gear, flaps, slats, and spoilers, all of which allow aircraft to land at lower speeds and to cruise at higher speeds. Adaptive materials have been used to reduce vibrations, eliminate noise, or control local air flow features such as separation. More recently, wings with the ability to drastically change planform area and shape have been proposed, and a few advanced concepts have been built. Adaptive materials are important elements of a morphing aircraft structure due to their ability to change or alter material properties and structural shapes using energy inputs such as light, heat, and electric or magnetic fields. Adaptive materials include heat-activated shape memory alloys like NiTiNOL; ceramics (e.g., lead zirconate titanate); photonically activated, lightweight, flexible shape memory polymers; electrically activated piezoelectrics; and magnetorheological fluids. These materials, some of which are self-sensing and self-actuating, can be developed into motors, combined with mechanisms, or distributed as actuators to produce highly efficient, lightweight airplanes. The success of morphing designs requires new adaptive materials and mechanisms, as well as innovative aircraft designs.

Morphing aircraft predate the field of adaptive structures. Fighter aircraft with variable swept wings first appeared in the 1960s. These aircraft were required to take off and land on aircraft carriers and achieve efficient supersonic speeds, yet they weighed less than comparable aircraft with fixed-wing designs. The field of adaptive structures evolved in the 1970s with work on vibration suppression for optical devices using piezoelectric materials and continued into flutter suppression work in the 1980s, some of which was conducted at NASA Langley. More recent design and testing of adaptive aeronautical structural components, such as rotor blades, have also demonstrated promise for allied areas such as vibration and noise control.

Recent morphing wing designs include planform area changes of up to 50 percent. These concepts, along with adaptive materials, expand possibilities for advanced aircraft. This includes providing new opportunities to expand wing area on demand, providing local flow control devices for drag reduction, and reducing vibrations and noise (inside the cabin and externally). These combined and synergistic developments allow innovative aircraft designs that operate efficiently over a wide range of speeds, for example, to land at very low speeds, loiter for long periods of time, and cruise efficiently at both subsonic and supersonic speeds.

Morphing structures with adaptive materials are not limited to wing surfaces. They could also be used to enable engine inlets that dramatically reshape or alter flow characteristics over a wide range of speeds. However, this would require development of high-temperature adaptive materials. These and other applications provide opportunities for revolutionary changes in future aircraft structures, although the costs of incorporating such technology have not yet been determined.

DARPA's Morphing Aircraft Structures Program has sponsored several wind tunnel experiments at NASA Langley Research Center's Transonic Dynamics Tunnel. These tests, which cost more than $35 million, produced a wide range of experience and identified innovative aircraft and rotorcraft concepts and critical materials technologies for future work. DARPA has also sponsored flight research on morphing technologies applied at the system level. Although the results were promising, these tests also showed that the technologies relevant to this Challenge are still immature and require additional research.

Adaptive materials have received a great deal of attention during the past decade, mostly directed toward the development of piezoelectric devices (Chaplya and Carman, 2002; Ritter et al., 2000), metallic shape memory alloys (Shin et al., 2004; Lim and McDowell, 1999), and ferromagnetic actuations (e.g., both magnetostrictive (Moffett et al., 1991) and ferromagnetic shape memory alloys (Ullakko et al., 1997)). However, to date, few viable aeronautical concepts have been created. The principal missing technologies are self-actuating adaptive materials with sufficient power output per unit mass and new materials that can be easily molded and remolded, stretched, or reshaped drastically. Current materials are not suitable for large actuation or elastic strains required for commercial aircraft. The requirements for load-carrying efficiency (high stiffness) and actuatable structures (low stiffness) have resulted in quick fixes using stiffened polymers as wing panels.

Morphing designs that change the aircraft shape require (1) development of new materials that can be turned on and off and (2) integration of these materials into novel mechanisms with embedded, distributed power sources that can activate the structures and move them efficiently from one point to another. These new materials will require lifetimes comparable to those of current materials.

Advanced aircraft concepts require designers to think differently about how aircraft and systems can be designed to demonstrate lower landing speeds, higher cruise speeds, and longer ranges than are possible today. A fundamental task is to characterize the mechanical response of these inherently nonlinear materials, including hysteresis, fatigue, long-term behavior, and damage behaviors. Analysis and design tools that accurately predict these responses will open the door to even more applications of these revolutionary adaptive structural concepts that could optimize performance and expand the flight envelope. Key milestones include

- Identify new morphing missions and designs for reconfigurable civil aircraft, including supersonic aircraft with low sonic boom.
- Develop the next generation of high-strain adaptive materials or devices that can be activated and deactivated for repositioning, with actuation deformation up to 100 percent.
- Develop novel integrated adaptive materials that allow wing surfaces and fuselages (including inlets) to rapidly change shape or alter load paths.
- Conduct scaled wind tunnel and flight tests on active, morphing aircraft to enable innovative, lightweight designs.
- Develop new, structurally integrated adaptive devices for flow control on a commercial aircraft to, for example, reduce drag and improve performance in off-design conditions.
- Develop analysis and design tools that account for and accurately predict nonlinear behaviors of adaptive materials and morphing structures.

Relevance to Strategic Objectives

Capacity (9): Adaptive materials and structures have already led to innovative solutions to aeronautical problems across a broad spectrum of requirements, including ice removal and vibration reduction. Morphing civil aircraft can adapt to a wider variety of landing and cruising speeds and may be able to fly at supersonic speeds over inhabited areas with minimal sonic boom.

Safety and Reliability (3): Adaptive materials are key elements of some aeronautical health monitoring concepts. Morphing airliners might also allow damaged aircraft to be reconfigured for safe flight.

Efficiency and Performance (9): Adaptive materials, as part of flow control devices and seamless control surfaces, can enhance wing lift. Morphing civil aircraft would be able to reconfigure for optimal performance at various altitudes, weights, and speeds.

Energy and the Environment (3): Adaptive materials reduce fuselage noise and drag, which indirectly reduces emissions. Active vibration suppression has been successful in reducing rotorcraft noise.

Synergies with National and Homeland Security (9): Adaptive materials and structures allow aircraft to perform tasks that are difficult or impossible for conventional aircraft, such as vibration control. Morphing aircraft are high on the DoD future concepts list because of their ability to adapt to unforeseen situations, to reduce the number of aircraft required for certain missions, and to operate over a wide range of conditions.

Support to Space (3): Adaptive materials have been used for vibration control for payloads delivered to space, for opening solar arrays from compact packages, and as adaptive optics.

Why NASA?

Supporting Infrastructure (9): NASA has historically supported this type of technology and has a strong cadre of researchers with capabilities in flow control devices and adaptive materials. NASA has flight test organizations and facilities that uniquely support this type of activity. NASA Langley also has personnel developing new adaptive materials, including high-deformation piezoelectric actuators and piezo-fiber composites. More recently NASA Glenn has been developing high-temperature adaptive materials (e.g., piezoelectric and ternary nickel-titanium shape memory alloys).

Mission Alignment (9): This Challenge is very relevant to NASA's mission.

Lack of Alternative Sponsors (1): Morphing structures and adaptive materials are currently being studied by academia and the DoD.

Appropriate Level of Risk (9): Some technologies related to this Challenge face very high risk, as evidenced by DARPA interest. However, recent government-sponsored research has reduced risk and clearly identified high-risk investments and issues in materials and structures that are appropriate for NASA to pursue.

C3 Multidisciplinary analysis, design, and optimization

Methods for simulation-based, multidisciplinary design and optimization (MDO) are at the very core of a philosophy that moves away from the build-test-build approach, which has proven to be expensive and ineffective in exploring the aeronautical design space. MDO processes develop synergistic benefits by integrating people, analytical tools, experimentation, and information to design complex structural components and systems that are characterized by strong interactions. These approaches allow for development of optimal configurations, topologies, and dimensions for structural members and components to achieve design objectives, and they permit designers to examine the myriad what-ifs that characterize sophisticated designs with interdisciplinary trade-offs (Sobieszczanksi-Sobieski and Haftka, 1997).

Tools and platforms to enable the effective integration of analysis methods for the study of multidisciplinary interactions in structural design are the focus of this Challenge. A systematic analysis of uncertainty in all aspects of the design process is also important, as is increasing the level of detail in representing the structure. Uncertainties surrounding the predictive capabilities of physics-based models used in design—and their propagation in coupled systems—must be systematically included in the design process. Such design tools would greatly facilitate the process of sorting through design options and help identify and exploit interactions among multiple disciplines involved in the design process. The availability of MDO processes and analytical techniques enables experienced designers to collaborate with skilled analysts to identify and create innovative structural designs.

The rapid development of aerospace technologies has resulted in many new possibilities for aircraft design. Unfortunately, the benefit of these technologies, whether applied to conventional or radically new designs, is often slow to appear. MDO processes allow for the rapid identification of game-changing designs and design features, with significant potential impact on structural and material issues related to the design of a new generation of aerospace vehicles. The systematic inclusion of the effects of uncertainty, whether in loading, material behavior, or mission requirements, yields a rational approach for quantifying the risk associated with a certain design and allow for meaningful study of trade-offs among competing concepts. Inclusion of risk and reliability analysis in the design provides a time-dependent description of risk associated with structural and material systems in service, facilitating decisions that enhance vehicle availability and reliability.

After almost two decades of R&D, MDO processes for conventional designs have reached a high level of sophistication. In structural designs where the topology or outer mold lines are defined, a high level of success can be achieved by coupling analytical methods such as the structural finite-element technique with similar tools for load assessment. However, for designs with a multiplicity of topologies, some of which are not well-defined, and for problems where a large number of design parameters and constraints must be considered in the early stages of the design process, MDO methodologies are still underdeveloped (Giesing and Barthelemy, 1998). Major effort must also be directed at including the effects of uncertainty in the design process, as well as increasing the level of detail in representing the structure. Risk- and reliability-based design has emerged as a key research area, with industry, government, and academia focusing on developing robust methods that transition deterministic design tools to a nondeterministic environment. This effort has been primarily confined to disciplinary design with a dual focus: (1) developing approaches for modeling uncertainty in problem parameters and (2) seeking an efficient adaptation of design tools that account for the modeled uncertainties in a formal design process. New ways of formulating problems that incorporate quantitative reliability measures to facilitate effective design decisions have already been considered in this context. The extension of these approaches to large-scale structural and material design problems represents an entirely different level of problem complexity.

Significant new developments are required in both the platforms and the embedded tools that constitute the MDO process. Efficiency and effectiveness of the search process continue to be a problem, particularly in large-dimensionality problems and multimodal or disjointed search spaces. Existing problem formulations and search processes do not naturally allow for the emergence of radical design concepts and the associated design constraints. Current platforms are ill equipped to efficiently parse the vast amounts of data associated with the design process. Furthermore, not all methods are ideal for all problems. The goal of this Challenge is not to generate one perfect, all-encompassing algorithm but to use the most efficient and effective method or combination of methods for each problem. Proper algorithm selection in itself is an important research topic. There is a marked need for developing analysis modules for the search process to query. Some structural design issues, such as manufacturability, cost, repair, and environmental impact, are seldom represented in the design process. These analysis tools must be developed at multiple levels of granularity and precision, to coincide with the appropriate stage of the design process. The numerical efficiency of these tools is paramount, and alternative paradigms that take advantage of a new generation of parallel computational hardware must be sought (Giesing and Barthelemy, 1998; Sobieszczanksi-Sobieski and Haftka, 1997). Furthermore, existing analysis tools lack quantitative measures of prediction uncertainty. Uncertainty modeling in a data-lean environment, specifically for new concepts, continues to be an issue in this regard. There is a similar dearth of computationally efficient methods for reliability assessment, particularly in situations where uncertainty distributions do not conform to standard forms or where components or elements exhibit discrete behavior. The propagation of uncertainty in complex and highly coupled multidisciplinary systems needs to modeled, and tools for design and optimization in a nondeterministic environment continue to be computationally intractable, especially when applied to design problems involving a large number of nondeterministic variables, parameters, and design constraints.

New search methods and platforms that allow for an effective integration of analysis tools are required for the design of the next generation of aerospace structural and material systems. These methods would incorporate effective and computationally efficient procedures and tools for quantifying risk and would be integrated with design optimization and decision-making tools and software at all levels of the design process. A systematic approach for modeling risk and uncertainty in complex coupled systems should be a key concern in this area of inquiry. Use of commercial tools in optimization is not enough to advance the state of the art in MDO. Optimization is only one piece of the analysis, design, and optimization triad. It is the tightly integrated development of analysis and optimization tools that furthers the potential of MDO methods. In the aerospace arena, such expertise is unique to NASA—additional gains can be realized with NASA working in close collaboration with researchers from academia and industry. A number of synergistic benefits could also be achieved by developing this approach in concert with health-monitoring technologies (see R&T Challenge C1). Key milestones include

- Develop multidisciplinary analysis tools that incorporate aerodynamics, structural dynamics, vibration, thermal response, and acoustic response with structural response to mechanical loads.
- Extend multidisciplinary tools to incorporate explicit mathematical modeling of design issues such as manufacturing processes, life-cycle cost, and repairability.
- Develop efficient approaches for multivariable optimization.
- Develop efficient and effective search processes for analysis of large complex systems.
- Develop approaches for modeling uncertainty in data-lean environments.
- Develop computationally efficient methods for reliability assessment.
- Develop systematic approach for modeling risk and uncertainty in complex coupled systems.

Relevance to Strategic Objectives

Capacity (9): This Challenge would support the development of new vehicle concepts and designs that are more efficient from the standpoint of speed, payload capacity, and fuel burn. The formal inclusion of multiple design objectives would permit the development of vehicles capable of operating in multiple operating environments and could alleviate traffic congestion in busy air corridors.

Safety and Reliability (3): Quantitative inclusion of risk and uncertainty in the design process could result in structural designs that promote vehicle safety and reliability. Temporal estimates of structural reliability might allow for improved practices in airframe maintenance and repair.

Efficiency and Performance (9): This Challenge will support development of structural components and systems that exploit synergistic benefits of multidisciplinary interactions to optimize explicit design criteria related to efficiency and performance.

Energy and the Environment (1): Lightweight structures would probably increase payload capacity but not necessarily reduce fuel burn of particular vehicles. MDO can increase understanding of structural acoustics and noise due to airflow, vibration, and structural dynamics but would not necessarily offer any solutions.

Synergies with National and Homeland Security (3): This Challenge would also improve the design and development of military aircraft.

Support to Space (3): The design of space structures has many of the same multidisciplinary elements as the design of advanced aircraft. The inclusion of quantitative measures of risk in the design decision process would be of particular relevance to the design of space structures.

Why NASA?

Supporting Infrastructure (9): MDO tools and processes have a history of healthy development at the NASA centers. The multidisciplinary expertise required for the development of relevant platforms and tools—disciplinary experts, experimental facilities, and information technology support—are all available at NASA. Furthermore, NASA is uniquely qualified to conduct MDO research with aeronautics applications in mind.

Mission Alignment (9): This Challenge is very relevant to NASA's mission; it is a cross-cutting, enabling technology that supports many NASA missions.

Lack of Alternative Sponsors (3): Some support for this activity exists within the DoD and the aerospace industry, but there is a lack of infrastructure or research leadership in this area at these other agencies and organizations. Arguably, commercial tool set providers may be able to better accomplish this challenge.

Appropriate Level of Risk (9): This Challenge faces moderate risk. The research tasks are all plausible but not routine and would require NASA researchers to collaborate with academia and industry.

C4 Next generation polymers and composites

Over the past 50 years, polymeric composites have revolutionized and improved the performance of aircraft structures. Future needs for enhanced structural performance, high-temperature capability, and durability can only be met by the next generation of high-temperature, polymer-based composites. Next-generation composites will take advantage of improved polymeric matrices, new reinforcement materials, hybrid reinforcement approaches, improved joining technology, and science-based manufacturing with controlled three-dimensional placement of reinforcements. They will potentially lead to significant improvements in structural efficiency, safety, and high-temperature performance, as well as a reduction in data scatter, increased damage tolerance (e.g., resistance to delamination), and improved manufacturability (elimination of hand lay-up). These composites will likely incorporate adaptive materials and multifunctional concepts, thus serving as enabling materials for visionary concepts in nacelle components, wing structures, and fuselage materials.

Relevant research is currently directed toward development of higher temperature matrix materials; nanoscale reinforcements such as nanofibers, nanoclays, and carbon nanotubes; composites with multiple, different reinforcement fibers, and integration of adaptive materials to increase functionality. These materials have potential application in propulsion systems and for supersonic aircraft. Multiscale modeling efforts are also being conducted to guide the design and development of these materials across the nano-micro-meso structural levels.

The development of next-generation composites requires three capabilities to gain a complete understanding of the potential advantages of these materials: multiscale modeling, science-based processing techniques, and structural and mechanical testing. Multiscale models that link nano- and microstructure to structural composite response as well as the introduction of hybrid and multifunctional models are a critical concern. Research should also target (1) development of science-based processing techniques that account for resin chemistry, cure kinetics, and flow physics in guiding placement and distribution of the different reinforcement phases and (2) optimization of the reinforcement–matrix interface. Finally, structural and mechanical testing capabilities are needed to evaluate both the design and processing parameters. These three capabilities will enable the creation of effective life-prediction models and thereby eliminate statistical variation and defects. Key milestones include

- Demonstrate fabrication of composites with multiple different reinforcement fibers.
- Integrate adaptive materials to increase functionality.
- Develop techniques for manufacturing, processing, and dispersion of nanoscale reinforcements.
- Develop a fundamental understanding of how different kinds of reinforcements (e.g., nano, functional, or hybrid additives) affect the performance of polymers and composites.
- Improve damage tolerance for high-temperature polymers.
- Develop effective life-prediction models for polymers and composites.
- Investigate environment-friendly end-of-life reuse or disposal strategies.

Relevance to Strategic Objectives

Capacity (9): This Challenge has the potential to reduce structural weight, increase strength, and enable higher temperature multifunctional structures that will enable new vehicle concepts and increase capacity.

Safety and Reliability (3): Advanced composites will have improved damage tolerance, reduced delamination, and improved crashworthiness, which will moderately improve overall aircraft safety and reliability.

Efficiency and Performance (9): This Challenge has the potential to increase structural efficiency, increase aircraft range, increase the use of polymers in engine applications, and facilitate the development of advanced aircraft, such as an economically viable commercial supersonic aircraft.

Energy and the Environment (1): This Challenge contributes to this objective only indirectly, via improved structural efficiency.

Synergies with National and Homeland Security (9): Polymers and composites continue to be important for many DoD and DHS applications.

Support to Space (3): High-temperature polymeric composites are important for some space applications.

Why NASA?

Supporting Infrastructure (9): NASA (particularly at Langley) has excellent infrastructure for composites research, including development of methodologies to deal with damage tolerance and fracture mechanics. These efforts are particularly important because composites do not conform to linear damage theory. Nanocomposite efforts at Glenn have been initiated and are focusing on high-temperature polymers.

Mission Alignment (9): This Challenge is very relevant to NASA's mission to develop new materials with improved structural performance and reduced weight.

Lack of Alternative Sponsors (1): DoD funded the first generation of composite materials, although it is not currently funding basic composites research. Industry has sponsored relevant research in recent years.

Appropriate Level of Risk (9): The level of risk is high. Fundamental research has been conducted at universities and government laboratories, but only at the small coupon level.

C5 Noise prediction and suppression

At this time, many passive and active control concepts are being pursued to control the interior and exterior noise of aircraft, including rotorcraft, over a wide range of flight regimes. However, efficient solutions have not yet been achieved. Local communities in this country and abroad are becoming extremely aggressive in passing stringent noise regulations. As a result, landings and take offs at many airports have been restricted. Complying with additional noise restrictions could impose an enormous weight penalty on many aircraft. Additionally, the flight envelope is often restricted to keep noise within acceptable limits. Regulations pertaining to noise levels have limited civil applications of helicopters even at conventional airports because rotorcraft are noisier than most commercial aircraft. In addition, lack of public acceptance has been a major barrier to the widespread commercial use of helicopters to serve off-airport locations. Efficient methods to reduce noise will help expand the use of rotorcraft and fuel-efficient prop-rotor aircraft in commercial operations and increase the capacity of the air transportation system.

Reliable noise predictions are required to design efficient passive and active noise suppression devices. The problem is multidisciplinary, but with an important structural component. Advanced tools are required to accurately predict and alleviate noise, especially the aeroacoustic noise of rotorcraft. Research is needed to better understand the basic mechanisms of exterior and interior aircraft noise generation for different flight conditions. Current prediction tools for structurally transmitted noise are either finite-element-based (and therefore applicable only in the lower frequency range) or are energy- or power-based (covering a wider frequency range but with limited accuracy). Effective noise control techniques must take into account multiple types of aerodynamic and acoustic excitations. Therefore, structural prediction tools must be integrated with computational aeroacoustic and fluid dynamic prediction tools for a fully coupled solution to the problem of structural noise.

Suppression of exterior aircraft noise using smart materials holds great potential, but much R&D is needed. Advanced materials for larger, stronger fan blades and higher-temperature turbine blades, together with the development of very-high-bypass-ratio engines, is the biggest single factor in reducing external noise produced by jet aircraft. Good payoff can be achieved by developing locally morphing structures that smartly deploy themselves as needed, to reduce the noise generated by the propulsion system as well as the airframe. Variable-geometry-chevron nozzles, which could be driven by the shape memory alloy NiTiNOL, have been demonstrated to provide reduced noise during takeoff and then reconfigure themselves to a more efficient shape for cruise (Calkins and Butler, 2004). Smart materials can also be used to make fan duct liners that can adapt themselves (by changing cavity depth, face sheet porosity, etc.) according to the fan operating conditions to maximize noise reduction. Similar concepts can be applied to airframe noise suppression, where smart morphing structures could be integrated with noise reduction devices installed on aircraft high-lift systems (flaps and slats). These devices would operate at normal landing and takeoff conditions for noise reduction but could rapidly retract (or change configuration) as needed to increase lift and power during an emergency to maximize aircraft performance.

Noise experienced by flight crew and passengers is due largely to the excitation of the fuselage by the exterior flow. The fuselage structural design plays a key role in determining the amount of add-on noise control treatment needed to meet interior noise goals. Major strides in controlling noise in the aircraft cabin can be achieved using advanced structures and materials techniques.

The use of lightweight composite structural designs in commercial aircraft has greatly increased over the last few years. Unlike metallic structures, composite structures are excellent radiators of noise. Structural tuning concepts such as those pioneered in NASA-funded research for the reduction of turboprop tones may provide new opportunities to reduce noise while maintaining the strength and weight benefits of composite material systems. Experiments on current composite fuselage designs show that they would benefit from composite material systems with higher intrinsic damping. Work is needed to balance the structural and noise reduction requirements of honeycomb structures. Experiments have shown that partially filling cells with small loose particles can increase the damping properties of honeycomb panels. Other promising approaches include tailored lay-up using high-damping composite materials; nanotechnology to enhance structural damping; new acoustic and thermal insulation; morphing or tailored structures for achieving laminar flow and noise control; multifunctional composite structures (which offer improved noise control, strength, health monitoring, thermal insulation, and so on); and smart materials employing nanobiotechnology that can sense and respond to acoustic, elastic, thermal, and chemical fields in a positive, human-like manner. Key milestones include

- Measure noise signatures in controlled environments such as anechoic wind tunnels, for a range of flight conditions.
- Predict noise signatures using advanced multidisciplinary methodologies, validating against test data for level and maneuvering flight modes.
- Develop efficient structural solutions for interior noise control, i.e., structural optimization.
- Design non-load-bearing passive noise control.
- Design active controls for interior and exterior noise through smart structures technology.
- Develop low-noise rotors.
- Selectively flight test full-scale systems with noise signature measurement.

Relevance to Strategic Objectives

Capacity (9): Efficient suppression of noise would help to expand the flight envelope, including forward speed, rate of climb, descent, and flight course. Additionally, it would help to develop advanced vehicle concepts and next-generation heavy-lift rotorcraft. This will be a key step toward runway-independent aircraft for civil operations.

Safety and Reliability (1): This Challenge will have little impact on this Objective.

Efficiency and Performance (3): Efficient passive and active noise suppression devices could help to increase the flight envelope, particularly in congested areas.

Energy and the Environment (9): Reducing exterior noise will reduce the environmental impact of aviation.

Synergies with National and Homeland Security (3): Noise suppression will help to develop stealth vehicles, which is a key mission of DoD. In addition, noise suppression will enable increased use of military aircraft on bases near population centers.[5]

Support to Space (1): This Challenge would provide limited benefits to space operations.

Why NASA?

Supporting Infrastructure (9): NASA has some unique experimental facilities to measure noise signatures, including the 20 × 24 × 30 ft Anechoic Quiet Flow Facility, the 27.5 × 27 × 24 ft Anechoic Noise Facility, and the 21 × 31 × 15 ft Reverberation Chamber at Langley, and the 40 × 80 ft/ 80 × 120 ft wind tunnel with acoustic lining at Ames.

Mission Alignment (9): This Challenge is very relevant to NASA's mission.

Lack of Alternative Sponsors (3): Prediction and measurement of noise is of low priority for industry. Also, industry infrastructure is inadequate for long-term systematic studies. There are some limited research activities in DoD laboratories.

Appropriate Level of Risk (9): This Challenge faces moderate to high risks.

C6a Innovative high-temperature metals and environmental coatings

The goal of this Challenge is to provide the underlying technologies for materials modeling tools that can predict properties of new high-temperature metallic materials and associated protective coatings. The effort would include generation of the necessary fundamental data, complemented by testing that simulates realistic jet engine operating conditions to validate the models. These tools would then be applied, in concert with industry, to the development of innovative propulsion materials. Typically, substrate materials are developed separately from environmental coatings and then integrated toward the end of the development program, or coating development follows substrate development. This stretches development time considerably, often to a decade or more. These modeling tools are expected to reduce development time for high-temperature materials and coatings by 50 percent (NRC, 2004).

Advanced high-temperature materials are critical to advancing the next generation of jet engines that will power subsonic and supersonic fixed-wing aircraft, while enabling reduced operating costs and improved engine safety and reliability. Improved metallic alloys are needed for high-temperature structural applications such as turbine disks, blades, and pressure cases. Increases in operating temperature of 50°C or more for jet engine parts is possible, but the length of time and cost to develop these materials, and the risk that success will not be achieved, have been a huge disincentive to aggressive development of innovative materials.

The abilities to estimate materials properties and model complex phase fields, microstructures, and materials processing are advancing through the use of new computational tools and powerful desktop computers. However, application of this work to guide materials research is still in the beginning stages. NASA has the opportunity to be a leader in this technology.

The key to improving engine efficiency lies in developing turbine materials systems (i.e., alloy substrates for the turbine blade, disk, and shroud, plus necessary environmental coatings) that possess structural performance at higher temperatures while maintaining stability for tens of thousands of operating hours within an environment that is highly oxidative, corrosive, and erosive. Long development times would be reduced by the ability to conduct many experiments computationally, analogous to what is currently done by the tools for computational fluid dynamics. Key milestones include

- Define required models and a model integration strategy to provide necessary functionality for simulations.
- Select models for further development, based in part on how well they are aligned with materials systems that provide greatest benefit for propulsion systems.
- Develop models for selected substrates and associated environmental coatings; determine all the physical parameters required by the models.
- Validate the models by applying them to the development of new materials that are selected in concert with industry.

Relevance to Strategic Objectives

Capacity (3): Advanced high-temperature alloys with environmental protective coatings are one enabler for supersonic flight. These materials will increase maintenance intervals and could contribute to more flexible flight operations. Materials modeling will enable these materials to be developed faster, at lower risk.

Safety and Reliability (9): This Challenge will resolve safety and reliability issues with engines that operate at higher temperatures by improving the inherent capability of materials to withstand degradation modes.

Efficiency and Performance (3): Higher-temperature materials will directly improve the aerothermodynamic efficiency of the engine, thereby reducing fuel burn. These materials also reduce maintenance costs. Higher temperature materials will enable supersonic propulsion systems. The

[5]R. Flater, American Helicopter Society, Testimony before the House Armed Services Committee, Subcommittee on Tactical Air and Land Forces, 108th Congress, March 12, 2003. Available at http://vtol.org/pdf/congr03.pdf.

next generation of a turbine materials system resulting from this Challenge is anticipated to increase operating temperatures by about 50°F within 5-10 years, resulting in only a moderate impact on this Objective.

Energy and the Environment (1): This Challenge has little impact on this Objective.

Synergies with National and Homeland Security (9): This Challenge would benefit efforts by DoD to develop improved, long-life propulsion systems for supersonic and hypersonic flight.

Support to Space (3): This Challenge would benefit efforts to develop air-breathing access-to-space turbine engines. Advanced modeling tools might be used to guide development of specialized alloys expressly tailored for space launch systems.

Why NASA?

Supporting Infrastructure (9): NASA Glenn Research Center has a cadre of experts in the modeling and development of advanced turbine alloys and coatings. NASA Glenn also posseses unique specialized testing capabilities that simulate severe engine conditions, such as foreign object damage, creep, fatigue, and various environmental conditions. NASA also has outstanding electron microscopy equipment and facilities for high-temperature materials characterization.

Mission Alignment (9): This Challenge is very relevant to NASA's mission. In the past, NASA contributed significantly to this field.

Lack of Alternative Sponsors (3): The Air Force, Navy, and DARPA have all conducted work related to this Challenge. However, the DoD effort is not aimed at commercial engine or rotorcraft applications and is only likely to fund point solutions to specific problems.

Appropriate Level of Risk (9): This Challenges faces high risk.

C6h Innovative load suppression, and vibration and aeromechanical stability control

Because the aerodynamic environment surrounding an aircraft is unsteady, the aircraft experiences significant vibratory loads that need to be either isolated or absorbed to minimize their impact on passengers and key structural components and instruments. Also, unsteady aerodynamic forces couple with structural and inertial forces, resulting in potentially catastrophic aeromechanical instabilities. This Challenge will minimize the impact of vibratory loads using innovative passive and active techniques. It will also examine innovative techniques to increase the aeromechanical stability margins of aircraft in all flight modes. Minimizing vibratory loads enhances ride quality, increases the structural life of components, and improves handling. Aeromechanical stability (aeroservothermoelasticity) is the key to expanding the flight envelope.

Current aircraft use numerous passive devices to isolate or absorb loads. Prediction of vibratory loads, especially in rotorcraft, is far from satisfactory. Mechanisms of vibration in maneuvering flight are not completely understood. Vibration is a nonlinear-coupled phenomena that involves unsteady aerodynamics and wakes; nonlinear structural deformations and inertial couplings; and interactions between the flow and the structure. Aeromechanical stability can be a major issue with new configurations and expansions to the flight envelope.

Advanced CFD methodology needs to be coordinated with (1) comprehensive structural analyses to predict aeromechanical stability, vibratory loads, and vibration signatures at different stations in the airframe and (2) systematic validation against test data. The development of comprehensive MDO studies focused on inherently stable, low-vibration aircraft is another area worthy of research. Experimental issues involve the performance of systematic wind tunnel and flight tests using dynamically scaled and full-scale models. Problems also stem from a fundamental mismatch in basic structural models and reduced-order control models. Key milestones include

- Predict vibration using advanced CFD methodologies and validate experimentally.
- Predict aeromechanical stability for advanced configurations and expanded flight envelope (including hypersonic flight) and validate experimentally.
- Measure vibratory loads and vibration signatures under controlled wind tunnel environments for a range of flight conditions.
- Develop novel techniques for control-oriented modeling.
- Selectively flight-test full-scale systems, measuring vibration signatures and damping levels at level and maneuvering flight conditions.
- Innovate and employ active or passive techniques to minimize vibration and increase stability margin.
- Develop MDO techniques to develop low-vibration, stable systems.

Relevance to Strategic Objectives

Capacity (3): This Challenge would help increase payload and expand the flight envelope.

Safety and Reliability (9): Minimizing vibratory loads will increase structural life of components and improve reliability of controls and instrumentations. Aeromechanical stability margins are important for flight safety.

Efficiency and Performance (3): Increased stability margins would result in moderate increases in airspeed, payload, and structural efficiency.

Energy and the Environment (1): Reducing vibratory loads might have some impact on noise, but the effect would be minimal.

Synergies with National and Homeland Security (9): Vibratory loads have direct impact on DoD missions, especially the accuracy of some aircraft weapons. This Challenge would also increase the life of structural components and would help increase the maneuverability of military aircraft.

Support to Space (3): Active and passive vibration suppression techniques and stability augmentations could also apply to vibration suppression for access-to-space vehicles.

Why NASA?

Supporting Infrastructure (9): NASA has unique experimental facilities that are important for full-scale and small-scale model testing. These include the 40 × 80 ft/80 × 120 ft Wind Tunnels at the National Full-Scale Aerodynamics Complex at Ames and the Transonic Dynamics Tunnel at Langley for Froude- and Mach-scale testing.

Mission Alignment (9): This Challenge is very relevant to NASA's mission to improve aircraft performance.

Lack of Alternative Sponsors (3): Prediction and understanding of vibration and aeromechanical stability is already of moderate priority to industry and DoD.

Appropriate Level of Risk (9): This Challenge faces moderate to high risk.

C8 Structural innovations for high-speed rotorcraft

Recently, NASA conducted a systematic preliminary design study (Johnson et al., 2006) on a high-speed, heavy-lift civil rotorcraft with a cruise speed of 250 knots (Mach 0.6). It was "neighborly" quiet, economically competitive with a Boeing 737 aircraft, flexible in cruise altitude, and runway independent. Potential rotorcraft configurations include tilt-rotor, tandem-rotor compound, and an advancing blade concept (compound coaxial rotor). The development of a next-generation, cost-effective, high-speed rotorcraft that can meet these technology goals, faces numerous difficulties and barriers, which include scaling effects, aeromechanical efficiency, power-train limitations, structural efficiency, and life-cycle cost. Overcoming this Challenge will be a giant step toward the development of a runway independent aircraft that can efficiently expand capacity.

For the development of a high-speed, heavy-lift, efficient rotorcraft, a multidisciplinary aeromechanics research program needs to be established. It should encompass innovative rotor designs, active vibration and load control, variable-speed rotor technologies, active noise control, rotor morphing, lightweight and crash-absorbing airframe technologies, advanced variable-speed transmission systems, diagnostics and prognostics of drive trains and rotor head systems, increased autonomy and maneuverability, and enhanced handling quality. Because of an enormous increase of dynamic loads at high-speed flight (due to increased aerodynamic asymmetry), advanced composite designs with high damage tolerance are key structural issues. This provides opportunities to incorporate many disruptive and nondisruptive technologies in rotorcraft design, with an enormous payoff in performance and life-cycle cost compared with existing helicopters. Key milestones include

- Develop comprehensive aeromechanic analyses for high-speed rotorcraft that include tilt-rotor, tandem-rotor compound, and compound coaxial rotors for level and maneuvering flight conditions.
- Develop aeromechanics and technology tools for the drive-train system and other key components necessary for variable-speed rotors.
- Design and develop lightweight, crash-absorbing com posite airframes.
- Develop technology for all-weather rotorcraft operation.
- Develop advanced composites with high damage tolerance for use in large dynamic structural components.
- Reduce required shaft power by 15 percent from current levels using elastically tailored composite blades, active structural and flow control, and advanced airfoils.
- Reduce life-cycle cost using health and usage management systems, low-cost tailored airframes, and lightweight low-vibration rotors.

Relevance to Strategic Objectives

Capacity (9): The performance of a next-generation, revolutionary civil rotorcraft greatly increases the capacity of the air transportation system.

Safety and Reliability (1): This Challenge involves the use of many disruptive technologies and is unlikely to significantly increase safety or reliability.

Efficiency and Performance (3): This Challenge would increase the speed, range, payload, and structural efficiency of rotorcraft.

Energy and the Environment (1): A next-generation rotorcraft should be designed to fly at an altitude of 15,000 ft or higher, to reduce noise on the ground. A variable-speed rotor could also be used to reduce external noise. However, although these efforts would reduce the noise produced by rotorcraft, it would only reduce them to levels comparable to fixed wing aircraft. Therefore this Challenge will not reduce the overall noise produced by the air transportation system.

Synergies with National and Homeland Security (9): High-lift, high-speed rotorcraft will have an enormous impact on emergency evacuation, heavy-lift operations, and other military applications.

Support to Space (1): This Challenge is not relevant to this Objective.

Why NASA?

Supporting Infrastructure (9): NASA has unique experimental facilities that are important to rotorcraft full-scale and small-scale testing. These include the 40 × 80 ft/80 × 120 ft

Wind Tunnels at the National Full-Scale Aerodynamics Complex at Ames, the Drop Test Facility and Transonic Dynamics Tunnel for Froude- and Mach-scale rotor testing at Langley, and the Power Train Full-Scale Test Facility at Glenn.

Mission Alignment (9): This Challenge is very relevant to NASA's mission to improve aircraft performance.

Lack of Alternative Sponsors (3): The U.S. Army is the primary supporter of research on high-speed rotorcraft, but the Army will not investigate issues related to civil applications.

Appropriate Level of Risk (9): This Challenge faces high risk.

C9 High-temperature ceramics and coatings

Advanced structural ceramics, including oxide-, carbide-, nitride-, and boride-based systems, are characterized by high strength, stiffness, hardness, corrosion resistance, and durability. Furthermore, they retain these properties at high temperatures, making them ideal for a wide range of demanding applications, including engine components for subsonic aircraft (combustor liners, exhaust-washed structures, high-temperature ducts, heat exchangers, and nacelle insulation) and airframe and propulsion systems for high-speed vehicles. Due to their inherent brittleness and low fracture toughness at ambient conditions, they are often combined with other materials to produce composite systems that take advantage of the high strength of small-diameter ceramic fiber reinforcements or used as nonstructural coatings for thermal management. In general, each class of material exhibits inherent benefits and limitations for particular use environments (NRC, 1998).

The primary benefit of structural ceramic materials is increased thermal capability, which improves propulsion system efficiency, increases lifetime, enables higher operating speeds, and expands the margin of safety in airframe applications. The selection of the ceramic system and its use temperature is dictated by the chemistry of the operating environment and the minimum acceptable lifetime of the component. Oxide composites with operating temperatures as high as 1200°C and lifetimes of thousands of hours in highly oxidizing combustion or reentry environments are very suitable for some engine components, warm structures, and thermal management components.

Nonoxide composites made of silicon carbide reinforced either with carbon fibers or a combination of carbon fibers and silicon carbide fibers are capable of operating temperatures of 1300°C-2500°C for short times in highly oxidizing environments or for much longer times near the lower end of the thermal range when protected with coatings. Furthermore, because nonoxide fibers exhibit higher strength and better strength retention than oxide fibers, they are being widely researched for applications in combustion environments as well as for hot structures of reentry or hypersonic vehicles.

Refractory metal (e.g., hafnium or zirconium) carbides and borides are capable of surviving thermal excursions up to 2000°C-2500°C for short times with little material recession, making them strong candidates (in either monolithic form or as a composite matrix) for the sharp leading edges of hypersonic vehicles.

During the last 10 years, significant progress has been made in the processing, development, and demonstration of many ceramic systems for specific applications. Oxide composites deriving damage tolerance from highly porous matrices have been commercialized, and other systems with novel fiber coatings have been demonstrated in subscale testing for reentry vehicle thermal protection systems. Silicon carbide matrix processing approaches have advanced significantly, with systems produced by chemical vapor infiltration, melt infiltration, and preceramic polymer infiltration all having been demonstrated in subscale testing for jet or rocket engine components. NASA Glenn has led efforts to fabricate and test jet engine components such as exhaust nozzle liners, combustor liners, and turbine airfoils with silicon carbide matrix composites. Rocket nozzles fabricated from silicon carbide materials have been rig tested, and NASA Ames has demonstrated the ability to reproducibly fabricate refractory metal carbide and boride systems.

Despite the above successes, component fabrication is not often taken much beyond the prototyping stage. Fabrication methods can significantly affect material properties, but it has not been possible to devote enough effort to understanding the effect of variations. In addition, new design rules and attachment methods must be developed to fully exploit the unique properties of these materials. There has not been sufficient component testing under actual operating conditions to validate design rules, nor are there sufficient mechanical and thermal performance data to fully characterize these ceramic systems. Therefore, there is currently low confidence in designing structural components with these material systems.

Advancing the state of the art for high-temperature ceramics suitable for aeronautical applications requires research in several key areas: fabrication and testing, modeling, and attachment methods. As mentioned above, insufficient fabrication and testing experience deprives designers of confidence in the long-term behavior of these materials and in the design rules for translating material characteristics into component designs. As a consequence, structures are typically overdesigned or constrained by existing designs for other classes of materials. At best, structures are rarely optimized; at worst, they fail when the material is inappropriately applied.

Additionally, modeling tools to predict the life of components made of these materials are inadequate. This leads to inaccurate performance and cost assessments and further limits the use of ceramic materials. Since these materials are only considered for niche applications, no economics of scale can be anticipated. This problem could be alleviated to a large extent through the development of better design tools, a more thorough understanding of the effects of process variations, and more efficient commercial fabrication approaches.

Work is also needed to develop robust methods for attaching hot components to warm and cool structures and to develop textile approaches for integrating complex component architectures with key features such as stiffeners, sensors, and cooling features. Near-term milestones include

- Generate material property databases appropriate for design of a high-temperature ceramic component.
- Complete full-scale testing of at least one ceramic composite component with improved performance for subsonic aircraft applications (e.g., fairing heat shields, combustor liner, or turbine airfoil).
- Develop models to optimize a structure for a new, rather than an existing, platform.
- Model crack growth under actual operating conditions.
- Develop advanced ceramic composites for large surfaces and leading-edge components for supersonic and hypersonic vehicles and complete relevant environmental testing of subcomponents.

Far-term milestones include

- Flight test at least one ceramic composite component for improved subsonic flight vehicles and transfer the technology to industry.
- Verify model predictions of performance using flight test data.
- Extend model predictions to new flight speed regimes to optimize supersonic and hypersonic vehicle designs for hot structures and engine components.
- Demonstrate, through full-scale testing, at least one ceramic composite component for a supersonic or hypersonic platform.

Relevance to Strategic Objectives

Capacity (3): Ceramics, ceramic composites, and ceramic coatings could improve the operating temperatures of aircraft engines, leading to increased aircraft speed.

Safety and Reliability (1): This Challenge would increase safety margins for aircraft components such as nacelle insulation and burn-through shields, but the overall effect would be minor.

Efficiency and Performance (9): This Challenge would increase the operating temperatures of aircraft engines, resulting in higher efficiencies. They are an enabling technology for airframes and engines for hypersonic aircraft and would facilitate the development of commercial supersonic aircraft.

Energy and the Environment (3): High-temperature materials would enable engines to run hotter and to potentially decrease NO_x emissions.

Synergies with National and Homeland Security (3): This Challenge is relevant to military supersonic and hypersonic aircraft.

Support to Space (9): This Challenge is very relevant to access-to-space vehicles, as evidenced by the thermal protection systems developed for the space shuttle and more recent X-vehicles.

Why NASA?

Supporting Infrastructure (9): NASA Glenn has materials expertise and unique facilities in which to test the performance of ceramic engine components under appropriate flow and pressure conditions for jet aircraft engines. It also has a full suite of materials characterization facilities, and facilities to evaluate component and material behavior in rocket engine environments. NASA Ames has unique environmental exposure facilities and arc jet tunnels capable of simulating hypersonic and reentry environments.

Mission Alignment (9): The development of high-temperature, ceramics-based systems is very relevant to NASA's space exploration mission.

Lack of Alternative Sponsors (3): DoD continues to support relevant research. However, NASA is still positioned to lead in the development of these materials for the most extreme conditions.

Appropriate Level of Risk (9): This Challenge faces high risk.

C10 Multifunctional materials

Materials that possess multifunctional behavior combine electronic, magnetic, chemical, thermal, and mechanical properties at the macro, micro, or atomic level. These materials present unique opportunities for integrating communication, actuation, sensing, self-healing, and energy-harvesting functionalities into lightweight, load-bearing structures. Multifunctional materials enable a wide range of benefits, including improved aircraft telecommunications (wired, wireless, and optical); enhanced potential capabilities and flexibility for electronic and optoelectronic platforms, such as agile phased-array and multifunctional radar systems; structural prognosis and nondestructive evaluation; self-sensing and self-repair; camouflage and avoidance; and local power generation though energy harvesting. The use of structural elements to provide new functions to aircraft platforms increases structural efficiency and enables new aircraft capabilities.

A number of multifunctional concepts have been demonstrated, such as the conformal load-bearing antennas implemented in military vehicles. Additional work on materials capable of self-healing and local power generation has been conducted and is being integrated into research aircraft. The first use of multifunctional materials in operational aircraft will likely occur in UAVs, where consequences of failure are low and issues such as energy harvesting are important. While the majority of research to date has been confined to coupled electromechanical domains, a much broader vision is possible. Recent discoveries of electrochromic, magnetoelectric, and thermomechanical materials show substantial promise for future multifunctional materials.

From a programmatic viewpoint, the research problems inherently involve multiple domains and require a multiphysics (electrical, magnetic, chemical, structural, thermal, and electromagnetic) approach. Additionally, facilities to fabricate and test either monolithic or composite multifunctional materials are rare. The main tasks are to develop new materials based on fundamentals or to design composite materials that couple functionality without disrupting structural performance. To date, little research has been conducted at the atomic or composite levels, with even less on fabrication and testing. Relevant research should be conducted in collaborative between universities and NASA. Key milestones include

- Develop a comprehensive analysis to predict the performance of selected monolithic and composite multifunctional materials.
- Use this analysis to guide parametric studies to explore and optimize material response with the goal of understanding the combined response of the multifunctional material.
- Fabricate materials according to model predictions.
- Evaluate material performance, both coupled and structural, and compare with analytical predictions.
- Integrate multifunctional materials into a structural component for benchtop verification.
- Conduct flight tests on a structural component.

Relevance to Strategic Objectives

Capacity (3): Multifunctional materials can increase the payload fraction of individual aircraft via improved structural efficiency, allowing a given airplane to carry more passengers or cargo.

Safety and Reliability (3): Multifunctional materials potentially increase aircraft safety and reliability. For example, self-healing structures provide the opportunity for in-flight repair.

Efficiency and Performance (9): Multifunctional materials can improve the efficiency of engines by evaluating the flow characteristics at or near the engine inlets and outlets and provide locally generated power by converting thermal energy into electrical energy.

Energy and the Environment (3): Multifunctional materials can harvest energy from the surrounding environment, by converting solar, mechanical, or thermal power into electrical power.

Synergies with National and Homeland Security (9): This Challenge is very relevant to remote observation, hypersonic flight, and aircraft protection (through radar absorption or cloaking).

Support to Space (9): Multifunctional materials support access to space by enabling structures with inherent sensing capabilities that can transmit data to a central location.

Why NASA?

Supporting Infrastructure (3): NASA facilities at both the Langley and Glenn research centers are very relevant to this Challenge. Langley has developed relevant expertise through a recent morphing program. However, academia and the DoD possess similar facilities.

Mission Alignment (9): This Challenge is very relevant to NASA's mission to conduct fundamental aeronautics research.

Lack of Alternative Sponsors (3): DoD is interested in relevant research but not in applications to civil aviation. University-funded research in this area has demonstrated the viability of these materials but has not yet explored their relationship to aircraft systems.

Appropriate Level of Risk (9): Relevant technologies are still very immature, and this Challenge faces high risk.

C11 Novel coatings

Exterior and interior coatings can be designed to provide novel, yet beneficial functionality through the use of nanoscale fillers or self-assembled monolayers. Advanced coatings are attractive since they can be applied to existing structural components as an add-on technology.

This is a broad Challenge that may enable visionary concepts in aircraft design and improve structural efficiency and safety. Potential benefits include self-assembled monolayers for corrosion protection, soft polymeric coatings for noise reduction, ultrahydrophobic surfaces for drag reduction, coatings that enable aircraft to shed or melt ice without the use of deicing fluids, nanoparticle-filled coatings for wear resistance, electrically conductive coatings, and self-sensing and self-repairing surfaces. Development of novel coatings is an active research area in both academia and industry, but there is not a lot of research targeted at aeronautics or aerospace applications. Many potential systems have been demonstrated in the laboratory environment, for example, self-assembled monolayers for corrosion protection—while many others, such as deicing coatings, are under more advanced development. Key barriers for these coatings include development of nanoscale fillers with the appropriate functionality, processing and dispersion of nanoscale fillers, and the high cost of many nanoscale fillers (e.g., carbon nanotubes). In addition, most of the coatings (e.g., self-assembled monolayers) need to scale up for use with larger structures such as an aircraft wing. Finally, most of the coatings are not yet durable enough to be used in aeronautical applications.

Novel coatings are an emerging field with significant opportunity for new materials to achieve the various functions described above. Key milestones include

- Develop more durable, environmentally stable formulations for novel coatings that can survive in an aircraft environment.
- Develop cost-effective methods of processing and applying novel coatings onto large aircraft structures.

Relevance to Strategic Objectives

Capacity (3): Novel coatings might reduce drag, which could increase speed and deicing or self-cleaning capabilities, which would, in turn, increase operating flexibility. In addition, these coatings might enable new aircraft concepts.

Safety and Reliability (9): Novel coatings may detect damage (e.g., color change), facilitate deicing, and protect against corrosion. In this way, they can lessen the risk of catastrophic malfunction, aircraft loss, and human injury.

Efficiency and Performance (3): These coatings can reduce drag, and they offer biomimetic functionalities that may enable new aircraft capabilities.

Energy and the Environment (3): These coatings could provide environmental protection, lead to noise reduction, and be used for harvesting energy. New corrosion-resistant coatings could replace the environmentally hazardous chromium and cadmium that is currently used.

Synergies with National and Homeland Security (1): Although the DoD sponsors similar research, civil aeronautic applications would have a different emphasis, and NASA's research would not be applicable.

Support to Space (1): This Challenge has little impact on this Objective.

Why NASA?

Supporting Infrastructure (3): NASA has infrastructure relevant to this Challenge but little ongoing research.

Mission Alignment (9): This Challenge is very relevant to NASA's mission of improving aircraft capabilities.

Lack of Alternative Sponsors (3): Novel coatings are currently being explored by industry, and there is some interest in this area from DoD, but neither places sufficient emphasis on civil aeronautics applications.

Appropriate Level of Risk (9): Basic research is being done at universities. High-risk research is necessary to mature and transition this research to aeronautic applications.

C12 Innovations in structural joining

Load transfer in airframe structures is accomplished by joining discrete structural members. These joints add significant weight to the airframe, thereby reducing its efficiency. The broad classes of joining in airframe structure are mechanical fastening, adhesive bonding, and welding (for metallic structures). Significant advances have been made over the past decades in semiempirical analysis and design methods for structural joints. However, to make substantial progress, such as a 50 percent reduction in structural weight fraction, and to enable the widespread use of new joining methods, an initiative to develop a rigorous, all-encompassing simulation of performance is needed.

Joints in airframe structures are weight and cost drivers that often call for specialized treatments at the edges being joined. Improved joining technology promises significant payoffs in structural efficiency. Leakproof joints will eliminate parasitic seals in structures that must hold fluids. Efficient joints will improve airframe repairability and arrest damage propagation.

Several NASA and DoD programs have yielded test data and semiempirical models to analyze and design mechanically fastened and adhesively bonded joints. Welding of metallic structures is a well-defined and -characterized process. With the advent of ultralightweight structures using foam or honeycomb core sandwich panels, however, new mechanical fastening concepts with accompanying analysis methods are required. Friction stir welding of aluminum-lithium (Al-Li) structures needs to be modeled and a methodology for simulating complex structural arrangements needs to be developed. For adhesively bonded joints, surface preparation techniques and damage propagation arrest features need to be developed to ensure flight safety.

Currently, a rational methodology is required for adhesive surface preparation to ensure consistent, high-strength, certifiably bonded joints. A design and analysis methodology is needed for mechanically fastened joints in extreme environments along with substantiating test data. A fatigue performance and design methodology to resist cracking is required for friction stir welded joints in airframe-grade materials. Reliable, nondestructive inspection and evaluation methods are required for adhesively bonded joints. Some key milestones in advancing this technology include

- Develop certification methodology and tools for bonded joints.
- Fully characterize friction stir welding processes for Al-Li structural materials.
- Develop nondestructive strength assessment techniques for bonded joints.
- Develop modeling and simulation capabilities for mechanically fastened joints in extreme environments.

Relevance to Strategic Objectives

Capacity (3): Although this Challenge would have few direct effects on capacity, advanced joining technology is an enabler for new aircraft concepts, increased aircraft size, and increased operating flexibility.

Safety and Reliability (3): The ability to rigorously model, simulate, test, and verify joint concept performance translates directly into increased reliability and, hence, safety. This is especially significant for pressurized airframe structures.

Efficiency and Performance (9): Improvement in joint efficiency will increase structural efficiency and, thereby, aircraft efficiency and performance.

Energy and the Environment (1): Joining would only have a small, indirect impact on this Objective via improved structural efficiency.

Synergies with National and Homeland Security (3): This Challenge might have some impact on military aircraft.

Support to Space (3): This Challenge might have some impact on access-to-space vehicles.

Why NASA?

Supporting Infrastructure (3): NASA Langley has some skills left over in this area, which it developed in the early 1980s in its Aircraft Energy Efficiency Program.

Mission Alignment (9): This technology is well aligned with NASA's goals for improving airframe structural efficiency.

Lack of Alternative Sponsors (3): DoD organizations support the goals of this Challenge, although in recent years DoD has not supported relevant research.

Appropriate Level of Risk (9): Previous research addressing the very difficult task of modeling real airframe joint behavior has reduced risk to the point where this Challenge faces moderate to high risk.

C13 Advanced airframe alloys

Significant portions of aircraft structures will continue to be designed using metallic materials. New alloys that possess higher strength, inherently high fracture toughness, very high resistance to fatigue crack growth, and significantly improved corrosion resistance will enable much lighter, more efficient airframe structures with increased durability and reliability. It is important that these materials be developed along with the manufacturing science required to turn them into viable structural elements. Improved aluminum alloys have a history of very rapid insertion into aircraft applications if they provide significant performance and cost benefits, and can be recycled at the end of the aircraft's life. New chemistries, an enhanced understanding of processing–microstructure–property relationships, and improvements in processing science are enabling the development of advanced aluminum alloys with lower density, higher strength, and greater stiffness. The entire metallurgy of titanium will be redone if meltless titanium processing is shown to be practical; progress to date is quite promising.

Materials modeling is just now becoming possible at the desktop, driven by new computational tools and ever-increasing desktop computer processing capability. The conventional approach for alloy development is highly sequential and typically occurs over a long period of time—a decade or more. Use of modern materials modeling tools would facilitate the development of advanced alloys, reducing time and effort by 50 percent or more. Applying modeling to guide the advancements of these materials is in the very early stages, and NASA has the opportunity to be the leader in this technology.

The ability to model complex phase fields, microstructures, and materials processing and to estimate the full gamut of materials properties will allow many alloy trials to be conducted by computer analysis. This model-based approach for designing materials will focus limited resources on the most promising new alloy candidates. Improved materials performance directly translates into higher structural efficiency and reduced product weight; reductions of up to 25 percent appear possible. Improved modeling could increase material reliability as well.

A key focus of this Challenge is developing an integrated set of physics-based models that accurately estimate material properties of new alloys, significantly accelerating the R&D of new aerostructural metals. This effort would include the generation of fundamental data, complemented by testing that simulates realistic jet engine operating conditions to validate the model.

Industry has recently made significant improvements in conventional aluminum alloys by incrementally improving chemistry and microstructural control. For example, alloy 2025-T3 has 15-20 percent better fracture toughness and twice the fatigue crack growth resistance of 2024-T3. New Al-Li alloys have lower density and higher fatigue resistance. For example, the Al-Li alloy 2097 that replaced 2124 in an F-16 bulkhead has three times better spectrum fatigue behavior, which allows approximately 5 percent higher spectrum fatigue stress. Materials modeling could leverage the results of this latest work.

In addition, the aluminum–magnesium–scandium family offers potential for high strength and outstanding corrosion resistance. Serious consideration of these alloys has been impeded by the high cost of scandium. However, a significant reduction in the price of scandium is anticipated because a very large new ore deposit and a new refining method are coming on line in Australia.

Other alloys under development are moderate-temperature, age-hardenable aluminum alloys that could retain their strength at operating temperatures up to 150°C and would therefore be useful for a Mach 2.4 aircraft. Laminated hybrids of aluminum sheet with fiber reinforcements have high fatigue resistance and the potential for significant weight saving. The prospect of innovative chemistries that cannot be produced by conventional melting routes offers exciting new titanium materials. Meltless titanium processing has recently been demonstrated using several different processes, spurred by a current DARPA program. Advanced titanium alloys can exploit the meltless processing approach to enable very fine-grained structure with superb fatigue strength. Key milestones include

- Develop databases of the physical and mechanical properties needed for design of materials.
- Develop physics-based models that accurately estimate the properties of new alloys.
- Validate material models through alloy trials and material testing.
- Optimize new metal-processing techniques and scale them up to production size.
- Characterize new alloy families based on new alloying elements.
- Optimize and scale up processing techniques for the most promising new alloys.

Relevance to Strategic Objectives

Capacity (9): Lighter weight structures made from improved metallic structural materials will allow higher payload fractions, increasing the amount of cargo or passengers an aircraft can carry. Improved corrosion resistance and damage tolerance will reduce the time required for maintenance, improving 24/7 operational flexibility.

Safety and Reliability (1): Although these new structural alloys could provide improved reliability by enhancing corrosion resistance, fatigue strength, and inherent toughness, the focus of this Challenge will be on improved performance and efficiency rather than safety and reliability.

Efficiency and Performance (9): Better corrosion resistance directly translates into reduced maintenance and inspection costs. Better fatigue resistance will give the airframe longer life, lowering operating cost. Supersonic airplanes will greatly benefit from having affordable aluminum and titanium alloys as design alternatives to composites.

Energy and the Environment (1): This Challenge is not relevant to this Objective.

Synergies with National and Homeland Security (3): The DoD would consider using advanced airframe alloys for a variety of aeronautical systems applications, including supersonic aircraft. The range of DoD applications for advanced titanium alloys based on meltless processing would be considerable and would go beyond aeronautics. Supersonic airframes would also directly benefit from this technology.

Support to Space (1): Although these new structural materials might be considered for structural applications in access-to-space vehicles and satellite structures, these applications would require considerable additional effort for evaluation, first, and then for the development of specialized manufacturing methods.

Why NASA?

Supporting Infrastructure (1): Within NASA, the core metallurgical expertise has not been refreshed as people left the organization and the emphasis shifted to composite airframe structural materials. Airframe alloy research can leverage the thermal structures research facility at Langley and the hypersonic materials environmental test facilities at Langley, Johnson, and Ames for evaluation of the higher temperature alloys.

Mission Alignment (3): Although NASA has contributed significantly to this field, it has since been adopted by industry, and the research is less aligned with NASA's mission.

Lack of Alternative Sponsors (1): If this work were not undertaken by NASA, it would be done by industry and by DoD. The former is already leading the development of new chemistries and processing for advanced airframe alloys.

Appropriate Level of Risk (9): This Challenge faces high risk.

C14 Next-generation nondestructive evaluation

NDE is an interdisciplinary field that is concerned with the development of analysis techniques and measurement technologies for the quantitative characterization of materials and structures by noninvasive means. Ultrasonic, radiographic, thermographic, electromagnetic, and optical methods are employed to probe interior microstructure and characterize subsurface features. Currently available NDE instruments are compact, rugged, and can acquire large amounts of wide-area multiphysics data via sensor arrays. Recent better-than-Moore's-law increases in computational hardware capabilities allow these data sets to be processed with compact, rugged, and inexpensive computers. To advance NDE capabilities beyond the paradigm of rendering high-quality imagery for humans to interpret, the missing piece, more and more often, is the understanding necessary to create multiphysics algorithms that would allow the enormously rich data sets to be automatically interpreted.

The primary goal of next-generation NDE is, therefore, to develop this enabling understanding and algorithms. In the short term, the goal is to create artificial intelligence that can provide a backup assessment, as is now done in x-ray mammography. In the medium term, the primary task of NDE technicians will be to transport and set up instruments; NDE measurements and interpretation will be fully automatic. In the long term, the instrumentation could be robotic, so that NDE inspections as well as interpretation would be automated.

Next-generation NDE improves safety and reliability by minimizing manufacturing defects and identifying in-service flaws before they cause malfunctions. It enables 24/7 operation by minimizing downtime due to faults and has the potential to make just-in-time maintenance feasible. Most next-generation NDE technologies will find application in DoD and access-to-space vehicles, as well as in a variety of important nonaerospace industries. Next-generation NDE is synergistic and closely allied with IVHM research; NDE would provide inputs to prognosis and life-prediction systems.

The hardware that acquires the NDE data is becoming less and less interesting from a scientific perspective, and it is not the appropriate focus for NASA. At the same time, brute force computer image processing, rendering, and visualization is not the focus either. The human visual system is set up to deal with surfaces, not volumes of data, so it can be argued that existing NDE systems already generate data and imagery in quantities that strain or exceed human limits. Accordingly, current work is directed at bringing an understanding of the instrumentation and measurement together with sophisticated, multiphysics models of the probing energy–material interaction that is taking place.

NDE for complex materials and structures includes developing and fusing multiple sensor techniques that provide orthogonal information on the state of a material or structure, as well as improving data reduction techniques for quantitatively mapping measurements to accurately characterize material and structural integrity. Computational NDE involves developing and validating accurate multiphysics simulations to reduce the cost of optimization and automation of NDE techniques. Autonomous NDE involves the development of techniques that accurately characterize materials and structures (including damage) with minimal or no human interaction, including techniques that are self-calibrating and methodologies for assuring proper instrument setup. Key milestones include

- Demonstrate successful, real-time, fully automatic interpretation for various individual NDE techniques in targeted families of applications in the laboratory.
- Fuse multiple orthogonal NDE techniques.
- Adapt laboratory-tested NDE techniques to field-portable configurations that can be demonstrated outside of the laboratory.
- Implement large-area, multiphysics NDE techniques and instrumentation robotically.
- Transfer techniques and algorithms to IVHM efforts, including optimization of the output of next-generation NDE technologies for input to research on prognosis and life prediction.

Relevance to Strategic Objectives

Capacity (3): Current NDE relates directly to enabling 24/7 operation by reducing downtime due to faults. Next-generation NDE will be faster and more automated, and it will help to enable just-in-time maintenance.

Safety and Reliability (9): NDE directly supports safety and reliability by minimizing manufacturing defects and identifying in-service flaws before they cause a malfunction. Next-generation NDE is synergistic and closely allied with IVHM efforts and provides input to prognosis and life-prediction systems.

Efficiency and Performance (1): Although next-generation NDE imparts enough confidence to permit adopting new material systems and structural concepts as well as to operate safely closer to performance margins, the improvement over current NDE will have less of an impact on this Objective than other technologies.

Energy and the Environment (1): This Challenge is not relevant to this Objective.

Synergies with National and Homeland Security (3): Some NDE technologies are applicable to security screening areas, in particular the automated real-time interpretation of multiphysics measurements. Most next-generation NDE technologies will also find application in DoD aircraft and other vehicles, with moderate impact expected.

Support to Space (1): Most next-generation NDE technologies will find some application in access-to-space vehicles, but the impact will be minimal because the small number of reusable vehicles means that most NDE applications for space will be for manufacturing quality control and related issues prior to flight.

Why NASA?

Supporting Infrastructure (3): NDE equipment and research facilities tend to be compact, relatively inexpensive, and broadly available, but NASA has some staff with unique expertise.

Mission Alignment (9): This Challenge is very relevant to NASA's mission in terms of (1) aviation safety (especially aging aircraft) and (2) facilitating development and insertion of new materials, structures, and processes by ensuring that they are manufactured according to specifications and are behaving as designed as they are put into service.

Lack of Alternative Sponsors (1): DoD and nonaerospace industries are supporting NDE research. Focusing on automated interpretation of multiphysics NDE measurement data would provide a niche where NASA could collaborate with academia without duplicating work by others.

Appropriate Level of Risk (3): Given that this Challenge encompasses many tasks that are standard industrial practice now, it faces low risk.

C15 Aircraft hardening

Improved aviation security requires that commercial aircraft be hardened against safety threats such as explosions and biological or chemical agents. Effective solutions will encompass detection, avoidance, and impact minimization of threats. The large number of potential threats and pathways for delivery dictates the development of multiple and varied technologies, ranging from sensors for threat identification and structural health monitoring to the development of highly durable blast containment systems and methods to accurately model structural damage due to blast events. Successfully hardening commercial aircraft increases their safety and reliability and contributes to on-ground safety by minimizing the use of aircraft as weapons.

Biological and chemical threat detection sensors and self-decontaminating coatings are in the early development stages. Hardening technologies developed for military aircraft such as fuel inerting and nonreflective, infrared-absorbing paint for signature reduction remain largely unimplemented for commercial flight because of their weight and/or cost. These considerations have also limited the use of blast containment technologies in commercial aircraft even though there has been a significant effort by the FAA and commercial companies to develop blast-resistant luggage containers, also known as hardened unit-loading devices.

Various schemes for improving the energy absorption capacity of the luggage containers have been investigated, including incorporation of honeycomb or foam elements. In addition, instead of using aluminum or fiberglass as the primary structural material, new containers could use (1) composites reinforced with fibers developed specifically for ballistic armor applications (e.g., Kevlar, Spectra, and Zylon) or (2) hybrid systems (e.g., a laminate composed of fiberglass and aluminum foil). FAA-certified designs have been developed and verified through blast testing. Inspection, maintenance, and repair methodologies, however, have not been adequately addressed. Similar concepts are currently under development for hardening overhead bin compartments.

Many of the impediments to incorporating threat-hardening technology in commercial aircraft stem from the constraints imposed by retrofitting existing aircraft and the difficulty of analyzing rapidly loaded, complex structures. The cost and weight of hardening existing aircraft is prohibitive. New aircraft designs will more efficiently incorporate features such as fuel protection filters; integrated threat detection (including biometric identification); health monitoring sensors; and highly durable, fire-resistant composite structures. Key milestones include

- Analyze, design, and test an optimized blast-resistant luggage container.
- Develop and validate accurate damage prediction models for blast events including shock overpressure and crack propagation due to hull pressurization.
- Integrate onboard biological and chemical sensors.
- Model and develop self-decontaminating coatings.

Relevance to Strategic Objectives

Capacity (1): Hardening of aircraft can reduce vehicle losses, but the weight increase associated with blast hardening in particular would decrease capacity.

Safety and Reliability (9): Hardening is a key strategy for improving aircraft safety by providing protection from onboard explosions and improving threat assessment.

Efficiency and Performance (1): Hardening would likely have a nominal or negative impact on this Objective due to increased weight.

Energy and the Environment (1): Hardening would likely increase structural weight and possibly volume, which could indirectly increase noise and emissions.

Synergies with National and Homeland Security (9): This Challenge is very relevant to DHS and DoD missions.

Support to Space (1): This Challenge has no impact on this Objective.

Why NASA?

Supporting Infrastructure (3): NASA supported relevant research under its Aviation Safety and Security Program, where the topics included control systems to detect and compensate for vehicle damage; fuel protection; fire-resistant, damage-tolerant composites; and sensing of onboard chemical and biological contamination.

Mission Alignment (3): Although NASA's 2003 Strategic Plan specifically discusses the need for NASA to "aggressively apply our expertise and technologies to improve homeland security," this Challenge is more closely related to the mission of DHS.

Lack of Alternative Sponsors (1): The FAA, DHS, DoD, and industry are all sponsoring research for hardening technology.

Appropriate Level of Risk (9): This Challenge faces moderate to high risk.

C16 Multiphysics and multiscale modeling and simulation

Multiscale and multiphysics modeling and simulation encompass computational modeling of interdisciplinary systems at multiple spatial and temporal scales (e.g., by finite-element methods, molecular dynamics, and ab initio methods). With advances in the understanding of material and system behavior at multiple spatial scales (from atomistic to continuum) and time scales (from the period of atomic vibration to structural lifetime) when subjected to multiple physical stimuli (mechanical, thermal, electromagnetic, chemical), the promise of designing new materials based on atomic characteristics is emerging. Given a specific requirement (e.g., strength, stiffness, or piezoelectric coefficient), it may become possible to design a new material from the atomic level up that will meet the requirement, replacing the make-it-and-break-it approach currently used to investigate new material systems.

Multiscale and multiphysics modeling has been examined for several years, but is still in its infancy. With recent advances in computational capabilities, there has been renewed interest in multiscale and multiphysics modeling. DoD has invested in this field through various programs, including the Multidisciplinary University Research Initiatives (MURIs), Materials Engineering for Affordable New Systems (MEANS) grants, and DARPA programs. Multiscale, multiphysics modeling has been identified as an integral component of the National Nanotechnology Initiative, which promises to invest close to $1 trillion per year in product development over the next 10 years. A number of leading universities are also beginning work in this area.

From a programmatic viewpoint, the research problems are inherently twofold. Overcoming this Challenge requires researchers familiar with multiple physical phenomona (mechanical, thermal, electromagnetic, chemical). In addition, it requires facilities with high-end computer facilities for storing, processing, and managing large data sets. Substantial computer power is required to construct even small-scale simulations of materials. Complex systems require tens to hundreds of simulations, requiring high-performance computing support (in general, hundreds to thousands of processors) to complete the simulations in a reasonable time frame. Similarly, each simulation can generate terabytes to petabytes of data. Thus, state-of-the-art visualization, data mining, and data analysis techniques are also critical to the success of this Challenge. Key milestones include

- Select an aeronautics-related material test problem
- Develop multiscale, multiphysics modeling software representative of the selected test problem.
- Procure necessary computer facilities for the modeling effort.
- Measure critical parameters necessary for formulating the models, including measurements made using electron-based optics with the ability to perform energy dispersive spectroscopy or electron energy loss spectroscopy.
- Complete a multiscale, multiphysics analysis of the selected test problem.
- Validate modeling by comparing the results to those of an experimental development program.

Relevance to Strategic Objectives

Capacity (3): Multiscale and multiphysics modeling will enable the development of revolutionary new material systems. These new materials will be lighter, stiffer, and stronger than current material systems, leading to increased payload fractions, so that an aircraft of given size will be able to carry more passengers or cargo.

Safety and Reliability (3): Multiscale and multiphysics modeling will allow a complete understanding of the manner in which materials fail. This will improve safety and reliability by improving the ability to predict and account for structural failures.

Efficiency and Performance (3): Multiscale and multiphysics modeling will lead to new, possibly more efficient materials for aircraft propulsion systems.

Energy and the Environment (3): Multiscale and multiphysics modeling will improve the understanding of acoustic dampening properties and interactions between airflow and structures. These improvements will aide in the development of quieter aircraft.

Synergies with National and Homeland Security (3): Multiscale and multiphysics modeling is going on throughout DoD—for example, to create new materials that are immune to radiation effects for use in nuclear reactors. Nationally, much of the research relevant to this Challenge (including DoD-funded research) is conducted by universities, and it is still at a low level of technology readiness. Therefore, the impact on national security, at least in the near term, will be limited.

Support to Space (1): Multiscale and multiphysics models applicable to civil aeronautics focus primarily on interactions between mechanical, material, and flow phenomena. Since high temperatures typical of space reentry vehicles are a minor consideration, most of the research relevant to this Challenge would not be applicable to space applications.

Why NASA?

Supporting Infrastructure (3): Facilities and personnel at NASA's Langley and Glenn Research Centers are supporting research relevant to this Challenge. Superior facilities and personnel may exist at many universities, however.

Mission Alignment (3): This Challenge is relevant to NASA's mission. However, the technology is in the early stages of development, and current research is rather generic. As these materials are advanced to the point where specialized research with a focus in aeronautical applications becomes necessary, the Challenge will become more relevant to NASA's mission.

Lack of Alternative Sponsors (3): DoD, DOE, and some universities are conducting research relevant to this Challenge. However, without NASA's attention, this research may never be used to develop materials and structure necessary for civilian aeronautical applications. Additionally, NASA's expertise in multidisciplinary design and optimization means that NASA is well qualified to implement much of the work done elsewhere.

Appropriate Level of Risk (3): Relevant technology is currently at a very early stage of development; this Challenge faces very high risk.

C17 Ultralight structures

The current state of the art in lightweight airframe structures is demonstrated by the Boeing 787, which makes extensive use of structural composites. Lightweight structures enable aircraft with longer range, more fuel efficiency, greater payload, and/or lower operating costs. Ultralight-weight airframe designs would increase these payoffs.

Ultralight structures programs have been characterized by highly innovative concepts that test the boundaries of the possible. The Gossamer Condor demonstrated human-powered flight. Several NASA and DoD ultralight airframe programs have produced prototype high-altitude, long-endurance UAVs, such as Helios, which demonstrated solar-powered flight. Typically these ultralightweight aircraft have been point-designed for specific flight conditions. To achieve their objectives, they disregarded usual aircraft design practices such as minimum skin thicknesses, redundant load paths, and damage-tolerant design criteria. The designs demonstrated in these programs are not robust enough to meet the strict certification requirements for commercial aircraft. However, they do motivate pragmatic adaptations of ultralightweight structural concepts suitable for commercial application. Promising concepts for ultralight airframes include the use of foam or honeycomb core-stiffened structures with integral, durable damage arrest features; high-performance fibers for increased strength; directional tailoring and unitized construction; and structural optimization methods. Including adaptive materials for variable camber morphing wings may reduce control surface weight. Embedding multifunctional technologies, such as integral antennas or new flexible polymer solar cells, could reduce subsystem weights. The details of many of these concepts have been discussed in other, more highly ranked R&T Challenges. A new initiative is needed to integrate these concepts and the associated design and analysis technologies, along with the substantiating test data, to enable robust, ultralight-weight airframe structures for commercial transport applications. Key milestones include

- Develop specific ultralightweight airframe concepts, leveraging lessons learned from experimental ultralight aircraft research to develop damage-tolerant, adaptive, and multifunctional materials.
- Develop a design and analysis methodology.
- Develop a structural optimization methodology.
- Perform verification testing.
- Demonstrate the potential of one or more concepts to reduce airframe weight by 40 percent.

Relevance to Strategic Objectives

Capacity (3): Reducing airframe weight can increase payload fraction, meaning that an aircraft of a given size will be able to carry more passengers or cargo. Ultralight airframes may enable new aircraft concepts and increased operating flexibility.

Safety and Reliability (1): This Challenge is not relevant to this objective.

Efficiency and Performance (3): Reducing airframe weight will help reduce aircraft operating costs and increase structural efficiency and performance.

Energy and the Environment (1): This Challenge would only have a small, indirect impact on this Objective via improved structural efficiency.

Synergies with National and Homeland Security (3): Some aspects of this technology, such as structures for micro-UAVs and long-endurance surveillance aircraft, are of great interest to the DoD and DHS.

Support to Space (3): This Challenge is relevant to the design of space structures.

Why NASA?

Supporting Infrastructure (3): NASA Langley has some relevant skills left over from the Advanced Composites Technology Program of the 1980s.

Mission Alignment (9): This Challenge is well-aligned with NASA's improved airframe efficiency goals.

Lack of Alternative Sponsors (1): DoD organizations support this goal, although in recent years there has been no investment in this area. Private entrepreneurs' investment in innovative ultralightweight concepts is frequently motivated by setting new aviation performance records (speed, altitude, endurance, etc.).

Appropriate Level of Risk (3): Incremental research to reduce the weight of airframe structures has been under way for a long time. Therefore, this Challenge faces low risk.

C18 Advanced functional polymers

Polymers that adapt their properties or alter their form in response to a change in their environment are known as functional or stimuli-responsive materials. Polymeric materials have demonstrated many responses coupled to a wide range of stimuli (temperature, pH, ionic strength, electrical potentials, and light). These polymers can provide unique functionality of great benefit to aeronautical and aerospace applications. They hold particular potential for achieving biomimetic functionality and sensing.

Potential benefits from this broad-based technology include new sensing capabilities, self-healing polymers for passive repair of damage, reversible liquid crystal adhesives, phase changing polymeric materials for managing interior temperatures, superabsorbent polymers for fire retardation, nanocomposite dispersions providing longer life and resistance to dirt (e.g., self-cleaning), ionic polymers for actuation, color change or other reaction to stress or an environmental threat, conductive polymers, and materials with energy-harvesting capabilities. Development of advanced functional polymers is an active research area in both academia and industry. However, many current applications are not necessarily targeted for aeronautics or aerospace. Many potential functionalities (e.g., self-healing and self-cleaning) have been demonstrated in the laboratory environment, but only in small samples have been generated. Field testing remains to be done. There is also a need to improve the durability and environmental stability of these materials so they can survive at very high or very low temperatures and in corrosive conditions. It would also be helpful to reduce the need for expensive catalysts or other additives, which many advanced functional polymers require.

Functional polymers are an emerging field with significant opportunity for synthesis and characterization of new polymers to achieve the varied applications described above. Key milestones include

- Demonstrate cost-effective methods for processing larger quantities of functional polymers.
- Transition new functional polymers from research laboratories to the sizes needed for aircraft component testing.
- Develop more robust, environmentally stable formulations that can survive in aircraft environments.

Relevance to Strategic Objectives

Capacity (1): Advanced functional polymers may increase operating flexibility, but the impact on aircraft size, new vehicle concepts, and speed will not be significant.

Safety and Reliability (3): Advanced functional polymers can enable self-repair, improve damage detection and fire retardation, and simplify maintenance.

Efficiency and Performance (1): This Challenge has no impact on this Objective.

Energy and the Environment (1): Advanced functional polymers may provide some energy-harvesting capabilities, but the impact on noise, emissions, environmental hazards, and development of alternative fuels is minimal.

Synergies with National and Homeland Security (3): Chameleon-like functionality, self-assessment, self-repair, and autonomous functions are of interest to DoD and DHS, which also fund research relevant to this Challenge.

Support to Space (1): This Challenge has no impact on this Objective.

Why NASA?

Supporting Infrastructure (9): NASA Langley Research Center has unique, relevant infrastructure for polymer research.

Mission Alignment (3): Development of functional polymers for aeronautics and aerospace applications is aligned with the NASA mission of increasing aircraft performance. However, the technology is in the early stages of development and current research is rather generic. As these materials are advanced to the point where specialized research with a focus in aeronautical applications becomes necessary, the Challenge will move into closer alignment with NASA's mission.

Lack of Alternative Sponsors (3): DoD sponsors work in this broad area, but the focus is on military applications.

Appropriate Level of Risk (3): At this stage, with basic research mainly done at universities, this Challenge faces low risk. However, when the technology advances to the point where research results are ready to be transitioned to aeronautical applications, the risk will increase.

C19 Advanced engine nacelle structures

Engine nacelles and pylons are critical portions of aircraft structure. Nacelles enclose the jet engine and pylons provide means for mounting the engine on the airframe. The front portion of a nacelle also directs air into the engine inlet and thus affects performance of the engine. For large commercial airplanes, nacelle designs have not appreciably advanced for a generation. Consequently, nacelle structures have not taken advantage of recent developments in materials and structures technology. The result is nacelles that weigh more than they should, which decreases range, payload, and airframe fatigue life.

New structural concepts take advantage of modern analytic design tools and advanced structural materials to significantly reduce the weight while maintaining structural integrity and improving engine efficiency.

Nacelles and pylons are critical structures that affect airworthiness, so that any change from current practice must be well understood, analyzed, and validated to assure that it gives at least the same level of safety as existing structures, which have served well in many airplane applications. Key milestones include

- Define the attributes of new design concepts for nacelles that reduce weight and engine noise based on input from large and small aircraft manufacturers as well as NASA experts.
- Perform multidisciplinary design analysis to identify new structural concepts for both large and small airplanes.
- Validate the analysis via testing of subscale models of these new concepts.
- Test design at full scale.

Relevance to Strategic Objectives

Capacity (3): Lighter weight nacelles would allow a higher payload fraction, meaning that an aircraft of a given size would be able to carry more passengers or cargo. However, this particular Challenge may have more overall benefit for small airplanes than for large ones.

Safety and Reliability (1): This Challenge would ensure that pylons and nacelles are as safe and reliable as current systems, but there would be no net improvement for aircraft or the airspace system as a whole.

Efficiency and Performance (3): This technology will leverage advanced analytic tools and advanced structural materials to significantly reduce the weight of these structures. As with capacity, effects may be more pronounced for smaller aircraft.

Energy and the Environment (1): This Challenge would offer little reduction in community noise.

Synergies with National and Homeland Security (1): DoD may gain the advantage of this technology once it becomes standard practice in the aeronautical industry. However, it is not likely to be active in advancing this Challenge. Furthermore, as noted before, larger aircraft, such as the tankers and transports the DoD would be interested in, would benefit less.

Support to Space (1): This Challenge has no impact on this Objective.

Why NASA?

Supporting Infrastructure (1): NASA seems to have little or no infrastructure relevant to this Challenge.

Mission Alignment (9): This Challenge is very relevant to NASA's mission to improve aircraft performance.

Lack of Alternative Sponsors (1): Although there is currently no known effort within government or industry, this is the sort of research that should be pursued by original equipment manufacturers. If they feel the payoff is great enough, they will pursue the technology themselves.

Appropriate Level of Risk (3): Although there is currently no known effort within government or industry, this is the sort of research that manufacturers should support. There, too, if they conclude the payoff is high enough, they will pursue the technology themselves.

C20 Repairability of structures

Modern airframes, whether composite or metallic, require repairs either to restore functionality or to extend their lives. To assess the repairability of structures and make correct repair-or-replace decisions, structural assessment methods and tools, tools for trade-off analyses, and repair technologies and processes are required. These methods, technologies, tools, processes, and analyses must be applicable to metallic, polymer composite, and ceramic composite structural elements.

The primary benefit of being able to assess repairability and repair airframe parts instead of replacing them is lower direct operating costs. Life extension via repair of aging aircraft also has a significant economic impact in the form of lower acquisition costs.

Several NASA and DoD programs and manufacturers' repair manuals for aircraft owners and operators provide up-to-date tools and processes for making airframe repairs. The DoD has published a repair design guide for combat and transport aircraft. However, new algorithms are needed that incorporate validated analysis of crack growth criticality in metals, defect or damage propagation in composites, and mathematical models for stress corrosion cracking. Further, these algorithms need to be integrated with repair integrity evaluation analyses to provide a complete modeling and simulation capability that assesses the economic effectiveness of the repair method.

Fatigue and fracture analysis and databases for metals and composites used in airframe construction need to be developed, along with software enabling complete modeling and simulation of repairs, including cost considerations, to identify realistic trade-off alternatives. Key milestones include

- Conduct damage and damage growth analyses for metallic and composite structures.
- Collect a compendium of repair processes for metallic and composite structures.
- Demonstrate computer codes to model and simulate repairs for decision making.

Relevance to Strategic Objectives

Capacity (3): Effective airframe repairs will increase operating aircraft availability.

Safety and Reliability (3): Improved quality of airframe structure repairs will reduce the likelihood of loss or human injury.

Efficiency and Performance (3): This Challenge reduces aircraft operating costs and postpones the need for new aircraft purchases.

Energy and the Environment (1): This Challenge would only have a small, indirect impact on this Objective by extending structural life.

Synergies with National and Homeland Security (1): The focus of this Challenge would be on the needs of civil aircraft.

Support to Space (1): This Challenge focuses on repairability issues for air breathing aircraft and would likely have little relevance for access-to-space vehicles, which operate in extreme environments.

Why NASA?

Supporting Infrastructure (3): NASA Langley has some capability relevant to this Challenge as a legacy of several fatigue and damage propagation research programs related to metallic and composite airframe structures during the 1970s and 1980s. These skills would need to be augmented and updated.

Mission Alignment (3): This Challenge is aligned with NASA's mission to improve aeronautical technology, but it addresses issues with operational aircraft that are also industry's responsibility.

Lack of Alternative Sponsors (1): This technology should be undertaken by industry; in the past, DoD has supported this goal as well.

Appropriate Level of Risk (3): This Challenge faces low to moderate risk.

REFERENCES

Calkins, F.T., and G.W. Butler. 2004. Subsonic Jet Noise Reduction Variable Geometry Chevron. AIAA 2004-190, 42nd AIAA Aerospace Sciences Meeting and Exhibit, Reno, Nev., January 5-8.

Carman, G.P., and G.P. Sendeckyi. 1995. Review of the mechanics of embedded optical sensor. Journal of Composite Technology and Research 17(3): 183-193.

Chaplya, P.M., and G.P. Carman. 2002. Compression of piezoelectric ceramic at constant electric field: Energy absorption through non-180 degrees domain-wall motion. Journal of Applied Physics 92(3): 1504-1510.

Giesing, J., and J.M. Barthelemy. 1998. A Summary of Industry MDO Applications and Needs. AIAA-1998-4737, 7th AIAA/USAF/NASA/ISSMO Symposium on Multidisciplinary Analysis and Optimization, St. Louis, Mo., September 2-4. Available online at <http://endo.sandia.gov/AIAA_MDOTC/sponsored/ summary_paper.pdf>.

Giurgiutiu, V., and A.N. Zagrai. 2002. Embedded self-sensing piezoelectric active sensors for on-line structural identification. Journal of Vibration and Acoustics-Transactions of the ASME 124(1): 116-125.

Johnson, W., G. Yamauchi, and M. Watts. 2006. Design and Technology Requirements for Civil Heavy Lift Rotorcraft. Proceedings of the American Helicopter Society Vertical Lift Aircraft Design Conference, San Francisco, Calif., January 18-20.

Lim, T.J., and D.L. McDowell. 1999. Mechanical behavior of an Ni-Ti shape memory alloy under axial-torsional proportional and nonproportional loading. Journal of Engineering Materials and Technology-Transactions of the ASME 121(1): 9-18.

Lin, C.Y., and E.F. Crawley. 1995. Design Considerations for a Strain Actuated Adaptive Wing for Aeroelastic Control. Journal of Intelligent Material Systems and Structures 6(3): 403-410.

Lin, M., and F.K. Chang. 2002. The manufacture of composite structures with a built-in network of piezoceramics. Composites Science and Technology 62(7-8): 919-939.

Moffett, M.B., A.E. Clark, M. Wun-Fogle, J. Linberg, J.P. Teter, and E.A. McLaughlin. 1991. Characterization of terfenol-D for magnetostrictive transducers. Journal of the Acoustical Society of America 89(3): 1448-1455.

National Research Council (NRC). 1998. Ceramic Fibers and Coatings: Advanced Materials for the 21st Century. Washington, D.C.: National Academy Press. Available online at <http://fermat.nap.edu/catalog/6042.html>.

NRC. 2004. Accelerating Technology Transition: Bridging the Valley of Death for Materials and Processes in Defense Systems. Washington, D.C.: The National Academies Press. Available online at <http://fermat.nap.edu/catalog/11108.html>.

Powrie, H.E.G., and C.E. Fisher. 1999. Engine health monitoring: Towards total prognostics. Proceedings of the 1999 IEEE Aerospace Conference 3: 11-20.

Ritter, T., X.C. Geng, K.K. Shung, P.D. Lopath, S.E. Park, and T.R. Shrout. 2000. Single crystal PZN/PT-polymer composites for ultrasound transducer applications. IEEE Transactions on Ultrasonics Ferroelectrics and Frequency Control 47(4): 792-800.

Shin, D.D., K.P. Mohanchandra, and G.P. Carman. 2004. High frequency actuation of thin film NiTi. Sensors and Actuators A-Physical 111(2-3): 166-171.

Simon, D.L. 2000. An Overview of the NASA Aviation Safety Program Propulsion Health Monitoring Element. 36th Joint Propulsion Conference and Exhibit, Huntsville, Ala., July 16-19.

Sobieszczanski-Sobieski, J., and R.T. Haftka. 1997. Multidisciplinary aerospace design optimization: Survey of recent developments. Structural Optimization 14(1): 1-23.

Stewart, A., G. Carman, and L. Richards. 2003. Nondestructive evaluation technique utilizing embedded thermal fiber optic sensors. Journal of Composite Materials 37(24): 2197-2206.

Ullakko, K., J.K. Huang, V.V. Kokorin, and R.C. O'Handley. 1997. Magnetically controlled shape memory effect in Ni_2MnGa intermetallics. Scripta Materialia 36(10): 1133-1138.

Wood, K., T. Brown, R. Rogowski, and B. Jensen. 2000. Fiber optic sensors for health monitoring of morphing airframes: I. Bragg grating strain and temperature sensor. Smart Materials and Structures 9(2): 163-169.

D

R&T Challenges for Dynamics, Navigation, and Control, and Avionics

A total of 14 R&T Challenges were prioritized in the guidance, navigation, and control, and avionics Area. Table D-1 shows the results. The R&T Challenges are listed in order of NASA priority. National priority scores are also shown.[1] This appendix contains a description of each R&T Challenge, including milestones and an item-by-item justification for each score that appears in Table D-1.[2]

D1 Advanced guidance systems

Advanced guidance systems consist of subsystems and processes (hardware and software) assembled for the purpose of providing an aircraft, spacecraft, or other dynamic system with desired state trajectories. These trajectories can be defined using either discrete or continuous data and can include information such as current velocity, acceleration, time of arrival, and desired position. The determination of the desired trajectory usually takes into account mission-dependent constraints, which can include obstacles (such as terrain, wake vortices, or other aircraft), hazards (such as weather), coordination with other aircraft (such as cooperative and multiaircraft guidance, formation flight, or swarming), and regulatory constraints (such as airspace class restrictions) (Doebbler et al., 2005).

State-of-the-art guidance systems enable aircraft to follow waypoints, perform automatic obstacle avoidance, and fly in formation with other aircraft (Schierman et al., 2004). Additional research is needed to develop guidance algorithms and mature them into flight-ready systems,[3] to develop improved reconfigurable and adaptive guidance systems, and to develop advanced guidance systems for UAVs. One concern, for example, is the need to develop improved technologies to avoid controlled flight into terrain, particularly in the case of all-weather operation of advanced rotorcraft. Some important research is inhibited by the limited number of programs and facilities capable of implementing and flying these systems on real aircraft. Also, certification and regulatory issues must be resolved so that the air transportation system can take advantage of the full capabilities of current and future guidance systems for piloted aircraft and UAVs.

Advanced guidance systems have the potential to greatly improve the capacity, safety, and efficiency of the air transportation system. In addition, they can enhance the performance of many existing and future military systems. Key milestones include

- Development of advanced algorithms and avionics for collision, terrain, and wake vortex avoidance; formation flight and cooperative and multiaircraft guidance; and ground operations guidance (taxi, takeoff, rollout, and turnoff).
- Expansion of facilities and programs capable of maturing the above technologies to flight-ready systems.
- Development and adoption of regulations for the certification and operation of autonomous UAVs in civil airspace.

Relevance to Strategic Objectives

Capacity (9): Advancing the state of the art in multiaircraft and cooperative guidance will allow more aircraft per unit time to move through the available airspace.

Safety and Reliability (9): Advanced guidance systems will allow aircraft to operate more safely in closer quarters than is currently possible both in the air and on the ground.

[1]The prioritization process is described in Chapter 2.

[2]The technical descriptions for the first 10 Challenges listed below are essentially the same as the technical descriptions for these Challenges as they appear in Chapter 3.

[3]R. Duren, associate professor, Baylor University, "Avionics research challenges," Presentation to Panel D on November 15, 2005.

TABLE D-1 Prioritization of R&T Challenges for Area D: Dynamics, Navigation, and Control, and Avionics

R&T Challenge		Strategic Objective						National Priority	Why NASA?				Why NASA Composite Score	NASA Priority Score
		Capacity	Safety and Reliability	Efficiency and Performance	Energy and the Environment	Synergies with Security	Support to Space		Supporting Infrastructure	Mission Alignment	Lack of Alternative Sponsors	Appropriate Level of Risk		
	Weight	5		3		1			1/4 each					
D1	Advanced guidance systems	9	9	9	3	3	3	**132**	9	9	3	9	**7.5**	**990**
D2	Distributed decision making, decision making under uncertainty, and flight-path planning and prediction	9	9	9	3	3	3	**132**	3	9	3	9	**6.0**	**792**
D3	Aerodynamics and vehicle dynamics via closed-loop flow control	1	9	9	3	3	3	**92**	9	9	3	9	**7.5**	**690**
D4	Intelligent and adaptive flight control techniques	3	9	9	3	3	9	**108**	3	9	3	9	**6.0**	**648**
D5	Fault-tolerant and integrated vehicle health management systems	3	9	3	1	3	9	**84**	9	9	3	9	**7.5**	**630**
D6	Improved onboard weather systems and tools	9	9	3	1	1	1	**104**	9	9	3	3	**6.0**	**624**
D7	Advanced communication, navigation, and surveillance technology	9	9	9	3	3	3	**132**	3	9	3	3	**4.5**	**594**
D8	Human–machine integration	3	9	9	1	3	3	**96**	3	9	3	9	**6.0**	**576**
D9	Synthetic and enhanced vision systems	3	9	3	1	1	3	**76**	9	9	3	3	**6.0**	**456**
D10	Safe operation of unmanned air vehicles in the national airspace	3	9	3	1	9	1	**82**	3	9	3	3	**4.5**	**369**
D11	Secure network-centric avionics architectures and systems to provide low-cost, efficient, fault-tolerant, onboard communications systems for data link and data transfer	9	9	9	1	9	3	**132**	3	3	1	3	**2.5**	**330**
D12	Smaller, lighter, and less expensive avionics	1	3	9	3	3	9	**68**	3	3	3	3	**3.0**	**204**
D13	More efficient certification processes for complex systems	3	9	9	1	1	3	**94**	3	1	1	3	**2.0**	**188**
D14	Design, development, and upgrade processes for complex, software-intensive systems, including tools for design, development, and validation and verification	3	9	3	1	1	3	**76**	1	3	1	1	**1.5**	**114**

Efficiency and Performance (9): The ability to safely operate aircraft closer to each other will allow more efficient use of the airspace and airport real estate.

Energy and the Environment (3): Advanced guidance systems can enable arrival and departure trajectories that reduce community noise. Also, multiaircraft guidance systems can increase fuel efficiency.

Synergies with National and Homeland Security (3): Cooperative and autonomous capabilities are also applicable to military aircraft.

Support to Space (3): Many multiaircraft guidance algorithms are applicable to satellite constellations and formations of mini- and microsatellites.

Why NASA?

Supporting Infrastructure (9): NASA has air transportation system simulation facilities and manned and unmanned aircraft simulation and flight test facilities. In addition, NASA has also been the primary facilitator of the unique and highly relevant Access 5 program.

Mission Alignment (9): This Challenge will have a broad benefit for aeronautics in general, and NASA has often done related research.

Lack of Alternative Sponsors (3): DoD is doing some military-specific work related to this Challenge. Relevant work by industry and academia is limited by certification and regulatory issues as well as the prohibitive cost of test facilities.

Appropriate Level of Risk (9): This Challenge faces moderate to high risk.

D2 Distributed decision making, decision making under uncertainty, and flight path planning and prediction

Improving the decision-making process used by pilots and aircraft systems, when coupled with improvements in flight-path planning and prediction, has been theorized as an effective approach to improving air transportation system capacity and safety. This Challenge has the potential to significantly improve the timeliness of real-time decisions to alter flight paths in the dynamic environment of congested airspace (Ding et al., 2004; Helbing et al., forthcoming; Rong et al., 2002). Coordinated decision making, which includes the direct exchange of data among different aircraft and the deconfliction of flight paths without the need to rely on ground-based controllers, addresses the inherent limitations of centralized ATC systems in terms of uncertainty and fault tolerance. A coordinated, distributed approach to decision making increases air transportation system reliability and safety by distributing control and mission management capabilities among multiple agents. It also allows for rapid response to changing dynamics and minimizes vulnerability to system failures.

Automated systems can help improve decision making and flight path planning. Levels of automation ranging from "pilot aid" (that is, systems that advise pilots to take specific action) to "fully autonomous" are achievable but have not yet been developed to the point where they can support high levels of automation for civil aircraft. Until now, coordinated distributed algorithms for constraint reasoning (for example, to optimize flight paths) have not been applied to the air transportation system because implementation with such a complex system would require aircraft to exchange a large number of messages, which raises substantive communications, bandwidth, and man–machine interface issues.

This Challenge should address the needs of a wide variety of conventional and unconventional aircraft types, including those with no distributed decision-making capability. Aircraft types of interest include commercial airliners, general aviation aircraft, civil helicopters, military aircraft, and UAVs.

This Challenge also has the potential to be of great benefit when applied to complex, nonaviation systems that operate in dynamically changing environments and require high-quality, real-time decision making. Key milestones include

- Develop fundamental system requirements, architectures, and system logic that are compatible with current and future regulatory requirements and ATM systems. This Challenge should include studies to determine the levels of automation appropriate to a wide range of decision-making applications.
- Develop simulation capabilities for evaluation and demonstration of certain high-performing strategies in the execution of realistic system architectures and applications.
- Develop a requirements flowdown to all affected aircraft systems, such as advanced communications, navigation, and surveillance (CNS) systems.
- Develop improved, automated logic and processes for contingency management.
- Develop a methodology to support verification and validation of future systems technologies developed by this Challenge.

Relevance to Strategic Objectives

Capacity (9): The capability to accomplish dynamic real-time flight path planning and replanning in the dynamic airspace environment will allow closer separations and increase capacity.

Safety and Reliability (9): The capability to accomplish dynamic high-quality flight path planning and replanning in the dynamic airspace environment will allow closer aircraft separations with increased safety.

Efficiency and Performance (9): The capability to accomplish automated and autonomous high-quality, real-time flight path planning and replanning in the dynamic airspace environment will reduce aircrew and controller workloads as well as the demands on the entire ATM system. Training requirements will also be reduced at all levels within the system.

Energy and the Environment (3): Dynamic flight path replanning can be used to improve mission efficiency and reduce fuel burn.

Synergy with National and Homeland Security (3): The capability to accomplish dynamic, high-quality, real-time decision making (e.g., planning and replanning) will have many applications in the national and homeland security environment, including the future integration of UAVs into the air transportation system.

Support to Space (3): Coordinated distributed decision making is very beneficial for spacecraft guidance, navigation, and control (GNC) tasks such as flight planning, rendezvous and docking, and reentry.

Why NASA?

Supporting Infrastructure (3): NASA has significant facilities and some experience relevant to this Challenge.

Mission Alignment (9): This Challenge will directly support multiple R&T Challenges and benefit military and civil aviation, as well as future applications in space.

Lack of Alternative Sponsors (3): Industry is developing technology relevant to this Challenge, but it is more geared to military systems with a focus on UAVs.

Appropriate Level of Risk (9): Mid- and long-term research can address issues related to this Challenge, and the results can be transferred to future civil and military applications.

D3 Aerodynamics and vehicle dynamics via closed-loop flow control

Closed-loop flow control appears to offer tremendous promise in improving aerodynamic performance. For example, active flow control approaches should allow the airfoil lift:drag (L/D) to remain high over large changes in angle of attack.[4] Flow control R&T could also be used to develop a spoiler-aileron to replace complex and heavy control surfaces and to reduce or eliminate turbulent flow over aircraft surfaces to reduce skin-friction drag. These applications could lead to new aircraft configurations (Chavez and Schmidt, 1994).

The mechanization of flow control systems may require a large number of distributed sensors measuring pressure or shear stress over the wing and changes in the boundary layer. Actuation might be accomplished by morphing the wing or introducing devices that induce sucking or blowing along the wing. These distributed sensors and actuators are coordinated so that control is obtained over large flight regimes, angles of attack, and attitudes.

Distributed sensing and actuation would also permit structures to be self-aware for health monitoring, thereby increasing system reliability. Airframe and engine structures could be monitored for changes in behavior.

Some of the techniques developed by this Challenge may also advance modeling and design capabilities applicable to morphing aircraft (Tandale et al., 2005; Valasek et al., 2005). Heretofore, aircraft have generally been fixed-frame structures. Morphing aircraft would be designed with distributed actuation and controls and with mechanization as an inherent property. They would lead to new capabilities and concepts in aircraft design. Examples include (1) biomorphic aircraft, such as ornithopters, that could maneuver robustly in complex environments and (2) hunter-killer aircraft that change shape to optimize performance for different tasks (e.g., surveillance, reconnaissance, and ground attack). Morphing technology might also enable aircraft capable of perching. Key milestones include

- Develop simpler representations of the aircraft system dynamics for control design.
- Develop distributed control algorithms and architectures.
- Demonstrate the ability to numerically solve distributed control algorithms at the Reynolds numbers associated with manned aircraft flight to demonstrate control performance.
- Implement integrated, distributed closed-loop flow control systems.
- Design and develop lightweight, mechanized, shape-changing structures.
- Experimentally verify the performance of shape-changing aerodynamic structures before flight testing.

Relevance to Strategic Objectives

Capacity (1): This Challenge has minimal application to this Objective, although networks of sensors and actuators could be utilized to monitor and maintain fleet readiness.

Safety and Reliability (9): Large arrays of sensors and actuators vastly improve system redundancy.

Efficiency and Performance (9): Adaptable flight characteristics will improve mission performance over a variety of conditions.

Energy and the Environment (3): Extremely high L/D reduces fuel usage, and adaptable engines could significantly reduce noise and emissions.

Synergies with National and Homeland Security (3): This Challenge will facilitate the development of UAVs with long endurance for surveillance.

[4]The flow over the specially shaped GLAS II airfoil remains naturally separated at the rear of its upper surface over a wide range of angles of incidence; in the absence of active control, its L/D does not exceed 25. At an incidence angle of 10 degrees, its L/D is nearly 500 (Glauert, 1945, 1948).

Support to Space (3): This Challenge could (1) improve the performance of aerodynamic first-stage launch vehicles, which would increase payload capacity, and (2) enhance the endurance of aircraft design to fly in the martian atmosphere. Shape control also allows for safer trans-atmospheric flight.

Why NASA?

Supporting Infrastructure (9): NASA has been a leader in computational fluid dynamics, producing the first direct numerical computation of the Navier-Stokes equations. Furthermore, Langley Research Center and Ames Research Center have conducted flow control R&T. Finally, existing NASA wind tunnels could be important for experimental verification of flow control algorithms.

Mission Alignment (9): This Challenge is very relevant to NASA's mission.

Lack of Alternative Sponsors (3): DoD is doing a lot of research relevant to this Challenge and has coupled it with funding to many universities.

Appropriate Level of Risk (9): This Challenge faces moderate to high risk.

D4 Intelligent and adaptive flight control techniques

The missions and capabilities of future aircraft, both manned and unmanned, will be more multifunctional than those of the current generation of specialized aircraft. Achieving aggressive performance targets in range, payload, reliability, safety, noise, and emissions will require a total system that is integrated to a far higher level than existing aircraft. R&D for military aircraft has been able to push the technological envelope associated with intelligent and adaptive flight control techniques farther than R&D for civil aircraft because of different safety limits. In the far term, as it advances, application of military technology to civil aircraft may be possible.

The vehicle management systems (VMS) paradigm offers the most promising path to realizing goals related to this Challenge. VMS takes a top-down systems approach to specifying, designing, and validating the aircraft as a single system with highly integrated inner and outer loops. It thereby unifies the traditionally separate fields of propulsion control, flight control, structural control, noise control, emissions control, and health monitoring. The current state of the art in VMS uses traditional feedback control, consisting of measurements of vehicle states such as airspeed, altitude, angle of attack, and linear and angular acceleration (Jaw and Garg, 2005). By incorporating an online learning capability to cope with new and unforeseen events and situations and nonlinear adaptive control, in which the controller self-tunes to maintain stability and tracking in the presence of disturbances and changing vehicle parameters, an intelligent and adaptive VMS can be developed with the promise of significant advances in capability, safety, and supportability (Tandale and Valasek, 2003).

Significant advances in the state of the art are required to develop an intelligent and adaptive VMS. Current nonlinear adaptive control approaches assume that (1) sensor information is reliable and (2) known nonlinearities can be modeled as slowly varying parameters that affect the system linearly. However, advanced actuators for flow control and structural control will have characteristics that are much more nonlinear than those of conventional control actuators. Control laws and control actuator allocation are currently treated as separate problems, such that optimization of the integrated control law is difficult or impossible. Finally, the problem of multiple correlated, simultaneous failures remains unsolved. Approaches that use analytic redundancy to finding failed sensors generally assume that aircraft dynamics have not changed, while adaptive or reconfigurable control approaches assume that sensor information is reliable. On an affordable aircraft with limited or no sensor redundancy, it is difficult or impossible to tell the difference between a degraded sensor and damage to the aircraft that changes the way it flies. Key milestones include

- Develop an adaptive, intelligent, fully integrated VMS that can operate safely without reliable sensor information.
- Demonstrate a mature methodology for designing and analyzing flight control laws for aircraft with large numbers of highly distributed control actuators and sensors—for example, shape memory alloys and piezoelectrics.
- Demonstrate a mature methodology for using information of different degrees of reliability without compromising flight safety (e.g., using data from what would traditionally be considered non-flight-critical systems within an inner control loop).
- Demonstrate long-term learning so that adaptation would only need to be used in novel situations. For example, following damage, the system adapts the first time it enters a particular part of the flight envelope but does not need to readapt if it leaves that part of the envelope and returns.
- Validate complex nonlinear systems to seek out worst-case scenarios that may not be identified with exhaustive testing.

Relevance to Strategic Objectives

Capacity (3): Advancing the state of the art in propulsion and flight control will allow manned and unmanned aircraft to operate more safely, thereby permitting UAV flight operations over highly populated areas and improving the ability of all aircraft to operate in poor weather conditions.

Safety and Reliability (9): Advanced propulsion and flight control will improve the ability of aircraft to continue operating in spite of control upsets, atmospheric disturbances such as gusts and turbulence, and damage (either natural or terrorist induced).

Efficiency and Performance (9): Advanced propulsion control increases engine efficiency, and flight control techniques such as relaxed static stability or de facto stability can improve range.

Energy and the Environment (3): Advanced propulsion control can reduce engine emissions.

Synergies with National and Homeland Security (3): Advanced propulsion and flight control can improve mission capability, safety, and supportability, which are important to military aircraft.

Support to Space (9): The intelligent and adaptive VMS for advanced propulsion and flight control directly applies to launch vehicles, spacecraft, and planetary landers and reentry vehicles.

Why NASA?

Supporting Infrastructure (3): NASA has a significant investment in high-quality vehicle system integration laboratories, flight simulators, and flight test facilities that are equal to any in DoD or industry and superior to any in academia.

Mission Alignment (9): This Challenge directly aligns with and enhances legacy NASA research in the stability and control of aircraft.

Lack of Alternative Sponsors (3): R&T related to advanced propulsion and flight control is being done by DoD, industry, and academia. However, these efforts are specialized to military aircraft and are not unified in objectives and scope.

Appropriate Level of Risk (9): This Challenge faces moderate to high risk.

D5 Fault-tolerant and integrated vehicle health management systems

Development of integrated vehicle health management (IVHM) system technologies is key to the acceptance of the automation needed in the transformation of the air transportation system. The technology provides an increased capability to accurately discover and assess system faults and reconfigure or recover from them. Although highly integrated, health management aspects consist of related components: fault detection and isolation, recovery and reconfiguration, and condition-based maintenance (CBM). In addition, modeling plays an important role in the development of these functions (Garg, 2005; Litt et al., 2005; Tandale and Valasek, 2006).

Fault detection, isolation, recovery, and reconfiguration

Fault detection, isolation, recovery, and reconfiguration involve processes and approaches that enable robust detection of faults from measured or estimated error residuals and isolation of faults with minimal latency in the presence of noise and environmental effects during aircraft operation. Fault detection, isolation, recovery, and reconfiguration are platform specific and should cover all flight regimes and mission types. Recovery and reconfiguration systems are developed with regard to the possibilities of faults, the nature of the latency of the fault detection and isolation system, and the controls available for recovery and reconfiguration. Redundancy management strategies for avionics and the airframe directly influence options for recovery and reconfiguration.

Condition-based maintenance

CBM involves maintenance processes and capabilities derived from real-time assessment of aircraft system conditions obtained by software from embedded and redundant sensors. The combination of software and sensors can create important communications and bandwidth challenges. More robust diagnostics and prognostics are needed to achieve the goal of CBM, which is to perform maintenance only on evidence of need, to prevent a failure that would reduce aircraft availability. In addition, CBM includes processes that couple real-time assessment of system and component performance with ground- and air-based logistics to improve aircraft system readiness and maintenance practices. CBM is a form of proactive equipment maintenance that forecasts incipient failures. CBM also aims to ensure safety, equipment reliability, and reduction of total ownership cost. Fault tolerance is achieved when CBM is married to decision strategies for safe and reliable operation of manned and unmanned aircraft.

Modeling

Physics-based models of sensors, actuators, avionics, components, and vehicle flight dynamics contribute to the development of methods for forecasting aircraft system performance, thereby helping to uncover faults. In addition, these models can be used for examining architectures and control strategies to reconfigure systems and ensure safety and reliability.

An aircraft is a very complex system. While individual fault-tolerant functions can be set up for each subsystem, the value of fault-tolerant designs is maximized when the system is modeled as a whole, since the behavior of each subsystem can influence that of other subsystems. The advantage of working with a total system model lies in the ability to discover a fault through its effects on other parts of the system before it is discovered in the individual subsystem itself. One primary thrust of fault-tolerant technology development is to identify system models that characterize the behavior of systems properly without developing an overly detailed and unnecessary representation. In other words, an optimum system is not a collection of optimized subsystems.

To advance the state of the art in fault-tolerant aircraft systems, fundamental R&T is required in the three topics above to develop a more robust image of the state, or health, of an aircraft in the presence of uncertainty. With a better model of itself, the aircraft can trace back system anomalies through the multitude of discrete state and mode changes to isolate aberrant behavior. Fault-tolerant systems combine simple rule-based reasoning, state charts, model-free monitoring of cross-correlations among state variables, and model-based representations of aircraft subsystems. Together, these models form a hybrid system model. Advances in computing resource technology have allowed hybrid system models to run in real time.

Fault-tolerant aircraft systems, coupled with CBM, may improve aircraft safety and reduce aircraft life-cycle maintenance and ownership costs. Critical research tasks include developing (1) robust and reliable hardware and software tools for monitoring components, detecting faults, and identifying anomalies; (2) prognosis analysis tools for predicting the remaining life of key components; (3) approaches for recovering from detected faults, including reconfiguration of the flight control system for in-flight failures of manned and unmanned aircraft; and (4) low-cost, lightweight, wireless, self-powered sensors with greater memory and processing capability. Key milestones include

- Specify nominal models and model behavior, interface, and test requirements for component and integrated system capability affected by degraded or failed operation of a representative subset of avionics and flight system components. Define suitable thresholds for levels of degraded and failed operation for component-level and system-level operations.
- Work with aircraft subsystem and flight system vendors to specify parameters that are candidates for maintenance logging. Develop models and compact representations that can incorporate measurements of these parameters in near real-time and develop thresholds that can be used for on-demand maintenance activities.
- Evaluate component capability in a simulated environment (ground test and hardware in the loop). That is, take a particular subsystem, such as a real landing gear system that has been represented by an appropriate behavioral model as specified and insert simulated faults to test for proper operation of the health monitoring system. Perform these tests for all representative subsystems that were specified above.
- Evaluate integrated system capability in a simulated environment. Take the subsystem health models previously specified and insert faults, preferably ones that were not detected as quickly as necessary by the individual component models that were evaluated in the above set of tests, and use the system models in order to evaluate the efficacy of their integrated operation.
- Test component and integrated system capability in a flight environment.

Relevance to Strategic Objectives

Capacity (3): Knowing the health of key components of an aircraft system reduces aircraft downtime and thus improves capacity. Fault-tolerant systems increase airport capacity through improved on-time dispatch of a flight.

Safety and Reliability (9): This Challenge addresses a primary component of safe and reliable operation. Monitoring the performance of key aircraft components significantly improves overall system safety and reliability and reduces uncontrolled flight into terrain for manned and unmanned aircraft.

Efficiency and Performance (3): With improved fault tolerance, overall flight performance is enhanced because faults can be isolated, thus ensuring robust operation of the aircraft. More efficient use of maintenance resources reduces aircraft downtime.

Energy and the Environment (1): Improved fault tolerance reduces life-cycle maintenance and operation costs and makes more efficient use of parts and supplies. This reduces environment effects, but only to a small degree.

Synergies with National and Homeland Security (3): Fault tolerance is important to military aircraft.

Support to Space (9): Fault-tolerant architectures for aircraft will be of great use to spacecraft systems, and fault tolerance has a major impact on space travel.

Why NASA?

Supporting Infrastructure (9): NASA has done R&T related to this Challenge and has a unique capability in applying fault tolerance to space applications. NASA has unique propulsion test facilities that would be critical for characterizing drive trains and engines for aircraft and rotorcraft. In addition, NASA has unique modeling and simulation capabilities that support fault-tolerant aircraft system modeling.

Mission Alignment (9): This research will benefit aeronautics in general, and NASA has often done similar research.

Lack of Alternative Sponsors (3): While industry and the military are looking at technology relevant to this Challenge, NASA has assembled a core competency that is unmatched for civil aircraft applications, especially for rotorcraft. NASA support is essential to address civil aeronautics applications.

Appropriate Level of Risk (9): More data are needed to determine how fault-tolerant systems impact the life-cycle costs of civil aircraft. Some information is available on how such systems can improve aircraft safety, especially for rotorcraft.

D6 Improved onboard weather systems and tools

Pilots—and the avionics software that provides in-flight, four-dimensional trajectory replanning and commands to the pilot or autopilot—require additional weather information to minimize the impact of weather on the control of flight in heavy traffic. Basic research is needed to determine the most cost-effective way of integrating real-time weather information into four-dimensional, integrated control of flight. This information might include information from data links with ground sites and other aircraft and weather video from ground stations and satellites (Bokadia and Valasek, 2001; Lampton and Valasek, 2005, 2006).

Other aircraft could provide information about geospatial position, wind, icing conditions, turbulence, lightning, and precipitation, as well as imagery from radars and other sensors. Data links with the ground could provide actual and forecast information on winds at different flight levels, pressure, icing potential, precipitation, ground-level temperatures, weather fronts, severe weather, airport surface conditions, and other information from significant meteorological information reports (SIGMETs), pilot reports (PIREPs), meteorological aviation reports (METARs), terminal area forecasts (TAFs), imagery from satellites and radars, and so on. Key milestones include

- Develop robust and reliable data links for collecting information from onboard sensors.
- Develop processes and tools for integrating weather information from onboard sensors and data links to the ground and other aircraft.
- Demonstrate effectiveness in practical decision-support applications relating to weather, with varying levels of information quality and uncertainty.

Relevance to Strategic Objectives

Capacity (9): Improving the quality and use of weather information will enable aircraft to avoid or fly through weather more effectively, which will reduce delays due to weather.

Safety and Reliability (9): Improving the quality and use of weather information, including information on runway conditions, will reduce the number of aircraft accidents caused by weather. Pilots will be less likely to fly into weather that they or their aircraft cannot handle.

Efficiency and Performance (3): Fuel consumption can be reduced by flight plans that incorporate real-time information on winds aloft to maintain the desired four-dimensional flight trajectory.

Energy and the Environment (1): This Challenge has little impact on this Objective. There might be some indirect benefit through the reduction of emissions or ground noise by optimizing flight plans, allowed by improved onboard weather systems and tools.

Synergy with National and Homeland Security (1): This Challenge would also benefit DoD and DHS flight operations involving civil airspace.

Support to Space (1): This Challenge has little or no application to this Objective.

Why NASA?

Supporting Infrastructure (9): NASA has an outstanding research facility for icing tests and evaluation and the infrastructure to develop and test weather-related tools.

Mission Alignment (9): This Challenge is central to NASA's safety and capacity mission.

Lack of Alternative Sponsors (3): Some airlines have invested in developing capabilities relevant to this Challenge, and the FAA and Air Force are also interested. However, these efforts are limited in comparison to what NASA could do.

Appropriate Level of Risk (3): Weather-related tools exist, but integration into other systems, especially onboard systems, is needed. This Challenge faces low risk, and transfer to industry is likely, because industry is developing some tools related to the integration of weather information.

D7 Advanced communication, navigation, and surveillance (CNS) technology

The capacity of the air transportation system is dependent on minimum spacing requirements for safe operation. Minimum spacing depends on many factors, including the capability of each aircraft to precisely fly a predetermined, geospatially time-referenced flight path.

Advanced, integrated, accurate, secure, and reliable CNS capabilities are required for network-centric operations, which can increase capacity in very high density airspace. Each aircraft may be considered a node in a network-centric, distributed, fault-tolerant ATM system. Communications between nodes (aircraft to aircraft, aircraft to ground, and aircraft to satellite to ground) must be highly reliable. (For example, the probability of a missed or incorrect message should be less than 10^{-7} per flight hour, depending on the consequence of the fault). Safe, secure, accurate, and certifiable CNS technologies that provide required capabilities are needed.

More precision aircraft navigation, coupled with the precise six-dimensional[5] guidance algorithms used in advanced flight management systems, will enable reduced spacing between aircraft operating en route and in the terminal airspace. CNS system functions must be tightly coupled in terms of information integrity, and they should allow pilots to operate cooperatively with ground systems without controllers continuously in the control loop. The CNS should transmit navigation, guidance, and other sensor data to other aircraft and ground operation centers via multichannel data links while, at essentially the same time, they receive similar information about other aircraft, the weather, airport conditions, etc. This information can prevent accidents by revealing the current and future status of other aircraft, weather

[5]The six dimensions refer to three position coordinates and three velocity vectors to define aircraft location, speed, and direction of motion.

phenomena, terrain, buildings, and vehicles on the ground at airports. This Challenge should also increase the affordability of onboard avionics to encourage aircraft owners and operators to procure more capable avionics. This Challenge encompasses the following CNS issues:

- Communications issues.
  —Fault-tolerant network connectivity and security.
  —Dynamic network control and reconfiguration.
  —Quality of service.
  —Spectrum allocation and usage.
  —Adequate communication bandwidth.
  —Required communications capability as a function of geospatial location and phase of flight.
- Navigation issues.
  —High-precision, six-dimensional estimate of aircraft state as a function of time.
  —Integration of satellite navigation with other navigation modes.
  —Navigation system capability, including reliability and quality, of input signals.
  —Functional integration of navigation system with guidance and flight control systems to ensure high-integrity, integrated control of flight during automatic and manual modes.
- Surveillance issues.
  —Capability of data links to provide accurate time-referenced data from navigation systems, guidance systems, and other sensors when interrogated by external systems or periodic broadcast.
  —Handling of multiple, simultaneous interrogations using multiple channels to provide high-integrity, secure data.
  —Processing and reacting to incoming data about other aircraft, hazardous weather, etc.
  —Continuous improvement in situational awareness through advanced sensors, communication links, and human–system interfaces.

Key milestones include

- Simulate avionics on an individual aircraft to determine the capability of each avionics function (communication, navigation, guidance, control, and surveillance).
- Demonstrate (1) fault-tolerant degradation of CNS capability (in terms of accuracy and availability of modes) and (2) processes needed to ensure that the individual aircraft can still transmit the needed aircraft state information and receive information and air traffic control commands with an extremely low probability of communication error.
- Evaluate different tracking and control algorithms with various faults that could occur in either the ATM system or airborne aircraft to determine whether the algorithms are able to detect the faults, identify them, and recover from them by reconfiguring the system in which the fault occurred as well as other systems to provide a satisfactory level of service.
- Document the feasibility of using space-based communications and surveillance as both a primary and backup means of ATM.
- Demonstrate modeling and real-time simulation using distributed control centers and different traffic levels, ranging from the current peak hourly load of about 6,000 airborne aircraft in the continental U.S. airspace to a predicted hourly load of 18,000 airborne aircraft, using current demand patterns. This effort is required to verify that the network of communication links, processing nodes in the network, and control algorithms provides the desired capacity while satisfying safety criteria.
- Demonstrate a means to provide seamless information flow between an aircraft's multiband antenna and the fiber-optic local area network that manages the information flow between aircraft systems and the radio channels.
- Demonstrate a robust IVHM system that detects permanent and transient onboard system faults and communicates system status to pilots and ground systems.
- For aircraft equipped with autothrottles, develop performance algorithms linked to aircraft dynamics to maintain the approved flight trajectory while minimizing fuel consumption. For aircraft that are not equipped with autothrottles, document the information required by the flight management system to generate speed commands to be displayed to pilots while minimizing pilot workload.
- Develop an air-ground communication protocol that (1) optimally allocates functions among pilots, avionics, air traffic controllers, and automated ground systems and (2) includes a means to alert ground systems and controllers that the data link or an onboard system has failed. This will require control algorithms that can handle multiple failures in terms of controlling the aircraft with the failures as well as adjacent traffic to minimize the impact on airspace capacity and efficiency.

Relevance to Strategic Objectives

Capacity (9): Tripling the number of aircraft in the airspace requires reducing the uncertainty of six-dimensional aircraft state (position and velocity) to less than one-third of the current required navigation precision of 0.1 nautical mile. Time-tagged state information must be broadcast so that adjacent aircraft as well as ground systems know the relative position and velocity of aircraft and each aircraft's deviation from its planned flight trajectory.

Safety and Reliability (9): This Challenge will provide fault-tolerant aircraft and ground systems that will permit safely reducing separation standards.

Efficiency and Performance (9): This Challenge will enable aircraft to fly flight trajectories that minimize fuel consumption.

Energy and the Environment (3): This Challenge will enable aircraft to fly flight trajectories that reduce noise and emissions.

Synergy with National and Homeland Security (3): This Challenge will benefit military aircraft that operate aircraft in civil airspace.

Support to Space (3): The systems developed for operation in Earth's airspace and atmosphere should have application to operations in a martian atmosphere.

Why NASA?

Supporting Infrastructure (3): NASA research centers have the engineering skills, computing and simulation facilities, and test aircraft to develop the technologies required to make advanced CNS a reality.

Mission Alignment (9): This Challenge is very relevant to NASA's mission.

Lack of Alternative Sponsors (3): Industry and federal government agencies are developing some CNS technologies, but they have lagged behind European development of some new technologies required to operate in the global airspace.

Appropriate Level of Risk (3): The individual technologies are low risk. The best methods of integrating these technologies remains to be developed and demonstrated.

D8 Human–machine integration

The ever-increasing demand for air transportation, combined with the rapid pace of technological change, poses significant challenges for effective integration of humans and automation. For the foreseeable future, humans will continue to play a central role in key decision-making tasks that directly influence the efficiency and safety of civil aviation. As technology evolves, it may be anticipated that the role of humans and the nature of their task will change accordingly. In order to maintain or improve upon existing standards of performance and safety, it is critical that the allocation of functions between humans and automation and the design of the human–machine interface be optimized based on a solid foundation of scientific principles that reflect our best understanding of human sensory, perceptual, and cognitive processes. Human–machine integration should remain an important element of NASA research directed toward civil aeronautics applications.[6] However, the emphasis should be shifted from development and testing of specific input and output devices toward more fundamental research involving modern instruments that measure brain physiology. Research should also include voice command and recognition technology, coupled with increased machine contextual understanding, to reduce workload. This will help define the future role of humans in complex, highly automated systems. Key milestones related to human–machine integration methods and tools include

- Develop improved system engineering processes and tools for determining optimum roles of humans and automation in complex systems and demonstrate the benefits of this improved methodology in a trial application. This milestone should include provisions for dynamic human–machine task allocation and monitoring of human performance by machines (e.g., automated terrain avoidance).
- Conduct fundamental research on the causes of human error and on human contributions to safety and document design guidelines that will (1) help minimize the potential for design-induced error and (2) facilitate positive human intervention in the event of system failures. Transfer these guidelines to government program offices and industry.
- Develop constructive models of human performance and decision making and validate model predictions against objective performance data acquired in high-fidelity human-in-the-loop flight simulation experiments.
- Develop and demonstrate rapid prototyping tools that enable comparative evaluations of alternative automation schemes early in system development.
- Develop and validate a technique for integrating human reliability estimates into system safety and reliability analyses.

Key milestones related to human–machine integration technologies for vehicle applications include

- Develop and test enabling technologies for pilot workload management and reduced crew operations (e.g., improved human–machine integration for a flight management system) while keeping pilot awareness at the proper level.
- Develop display concepts for maintaining operator situational awareness while monitoring highly automated processes. Demonstrate the ability of operators to rapidly and accurately intervene in the event of system failures.
- Develop technologies and/or display concepts enabling effective fusion of information from multiple sources, including real-world and synthetic imagery (i.e., augmented reality). Demonstrate the effectiveness of these concepts in practical decision support applications with varying levels of information quality and uncertainty (in terms of accuracy, timeliness, etc.).
- Develop and demonstrate technologies for machine vision (image-based object detection).

[6] J. Vagners, professor emeritus, aeronautics and astronautics, University of Washington, Presentation to Panel D on November 15, 2005.

- Develop tools and metrics to compare effectiveness of machine and human operators in see-and-avoid tasks to improve machine performance.

Relevance to Strategic Objectives

Capacity (3): This Challenge will address human performance limits that constrain overall performance of the air transportation system, such as aircraft separation, wake vortex avoidance, operations in reduced visibility, high-speed turnoffs, baggage and cargo handling, aircraft maintenance and servicing, etc.

Safety and Reliability (9): Human factors are the predominant cause of accidents and incidents in civil and military aircraft operations. Operational safety statistics show that 65 to 75 percent of mishaps are attributable to human error. This Challenge will reduce human error.

Efficiency and Performance (9): Reducing the workload of pilots and controllers would make more efficient use of human and automation resources.

Energy and the Environment (1): This Challenge has little or no impact on this Objective.

Synergies with National and Homeland Security (3): This Challenge will address issues related to command and control of complex, highly automated systems, situation awareness, and safety of flight operations, all of which are of interest to DoD. Elements of this Challenge that address information management and decision support systems (e.g., data mining, decision making under uncertainty, modeling and prediction of human behavior, etc.) are also relevant to DHS.

Support to Space (3): Elements of this Challenge directed toward ATM improvements may have applicability to ground-based control of space systems. Some research related to advanced human–machine integration devices (e.g., synthetic vision) may also be beneficial to ongoing manned spaceflight programs.

Why NASA?

Supporting Infrastructure (3): NASA has substantial capability (personnel and facilities) to do world-class research in human–machine integration. The facilities at NASA Ames and Langley Research Centers provide a unique environment for integrated, human-in-the-loop simulation of both flight deck and ATM technologies. These facilities have been specifically designed and instrumented for evaluation of advanced concepts for human–machine integration in a high-fidelity operational environment.

Mission Alignment (9): Advanced human–machine integration research is a mainstream activity for NASA Ames and Langley Research Centers and is very relevant to NASA's mission.

Lack of Alternative Sponsors (3): Some human–machine integration issues might be addressed by DoD labs, industry, or academia, if not addressed by NASA. However, research on fundamental human–machine integration principles and research directed specifically at civil aeronautics would probably be neglected without NASA leadership. The FAA historically looks to NASA to perform human–machine integration research, particularly as it relates to ATM.

Appropriate Level of Risk (9): Human–machine integration represents a broad spectrum of moderate- to high-risk technical challenges with both near- and far-term implications.

D9 Synthetic and enhanced vision systems

Synthetic and enhanced vision systems provide an out-the-window view of terrain, obstacles, and traffic. These systems can also be used as flight crew interfaces for flight trajectory and planning operations (Kelly et al., 2005). The synthetic vision systems that use databases to generate terrain and obstacles require high-fidelity, high-integrity information and a self-healing capability. Enhanced vision systems use forward-looking sensors such as infrared, radar, and laser ranging to allow the flight crew to visualize the real world when visibility is hindered. Currently, vision systems are limited by weather, human factors issues, and other issues. New sensors and improved sensor fusion are needed.

A combined synthetic and enhanced vision system has future potential as a navigation, approach, and landing sensor. The ability to "see" the airport in poor weather has the potential to reduce the likelihood of a go-around. Information fusion that exploits the capabilities of sensors and compensates for their deficiencies is needed, and the immature state of this art represents the most difficult obstacle to achieving these benefits.

Synthetic and enhanced vision systems are also intended to aid airport surface operations in poor weather, reducing runway occupancy and taxiing errors and reducing gate-to-gate travel time. Research topics of interest are as follows:

- Database integrity and quality
- Information fusion
- Object detection and avoidance
- Human–machine interface issues
- Verification of accuracy, fault tolerance, and reliability

Key milestones include

- Prepare an accurate and complete terrain and obstacles database and demonstrate real-time database monitoring and error correction.
- Develop procedures and rules for fusing image information from multiple imaging sensors as well as stored terrain data and traffic; identify common viewing parameters; and determine what role enhanced vision systems and synthetic vision systems should play in an integrated system.

- Demonstrate increased situational awareness and alerting to avoid air traffic, airport surface traffic, wires, and cables.
- Demonstrate displays that (1) eliminate image fusion artifacts that lead to misleading information and (2) present conformal information to pilots in a way that facilitates its transition to the outside world.
- Demonstrate tools for verifying database accuracy, fault tolerance, reliability, and overall system accuracy.

Relevance to Strategic Objectives

Capacity (3): This Challenge increases capacity due to better area navigation performance in the terminal area and improved surface operations, which increase capacity by compensating for the effects of bad weather and night vision constraints.

Safety and Reliability (9): Accurate synthetic and enhanced vision systems can increase safety dramatically during approach, landing, and ground operations. The intuitive information provided on advanced display is more easily understood than needles and gauges used in nonglass cockpits, especially with inexperienced pilots.

Efficiency and Performance (3): Terrain and obstacle information, combined with flight trajectory information, can result in more efficient flight paths.

Energy and the Environment (1): This Challenge has no impact on this Objective.

Synergy with National and Homeland Security (1): This Challenge does not apply to this Objective.

Support to Space (3): The technology developed for this Challenge can also be used by vehicles landing on other planets.

Why NASA?

Supporting Infrastructure (9): NASA has supported R&T relevant to this Challenge and has relevant expertise, aircraft platforms, and other test and evaluation facilities.

Mission Alignment (9): This Challenge is consistent with NASA's commitment to aviation safety.

Lack of Alternative Sponsors (3): Industry and DoD support R&T relevant to this Challenge. NASA's involvement is needed to provide overall leadership and to continue to push the envelope.

Appropriate Level of Risk (3): This Challenge faces low risk. Some commercial development has already occurred.

D10 Safe operation of unmanned aerial vehicles in the national airspace

The use of UAVs for a variety of civil applications (e.g., farming, communications relays, border monitoring, power line and pipeline monitoring, and firefighting) will continue to increase. Flight operations of military UAVs in civil airspace is also expected to increase. To facilitate these operations, UAVs should be integrated into the air transportation system. This requires them to be at least as safe as manned aircraft.

Most UAV technologies, capabilities, and processes are shared with manned aircraft and require research in several key topics, including the following four:

- *Aircraft.* Automation, system upgrade issues, and communications systems, all of which are distinct from those for manned aircraft.
- *Human–machine interaction.* Function allocation, human interface design, situational awareness, training, and required level of proficiency in the remote operation of the aircraft.
- *Maintenance and support.* In matters where UAVs differ distinctly from traditional aircraft.
- *Flight operations.* Sense- or see-and-avoid issues, person-to-person interfaces between operators and controllers, assurance of positive control of the aircraft (especially with highly automated UAVs that are not directly controlled by ground-based operators in real time), and automated contingency management.

Key milestones include

- Develop and demonstrate secure, reliable communications as well as procedures for interaction between UAVs and air traffic controllers.
- Design, develop, and demonstrate human interfaces for remote UAV operators under conditions extant in the air transportation system.
- Develop and test training programs for remote UAV operators.
- Develop and demonstrate sense-and-avoid technologies for UAVs.
- Demonstrate technologies for maintaining positive control of UAVs under adverse conditions.
- Develop and demonstrate automated contingency management for control of UAVs.

Relevance to Strategic Objectives

Capacity (3): This Challenge will enable more UAV flight operations in civil airspace. There is little impact on the movement of people but might be a significant impact on freight movement in the future.

Safety and Reliability (9): This Challenge is essential for safe and reliable operation of UAVs in civil airspace, both controlled and uncontrolled.

Efficiency and Performance (3): It takes months to obtain a waiver to allow UAV operations in civil airspace. This Challenge should help alleviate this situation and facilitate efficient commercial UAV operations.

Energy and the Environment (1): This Challenge has little or no impact on this Objective.

Synergies with National and Homeland Security (9): The ability to routinely operate UAVs in civil airspace will enhance national defense and homeland security operations.

Support to Space (1): This Challenge has little application to this Objective, although some synergies may exist with regard to automation and communications for spacecraft vehicle control.

Why NASA?

Supporting Infrastructure (3): NASA has more experience than any other entity in the test and evaluation of UAV systems.

Mission Alignment (9): This Challenge is very relevant to NASA's mission.

Lack of Alternative Sponsors (3): Industry is addressing DoD-related issues regarding airframes, operators, and maintenance. The commercial and government applications arena is hindered by operational issues that NASA can effectively address.

Appropriate Level of Risk (3): This Challenge faces low risk.

D11 Secure network-centric avionics architectures and systems to provide low-cost, efficient, fault-tolerant, onboard communications systems for data link and data transfer

As NASA moves into the network-centric vision of the future, data link assurance will become increasingly important. Threats to the integrity of information can be grouped into two categories: natural threats and malicious threats. Natural threats are associated with unintended system failures and include hardware and software flaws, lightning strikes, cosmic rays, and human error. Malicious threats are intelligent directed attacks. Historically, the former have posed the greater threat. However, as future aircraft become more network-centric, a new class of malicious threats could become increasingly destructive.

Data must be considered a valuable and critically important asset. Data loss and corrupt data can cause significant problems, especially in cooperative networked systems. In addition, data separation is required to protect International Traffic in Arms Regulations (ITAR) data from unauthorized disclosure. Considerations such as ITAR are increasing the need for avionics architectures that allow different nodes to communicate with each other over encrypted data links at different levels of security. Traditional approaches for assuring the security of networked information, such as the Transmission Control Protocol (TCP), have problems in high-latency environments such as deep space and parts of the air transportation system. DARPA has been developing a trusted key distribution mechanism, but it is unknown if this mechanism has similar problems with operating in such an environment. It would allow for NASA to address the needs of both legacy and new systems to function in a network-centric manner. Key milestones include

- Develop fundamental system requirements, architecture, and system logic that are compatible with current and future network-centric system requirements.
- Complete fundamental research necessary to develop and incorporate encryption techniques, threat detection, and counterthreat strategies.
- Develop simulation capabilities for evaluation and demonstration of certain high-performing strategies in the execution of realistic system architectures and applications.
- Develop requirements flowdown to all affected aircraft systems, such as advanced CNS.
- Develop requirements and methodology to support verification, validation, and certification of future systems incorporating this technology.

Relevance to Strategic Objectives

Capacity (9): This Challenge will increase capacity by developing key technologies and capabilities related to secure network communications, cooperative distributed networking systems, networked weather systems, and the like.

Safety and Reliability (9): Unreliable and insecure data links increase the likelihood of catastrophic failure. Secure data links prevent malicious and accidental corruption of data.

Efficiency and Performance (9): Improved communications security allows full exploitation of available communication infrastructure.

Energy and the Environment (1): This Challenge has little or no impact on this Objective.

Synergy with National and Homeland Security (9): Reliable and secure data links are required for national and homeland security to prevent both malicious and/or accidental corruption of data, especially with missions involving UAVs.

Support to Space (3): Reliable and secure data links are required for space missions to prevent both malicious and/or accidental corruption of data.

Why NASA?

Supporting Infrastructure (3): NASA has strong credentials in avionics architectures, network architectures, and encryption technology. NASA also has domain expertise to evaluate the applicability of these technologies.

Mission Alignment (3): This research will directly support multiple R&T Challenges and benefit military and civil aviation, as well as future applications in space.

Lack of Alternative Sponsors (1): Industry and DoD labs are developing technology related to this Challenge, but it is predominately focused on military systems. The commercial information technology business sector is very active in commercial encryption and networking technologies.

Appropriate Level of Risk (3): Near-term research can address issues related to this Challenge, and the results can be transferred to future civil and military applications.

D12 Smaller, lighter, and less expensive avionics

Today's commercial and military aircraft have benefited from a modest size, weight, and cost savings as a result of integrated avionics systems and fly-by-wire flight control systems, smaller antennas, smaller sensors, and digital data buses.

The expansion of smaller and lighter aircraft (in particular UAVs) resulted in development of significantly smaller and lighter avionics. While not common, entire systems weighing less than 1 pound have demonstrated that basic avionic functionality (navigation, communication, and autoflight) can be fit into a package (or set of packages) that are much smaller than and very different from conventional commercial and military avionics.

To maximize the efficiency and performance of future commercial and military aircraft, new technology is needed that significantly reduces the cost, size, and weight of current avionics as well as their supporting installation infrastructure (by minimizing or eliminating equipment mounting hardware and aircraft wiring). Key milestones include

- Demonstrate technologies that provide wireless onboard communications.
- Demonstrate methods to reduce processing requirements and power requirements.
- Demonstrate methods to improve avionics system capability, integrity, and reliability using low-cost components/sensors, including greater use of commercial off-the-shelf (COTS) components.
- Demonstrate greater use of microcontrollers as main processors and in distributed processing.
- Document common data standards that have broad application and usage.

Relevance to Strategic Objectives

Capacity (1): This Challenge could prompt some aircraft owners (especially in general aviation) to install more avionics, which would increase aircraft capability and indirectly increase capacity, although this effect would likely be small.

Safety and Reliability (3): Low-cost avionics could prompt some aircraft owners (especially in general aviation) to install more avionics, which would increase safety. However, the incorporation of COTS technologies into modern avionics systems has met greater scrutiny from regulatory agencies than ever before. The industry needs an affordable means to improve the performance, quality, and safety of increasingly complex avionics systems. These advanced avionics should come with or enable inherently fewer failure modes.

Efficiency and Performance (9): One pound of dead weight on a commercial transport increases fuel costs on the order of $100 per aircraft per year. Transport aircraft typically carry 200 pounds of avionics, not counting associated wiring. Smaller aircraft have a larger proportion of their weight in avionics, so the savings in fuel costs should be a larger proportion of total fuel costs. In addition, ultralight aircraft can have longer mission times if they can carry less weight. Common standards that permit cross-product strategies, such as wireless data protocols and transfer media reduce costs. ARINC standards are good examples of common protocols, but generally they focus on airline applications.

Energy and the Environment (3): Avionics weight savings reduce fuel consumption and will have a positive effect on the environment.

Synergies with National and Homeland Security (3): Technologies that enable weight and size savings, specifically wireless technologies that interconnect onboard systems and components, will need to be sufficiently secure and reliable to maintain aircraft safety. New systems that meet these needs may result in by-products that can be used in ground-based vehicles and fixed-base stations.

Support to Space (9): Smaller and lighter avionics are enablers for future space flight, especially for missions that would benefit from significantly reduced avionics power requirements.

Why NASA?

Supporting Infrastructure (3): The Jet Propulsion Laboratory (JPL) conducts relevant research, but its focus is space-based. This Challenge would use technologies developed for space to support aeronautics.

Mission Alignment (3): This Challenge will benefit the complete civil aviation community and have application in future space travel.

Lack of Alternative Sponsors (3): Industry is developing smaller and less expensive avionics, but they are geared toward unique applications like UAVs. In general, manufacturers of commercial and military aircraft are slow to make revolutionary changes in technology without a clearly understood business case or a clear assessment of risk; regulatory constraints are a major factor.

Appropriate Level of Risk (3): This Challenge faces low risk.

D13 More efficient certification processes for complex systems

Certification of aircraft and aircraft systems has focused on airworthiness using process-based standards. Products for which the processes have been followed are assumed to have acceptable quality. Many of the required processes in the standards, however, are expensive and time-consuming. In addition, following a good process does not necessarily imply the product is safe.

An alternative approach to process-based certification is to certify the product itself. Product-based certification and more efficient ways to do process-based certification could greatly reduce the time and cost of creating aircraft systems while potentially increasing safety. How to do such certification, particularly for software-intensive systems, is unknown, and research that could provide a valid and demonstrated basis for product-based certification could increase quality assurance while decreasing costs and could significantly influence industry standards and FAA advisory circulars.

Recently, a third, performance-based approach to specifying certification requirements has been applied to some systems. In this approach, a required performance level is specified rather than the process required to produce it or ways to evaluate the product. An example is required navigation performance (RNP), in which the performance level of certain navigational systems is specified. R&T is still needed, however, to demonstrate that the system will satisfy the performance requirement.

A final problem occurs when completely new technologies or procedures, such as reduced aircraft separation, are introduced in the air transportation system. Changes foreseen as necessary to transform the system and to solve critical problems in capacity involve significant new technology and operational procedures. Assurance of safety in past systems has relied on making few major changes and relying on historical data and experience, which will not be available when major changes are implemented over a short period of time. New, revolutionary approaches will be required to provide the necessary level of confidence in these new systems. Key milestones include

- Validate that following specific processes will produce required assurance levels.
- Demonstrate processes for certifying products such as software, where testing is unable to provide required levels of confidence.
- Demonstrate approaches to assuring that required performance levels will be achieved in complex systems, despite failures or environmental disturbances.
- Demonstrate approaches to ensuring the safety of proposed changes to the national airspace system for which historical data and prior experience are not available.

Relevance to Strategic Objectives

Capacity (3): New enhancements to the air transportation system that increase capacity will need to be certified and shown to be safe.

Safety and Reliability (9): Because the goal of certification is to ensure safety, more efficient and effective certification approaches could have a major impact on safety.

Efficiency and Performance (9): More efficient certification processes should reduce the number of resources required to develop and certify new aircraft capabilities, systems, and products.

Energy and the Environment (1): This Challenge has no impact on this Objective.

Synergies with National and Homeland Security (1): This Challenge has no impact on this Objective.

Support to Space (3): Some of the safety and reliability technologies developed by this Challenge will apply to space systems.

Why NASA?

Supporting Infrastructure (3): Some NASA infrastructure could be used in the evaluation of new certification approaches.

Mission Alignment (1): Certification is the responsibility of the FAA and the aircraft manufacturers, not NASA. NASA has capabilities, however, that could effectively help the FAA and industry address certification issues.

Lack of Alternative Sponsors (1): The FAA and other certification agencies have the responsibility for certification and thus interest in improvement. Aircraft and aircraft system manufacturers have great interest in this topic.

Appropriate Level of Risk (3): New certification approaches (including capability-based approaches) are feasible and are being recommended and used (e.g., RNP), so this Challenge faces low risk, but more research is needed into the effectiveness of the new and proposed approaches and how to implement them.

D14 Design, development, and upgrade processes for complex, software-intensive systems, including tools for design, development, and validation and verification

The introduction of software and digital components has allowed the development of increasingly complex systems but at the same time has required changes and additions to basic engineering approaches and methodologies. Basic research in new tools and techniques is needed for designing and testing software-intensive systems and for maintaining and upgrading them over time. For example, most software cannot be exhaustively tested, and the verification difficulties become even greater when nondeterministic artificial intelligence techniques are employed. There is a need for new ways to provide assurance, particularly for critical systems.

One technology relevant to this Challenge is model-based development, whereby models that can be executed and analyzed are constructed prior to system implementation and construction. Model-based development potentially can augment the detection of conceptual design errors early in development, when they are much less expensive to correct; decrease development time and risk; and allow for greater reuse of system engineering effort. Research should strive to improve analysis and specification of design rationale, visualization and simulation tools, and so forth. Key milestones include

- Develop easily used and reviewed modeling languages.
- Develop automated design and code generation.

Relevance to Strategic Objectives

Capacity (3): The development of new aircraft capabilities is increasingly being driven by the requirement to develop and upgrade software. Software development is a component of most approaches to increasing capacity, but it is not the most critical component.

Safety and Reliability (9): Because the behavior of complex systems is increasingly controlled by software, software can have a significant impact on safety and reliability.

Efficiency and Performance (3): Software costs are a driving factor in design and development. Improving design, development, and upgrade processes would increase efficiency and performance.

Energy and the Environment (1): This Challenge has relatively little effect on energy use and the environment, and the effect is indirect.

Synergies with National and Homeland Security (1): Software-intensive systems might be used to implement some new technologies, but the benefit to DoD and DHS would be indirect.

Support to Space (3): Some of the R&T relevant to this Challenge would apply to space control systems, but design requirements are very different.

Why NASA?

Supporting Infrastructure (1): NASA has some capable researchers in fields related to this Challenge, but NASA has outsourced most R&T relevant to this Challenge.

Mission Alignment (3): This Challenge is not well aligned with NASA's mission. Relevant tools and methodologies are applicable to any complex system, not just aerospace.

Lack of Alternative Sponsors (1): Industry, DoD, and academia are developing many tools relevant to this Challenge.

Appropriate Level of Risk (1): This Challenge faces very low risk. Many relevant tools and techniques exist and much of the basic research has been done.

REFERENCES

Bokadia, S., and J. Valasek. 2001. Severe Weather Avoidance Using Informed Heuristic Search. AIAA-2001-4232, AIAA Guidance, Navigation, and Control Conference, Montreal, Canada, August 6-9.

Chavez, F.R., and D.K. Schmidt. 1994. Analytical aeropropulsive-aeroelastic hypersonic-vehicle model with dynamic analysis. Journal of Guidance, Control, and Dynamics 17(6). 1308-1319.

Ding, Y., J. Rong, and J. Valasek. 2004. Feasibility Analysis of Aircraft Landing Scheduling for Non-Controlled Airports. AIAA-2004-5241, Proceedings of the AIAA Guidance, Navigation, and Control Conference, Providence, R.I., August 16-19.

Doebbler, J., P. Gesting, and J. Valasek. 2005. Real-Time Path Planning and Terrain Obstacle Avoidance for Aircraft. AIAA-2005-5825, Proceedings of the AIAA Guidance, Navigation, and Control Conference, San Francisco, Calif., August 15-18.

Garg, S. 2005. NASA Glenn Research in Controls and Diagnostics for Intelligent Aerospace Propulsion Systems. NASA/TM—2005-214036, Glenn Research Center, Cleveland, Ohio, December.

Glauert, M.B. 1945. The design of suction aerofoils with very large C_L-range. Aeronautics Research Council Reports and Memoranda 2111, National Research Establishment, November.

Glauert, M.B., W.S. Walker, W.G. Raymer, and N. Gregory. 1948. Wind-tunnel tests on a thick suction aerofoil with a single slot. Aeronautics Research Council Reports and Memoranda 2646, National Research Establishment, October.

Helbing, K., T. Spaeth, and J. Valasek. Forthcoming. Improving aircraft sequencing and separation at a small aircraft transportation system airport. Journal of Aircraft.

Jaw, L.C., and S. Garg. 2005. Propulsion Control Technology Development in the United States: A Historical Perspective. NASA/TM-2005-213978, October.

Kelly, W., J. Valasek, D.Wilt, J. Deaton, K. Alter, and R. Davis. 2005. The Design and Evaluation of a Traffic Situation Display for a SATS Self Controlled Area. Proceedings of the 24th Digital Avionics Systems Conference, Washington, D.C., October 30-November 3.

Lampton, A., and J. Valasek. 2005. Prediction of Icing Effects on the Stability and Control of Light Airplanes. AIAA-2005-6219, Proceedings of the AIAA Atmospheric Flight Mechanics Conference, San Francisco, Calif., August 15-18.

Lampton, A., and J. Valasek. 2006. Prediction of Icing Effects on the Lateral/Directional Stability and Control of Light Airplanes. AIAA-2006-6487, Proceedings of the AIAA Atmospheric Flight Mechanics Conference, Keystone, Colo., August 21-24.

Litt, J.S., D.L. Simon, S. Garg, T. Guo, C. Mercer, R. Millar, A. Behbahani, A. Bajwa, and D.T. Jensen. 2005. A Survey of Intelligent Control and Health Management Technologies for Aircraft Propulsion Systems. NASA/TM-2005-213622, ARL-TR-3413, May.

Rong, J., S. Geng, J. Valasek, and T. Ioerger. 2002. Air Traffic Conflict Negotiation and Resolution Using An Onboard Multi-Agent System. DASC-345, 21st Digital Avionics Systems Conference (DASC) on Air Traffic Management for Commercial and Military Systems, Irvine, Calif., October 22.

Schierman, J.D., D.G. Ward, J.R. Hull, N. Gandhi, M.W. Oppenheimer, and D.B. Doman. 2004. An approach to integrated adaptive guidance and control with flight test results. Journal of Guidance, Control and Dynamics 27(6).

Tandale, M.D., and J. Valasek. 2003. Structured Adaptive Model Inversion Control to Simultaneously Handle Actuator Failure and Actuator Saturation. AIAA-2003-5325, AIAA Guidance, Navigation, and Control Conference, Austin, Tex., August 11-14.

Tandale, M.D., and J. Valasek. 2006. Fault tolerant structured model inversion control. Journal of Guidance, Control, and Dynamics 29(3).

Tandale, M.D., J. Valasek, J. Doebbler, and A.J. Meade. 2005. Improved Adaptive-Reinforcement Learning Control for Morphing Unmanned Air Vehicles. AIAA-2005-7159, Proceedings of the AIAA Infotech@Aerospace Conference, Arlington, Va., September 26-29.

Valasek, J., M.D. Tandale, and J. Rong. 2005. A reinforcement learning-adaptive control architecture for morphing. Journal of Aerospace Computing, Information, and Communication 2(5): 174-195.

E

R&T Challenges for Intelligent and Autonomous Systems, Operations and Decision Making, Human Integrated Systems, and Networking and Communications

A total of 20 R&T Challenges were prioritized in the Area of intelligent and autonomous systems, operations and decision making, human integrated systems, and networking and communications. Table E-1 shows the results. The R&T Challenges are listed in order of NASA priority. National priority scores are also shown.[1] This appendix contains a description of each R&T Challenge, including milestones and an item-by-item justification for each score that appears in Table E-1.[2]

E1 Methodologies, tools, and simulation and modeling capabilities to design and evaluate complex interactive systems

The U.S. air transportation system is a complex interactive system whose behavior is difficult to simulate with currently available models. Methodologies, tools, and simulation and modeling capabilities suited for the design and integration of complex interactive systems are needed to understand the air transportation system as an integrated, adaptive, distributed system that includes aircraft, ATM facilities, and airports, each with its own complex systems, all of which interact with one another, the environment, and human operators. Simulations and models for complex interactive systems are needed to accurately estimate system performance, to properly allocate resources, and to select appropriate design parameters. Additionally, the large number of possible future system designs requires models that can be reconfigured to model a wide range of design parameters.

One key barrier to developing integrated aviation systems is the lack of basic research that regulators can use to develop new certification standards and testing methodologies. Tools and methodologies that can assess the reliability and effectiveness of complex, nondeterministic, software-intensive future systems need to be developed. In some cases, this will also require changes to FAA regulations and certification standards (Aerospace Commission, 2002, pp. 2-9). This Challenge will help ensure that the right architecture and design decisions can be made in developing the air transportation system of the future. Key milestones include

- Demonstrate methodologies and tools for the design, test, and certification of a flexible, robust, safe air transportation system that is readily adaptable to changing operational paradigms suited to new and different vehicles, including unmanned air vehicles (UAVs), very light jets (VLJs), and spacecraft operating in civil airspace; communications, navigation, and surveillance capabilities; and optimization techniques.
- Demonstrate a flexible ATM model that incorporates the performance characteristics and limitations of the wide mix of present and future aircraft arriving, departing, and operating within airspace surrounding major hub airports. This model should be capable of analyzing the impacts of (1) aircraft mix and (2) operator and controller decision making and actions on system efficiency and capacity.
- Demonstrate the ability of an enhanced version of the model to assess the impact of regional weather phenomena, such as convective activity, snow, and high winds.
- Demonstrate the capability to test and certify nondeterministic systems.
- Demonstrate the ability of an enhanced version of the ATM model to assess impacts of aircraft mix and operator and controller decision making.

Relevance to Strategic Objectives

Capacity (9): The capacity of the air transportation system must double or triple over the next 20 years to keep up

[1]The prioritization process is described in Chapter 2.

[2]The technical descriptions for the first 10 Challenges listed below are the same as the technical descriptions for these Challenges as they appear in Chapter 3.

TABLE E-1 Prioritization of R&T Challenges for Area E: Intelligent and Autonomous Systems, Operations and Decision Making, Human Integrated Systems, and Networking and Communications

		Strategic Objective							Why NASA?					
R&T Challenge		Capacity	Safety and Reliability	Efficiency and Performance	Energy and the Environment	Synergies with Security	Support to Space	National Priority	Supporting Infrastructure	Mission Alignment	Lack of Alternative Sponsors	Appropriate Level of Risk	Why NASA Composite Score	NASA Priority Score
	Weight	5		3		1			$^1/_4$ each					
E1	Methodologies, tools, and simulation and modeling capabilities to design and evaluate complex interactive systems	9	9	9	9	9	3	**156**	3	9	3	9	**6.0**	**936**
E2	New concepts and methods of separating, spacing, and sequencing aircraft	9	9	9	3	3	1	**130**	3	9	3	9	**6.0**	**780**
E3	Appropriate roles of humans and automated systems for separation assurance, including the feasibility and merits of highly automated separation assurance systems	9	9	9	1	3	1	**124**	3	9	3	9	**6.0**	**744**
E4	Affordable new sensors, system technologies, and procedures to improve the prediction and measurement of wake turbulence	9	9	3	1	1	1	**104**	3	9	3	9	**6.0**	**624**
E5	Interfaces that ensure effective information sharing and coordination among ground-based and airborne human and machine agents	3	9	9	1	9	3	**102**	3	9	3	9	**6.0**	**612**
E6	Vulnerability analysis as an integral element in the architecture design and simulations of the air transportation system	3	9	9	1	9	1	**100**	3	9	3	9	**6.0**	**600**
E7	Adaptive ATM techniques to minimize the impact of weather by taking better advantage of improved probabilistic forecasts	9	3	9	3	1	1	**98**	3	9	3	9	**6.0**	**588**
E8a	Transparent and collaborative decision support systems	3	9	9	1	3	3	**96**	3	9	3	9	**6.0**	**576**
E8b	Using operational and maintenance data to assess leading indicators of safety	3	9	9	1	3	3	**96**	3	9	3	9	**6.0**	**576**
E8c	Interfaces and procedures that support human operators in effective task and attention management	3	9	9	1	3	3	**96**	3	9	3	9	**6.0**	**576**
E11	Automated systems and dynamic strategies to facilitate allocation of airspace and airport resources	9	3	9	3	3	1	**100**	3	9	1	9	**5.5**	**550**
E12	Autonomous flight monitoring of manned and unmanned aircraft	3	9	3	1	9	1	**82**	3	9	3	9	**6.0**	**492**
E13	Feasibility of deploying an affordable broad-area, precision-navigation capability compatible with international standards	9	9	3	1	3	1	**106**	3	3	3	9	**4.5**	**477**
E14	Advanced spacecraft weather imagery and aircraft data for more accurate forecasts	3	3	9	3	1	1	**68**	3	9	3	9	**6.0**	**408**
E15	Technologies to enable refuse-to-crash and emergency autoland systems	1	9	1	1	3	1	**60**	3	9	3	9	**6.0**	**360**
E16	Appropriate metrics to facilitate analysis and design of the current and future air transportation system and operating concepts	3	3	9	3	3	1	**70**	3	9	3	3	**4.5**	**315**
E17	Change management techniques applicable to the U.S. air transportation system	9	9	9	1	3	1	**124**	1	3	3	3	**2.5**	**310**
E18	Certifiable information-sharing protocols that enable exchange of contextual information and coordination of intent and activity among automated systems	3	1	9	1	9	1	**60**	3	9	3	3	**4.5**	**270**
E19	Provably correct protocols for fault-tolerant aviation communications systems	3	9	3	1	3	1	**76**	3	3	1	1	**2.0**	**152**
E20	Comprehensive models and standards for designing and certifying aviation networking and communications systems	3	9	3	1	1	1	**74**	3	3	1	1	**2.0**	**148**

with demand. The capacity of the air transportation system is not the sum of the capacities of system components, because interactions among components are complex and interactions among system components may not be synergistic. The most effective way to estimate the capacity of the many options for the future U.S. air transportation system, and thereby identify the option that best meets future capacity needs, is to use system-of-system models that capture the functional relationships within the air transportation system.

Safety and Reliability (9): Models suited to complex interactive systems are needed to understand the complex behavior of the air transportation system and to quantify safety for the current system and proposed changes to the system. These models would provide a unique and vital capability to identify safety issues, including unintended consequences, especially for radically different system configurations.

Efficiency and Performance (9): Significant increases in efficiency will also be required to satisfy projected increases in demand. Models suited to complex interactive systems are needed to understand the behavior of the air transportation system and to quantify levels of efficiency and performance for the current system and proposed changes to the system.

Energy and the Environment (9): Environment considerations (noise and emissions) are limiting the growth in air transportation. Thus, increased demand will be satisfied only if there is a good understanding of the magnitude and location of environmental impacts. Models suited to complex interactive systems would help design an air transportation system with improved performance in terms of energy and the environment.

Synergies with National and Homeland Security (9): The DoD uses simulations of complex interactive systems to evaluate battlefield strategy and tactics. Thus, there are potential synergies in terms of using the simulation techniques that the DoD has developed to help evaluate commercial and private operations in the air transportation system. Additionally, the DoD and DHS would be interested in this Challenge because safety is also important for their operations, and the ability to distinguish between a component failure and an attack is predicated on the ability to predict and model failure modes.

Support to Space (3): This Challenge would facilitate space launch operations through civil airspace.

Why NASA?

Supporting Infrastructure (3): NASA has highly capable facilities (such as the Future Flight Central simulator and easily configurable full-motion cockpit simulators that can be integrated with other simulation facilities) that contribute to meeting this Challenge. The DoD, FAA, academia, and industry also have facilities and expertise that would help meet this Challenge.

Mission Alignment (9): This Challenge would directly contribute to the usefulness, performance, speed, safety, and efficiency of aircraft and the air transportation system, and it encompasses the type of long-term research that must occur before industry and operational agencies begin to develop specific components or synthesize components into system prototypes. Thus, this Challenge is well aligned with the NASA mission.

Lack of Alternative Sponsors (3): The capabilities that this Challenge would provide are essential. The DoD is sponsoring related research, but it is not focused on civil aviation applications.

Appropriate Level of Risk (9): This Challenge involves moderate risk.

E2 New concepts and methods of separating, spacing, and sequencing aircraft

Expected growth in the demand for air transportation will require efficient, denser en route and terminal area operations. This necessitates procedures that reduce minimum spacing requirements during all phases of flight and in all weather conditions, through an integrated approach that leverages a suite of emerging technologies such as required navigation performance and automatic dependent surveillance broadcast (ADS-B). The objective of this Challenge is to efficiently accommodate a large number and wide range of aircraft, including UAVs, through spacing and sequencing based on aircraft type and equipment rather than a common worst-case standard. Several concepts of operation should be systematically compared in terms of their technological, business, and human factors issues as well as their impact on capacity, safety, and the environment. This Challenge will study reduced separation operations within the context of existing ATM protocols and revolutionary paradigms that could significantly increase capacity, although the latter would involve a much more complicated transition process.

Integration of UAVs into the air transportation system will require procedures that can safely manage aircraft with diverse performance characteristics and highly automated onboard flight management systems (Sabatini, 2006). Safe, high-capacity operations in a complex future airspace environment will require fundamental research into alternative ATM paradigms such as simultaneous noninterfering operations (Xue and Atkins, 2006) in which general aviation, rotorcraft, and UAV traffic are threaded through airspace unused by commercial air traffic. As onboard automation and cooperative control algorithms are matured (McLain and Beard, 2005), UAV traffic might also be efficiently managed using formations of UAVs that are coordinated locally but treated as a single entity by air traffic controllers and pilots of nearby aircraft. Key milestones include

- Demonstrate high-efficiency airspace and airway structures that can be effectively managed and understood.

- Design and evaluate separation, spacing, and sequencing procedures for UAVs operating in civilian airspace and assess their impact on commercial aircraft capacity and safety.
- Extend models and simulation tools to enable accurate evaluation of emerging technologies (e.g., ADS-B) in all weather conditions and during all phases of flight.
- Complete an in-depth examination of the ability of concepts such as runway-independent aircraft and UAV formations or swarms to safely increase capacity and accommodate nontraditional aircraft operations.
- Demonstrate advanced, autonomous collision avoidance technologies and protocols.

Relevance to Strategic Objectives

Capacity (9): New methods for managing separation, spacing, and sequencing are key enablers to increase capacity in both en route and terminal area airspace.

Safety and Reliability (9): The air transportation system can safely manage growth only through a fundamental understanding of system behaviors and associated constraints. To be certified, new separation, spacing, and sequencing methods and associated procedures must prove they provide levels of safety that are comparable to or better than current operational procedures, even with much higher traffic density.

Efficiency and Performance (9): The current air transportation system operates near saturation, resulting in significant travel delays during periods of increased demand or adverse weather conditions. Alternative spacing and sequencing methods are of paramount importance to alleviate these delays, resulting in more efficient travel despite system growth.

Energy and the Environment (3): New methods to more efficiently space and sequence traffic will have a modest impact on energy use and the environment through reduced holding times and less circuitous flight paths.

Synergies with National and Homeland Security (3): Although primarily directed toward transport operations, new methods for traffic separation, spacing, and sequencing must also take into account surveillance and UAV traffic.

Support to Space (1): Given the relative infrequency of space launches, traffic density is not a factor. Therefore, this Challenge would have little or no relevance to the space program.

Why NASA?

Supporting Infrastructure (3): NASA has established and maintained a research group that studies alternative traffic management concepts in en route and terminal airspace. The FAA and industry also possess relevant expertise and facilities.

Mission Alignment (9): Research associated with this Challenge will greatly benefit the aeronautics community and broaden our fundamental understanding of highly dense flight operations. This Challenge would benefit the air transportation system in general and general aviation in particular.

Lack of Alternative Sponsors (3): NASA has a respected ATM research program capable of pursuing this Challenge. The FAA and private industry also are capable of performing related research—and leveraging work being done in Europe—but it is unclear that sufficient resources will be available from non-NASA sources, especially with regard to the development of revolutionary concepts with long-term application.

Appropriate Level of Risk (9): The development and validation of new methods capable of having a significant impact on capacity will require substantial research, but the goal can be attained with reasonable effort.

E3 Appropriate roles of humans and automated systems for separation assurance, including the feasibility and merits of highly automated separation assurance systems

Air traffic control is currently a labor-intensive process. FAA controllers—aided by radar, weather displays, and procedures—maintain traffic flow and assure separation by communicating instructions to aircraft in their sector of responsibility. Limitations to this traditional paradigm are, in some areas, constraining the capacity of the air transportation system. For example, the FAA required airlines serving the Chicago O'Hare airport to reduce some of their flights during 2005 because of congestion-related delays. A recent study of en route sector congestion suggested that capacity could be increased by a factor of two or more while maintaining existing spacing, by developing new systems that merge human and computer decision making and automate time-critical separation assurance tasks (Andrews et al., 2005).

Initiatives to reduce aircraft separation by providing automated advisories to air traffic controllers and flight crews have not lived up to expectations, because of controller workload concerns, institutional resistance, and other factors. The advent of UAVs has caused additional concern because it may not be feasible for UAVs with human-in-the-loop collision avoidance schemes to act in time to prevent midair collisions. This has led to interest in determining whether automating aircraft separation, whereby the controller is neither in the loop nor responsible for separation, is feasible and desirable. However, changing the role of the controller from tactical separation to traffic flow management and trusting automated systems to manage the tactical separation of aircraft would require resolution of major human factors, safety, and institutional issues (Wickens et al., 1998; Woods and Hollnagel, 2006). Collisions could occur if a UAV fails to respond or the automated traffic separation system fails and if human intervention is not effective. This Challenge would determine the appropriate roles of humans and automated systems to assure separation in high-density airspace during nominal and off-nominal operations. As part of this challenge, NASA should assess the feasibility and merits of highly automated separation assurance systems. Key milestones include

- Complete basic research necessary to determine the most appropriate separation assurance roles for humans and automation, for ground-centered and aircraft-centered designs.
- Complete the development of the NASA Ames Advanced Airspace Concept, an automated ground-based separation assurance system, for the en route domain.
- Determine how humans interact with the Advanced Airspace Concept and other automation designs.
- Determine how the Advanced Airspace Concept and other designs respond to air and/or ground automation failures, or when the flight crew fails to respond to automated directives.
- Develop an adaptation of the Advanced Airspace Concept or other designs for UAVs, and determine its performance.
- Determine through analysis and simulation the safety of the Advanced Airspace Concept and other designs.

Relevance to Strategic Objectives

Capacity (9): Recent research indicates that automated separation methods could enable significant en route and terminal capacity growth.

Safety and Reliability (9): The safe use of automated separation must be assured through independent monitoring. Automated separation will reduce traffic delays if it can better adapt to disruptions caused by adverse weather and congestion.

Efficiency and Performance (9): Automated separation could improve system efficiency and performance by reducing spacing requirements.

Energy and the Environment (1): Automation may reduce unnecessary variations in flight times due to holding and thus reduce fuel consumption.

Synergies with National and Homeland Security (3): The advent of automated separation may improve the ability to detect aircraft that deviate from approved flight paths because of terrorist activity or pilot error.

Support to Space (1): This Challenge is likely to have little relevance to the space program.

Why NASA?

Supporting Infrastructure (3): NASA, the FAA, and industry have the facilities and expertise to conduct research on automated separation.

Mission Alignment (9): This Challenge would have a broad benefit for aeronautics in general and civil aviation in particular.

Lack of Alternative Sponsors (3): Research related to this Challenge (by the FAA and foreign research agencies) would likely proceed even without NASA support, though it would be significantly diminished.

Appropriate Level of Risk (9): Developing fully automated separation systems would be very challenging, but a concerted effort is likely to produce worthwhile results and facilitate increases in system capacity and safety.

E4 Affordable new sensors, system technologies, and procedures to improve the prediction and measurement of wake turbulence

Existing wake vortex separation standards reduce system capacity during takeoff and landing operations and instrument approaches. Encounters with a wake vortex are also a growing concern in en route Reduced Vertical Separation Minima (RVSM) airspace (Reynolds and Hansman, 2001).[3]

Current research by the FAA and NASA is focused on procedural enhancements that take advantage of wake transport by winds (Mundra, 2001). For example, the capacity of San Francisco International Airport is expected to improve by using this approach to enable arrivals on both closely spaced parallel runways during low visibility weather. However, the relaxation of in-trail wake separation standards awaits improved measurement and prediction of wake behavior.

Existing sensors and models do not adequately characterize wake decay phenomena, especially at typical final approach altitudes. Improved sensors, including coherent pulsed lidars, capable of directly measuring wake rotational momentum, are needed to support phenomenological studies and enable more accurate predictions of wake magnitude and decay in various atmospheric conditions. Those predictions, combined with models of aircraft upset risk, should allow reduced wake separation standards without degrading safety.

R&T Challenge A10 will conduct research to improve techniques for predicting and measuring the formation, trajectory, and decay of vortices, including methods to accurately predict wingtip vortex formation and define changes in aircraft design to mitigate the strength of the vortices. This Challenge would complement that work by developing affordable new sensors, system technologies, and procedures to improve prediction and measurement of wake strength, location, motion, and aircraft upset risk in terminal and en route airspace. Together, Challenges A10 and E4 will enable safe flight with reduced in-trail wake separation. Key milestones include

[3]Reduced Vertical Separation Minima apply to the airspace from flight levels 290 to 410 (which is equivalent to altitudes of approximately 29,000 feet to 41,000 feet) and create twice as many usable flight levels, decreasing the vertical separation between aircraft from 2,000 to 1,000 feet. While increasing capacity, this also could exacerbate the effects of wake turbulence.

- Demonstrate new sensors, including a scientific, coherent lidar capable of accurate wake velocity strength measurements.
- Conduct phenomenological studies of wake behavior supported by field experiments using ground-based sensor(s) that measure wake decay and atmospheric conditions at altitudes up to 8,000 feet above the ground.
- Determine aircraft upset risks from wake vortices encounters, taking advantage of existing models and enhancing them where needed with field data.
- Demonstrate procedures, monitoring equipment, and other systems to safely reduce wake separation.
- Demonstrate an airborne means to sense and quantify the intensity of hazardous wakes en route in time for aircraft to evade them.

Relevance to Strategic Objectives

Capacity (9): Reduced wake vortex spacing requirements near airports would typically decrease arrival and departure aircraft spacing.

Safety and Reliability (9): Improved wake detection and avoidance systems and procedures will improve the ability to avoid accidents associated with wake vortices, even with higher traffic density.

Efficiency and Performance (3): Aircraft arrival and departure rates will be improved if wake vortex spacing can be safely reduced.

Energy and the Environment (1): Reductions in arrival and departure spacing may reduce airport and airborne congestion and could provide some reduction in fuel burn; however, the amount is difficult to calculate due to the lack of empirical data.

Synergies with National and Homeland Security (1): This Challenge would have little relevance to national and homeland security.

Support to Space (1): This Challenge would have little relevance to the space program.

Why NASA?

Supporting Infrastructure (3): NASA is currently leading a joint program with the FAA in wake vortex procedures and technologies. NASA has also sponsored research in wake upset risks.

Mission Alignment (9): This Challenge would have a broad benefit for aeronautics in general.

Lack of Alternative Sponsors (3): Industry is not supporting wake vortex research because it lacks facilities, and past efforts made little progress. Universities are not engaged due to the high costs of experimentation. Long-term research by NASA is needed to supplement near-term developments being made by the FAA.

Appropriate Level of Risk (9): This Challenge has moderate to high risks. Development of sensors and systems is challenging but doable.

E5 Interfaces that ensure effective information sharing and coordination among ground-based and airborne human and machine agents

The potential for sharing a wide range of information within the air transportation system raises additional questions about how multiple agents (pilots, controllers, other system users, and automated system elements) can coordinate and share information given their disparate viewpoints and contexts. For information sharing to be effective, information must be provided to the right agents, at the right time, and in a fashion that facilitates accurate interpretation regardless of the source of the information. Some of the shared information may be factual (e.g., aircraft position, speed, heading, altitude, and flight plan), while some of it may be less tangible (e.g., potential responses to disruptions). The information elements will also likely vary in their timeliness and accuracy, and access to some information will be restricted for security and business reasons. Developing appropriate interfaces (in terms of information-sharing protocols, as well as display and visualization technology) is a nontrivial challenge, because agents can be easily overwhelmed by too much information or by the need to translate and analyze the information relative to their own situation and goals (Woods et al., 2002). Interfaces for human agents, in particular, will need to include methods for visualizing and interpreting operational situations to facilitate effective judgments and decisions. In addition, information-sharing and decision-making processes will often be conducted collaboratively by multiple agents. Therefore, they will require knowledge of both individual human cognition and of collaborative work among agents with potentially conflicting goals and different representations of the immediate situation (Brennan, 1998; Olson et al., 2001). Information-sharing protocols become exceptionally critical during crises, such as 9/11, when control of the national airspace was transferred to the military. Communications and decision-making protocols were fragmented. Research related to this Challenge must be coordinated with DoD and DHS to avoid a recurrence of such problems. The Challenge should also capitalize on technologies pioneered in the telecommunications industry that would facilitate the transfer of diverse information through dynamically reconfigured networks using thousands of disparate nodes. Key milestones include

- Document improved understanding of human cognitive control, judgment, and decision making in a variety of contexts and under a variety of stressors.
- Document improved understanding of organizational dynamics and business concerns associated with information sharing.

Relevance to Strategic Objectives

Capacity (3): Making it possible for systems agents to share more information and make better use of it will enable the air transportation system to safely handle the expected growth in demand.

Safety and Reliability (9): Effective information sharing is vital, especially during emergencies, when operations are disturbed from established structures and real-time information is central to fluidly developing a course of action involving multiple agents.

Efficiency and Performance (9): Taking better advantage of information sharing will reduce unnecessary flight delays due to uninformed management of system resources and will also allow local entities (e.g., aircraft operators) to make better decisions relative to their individual goals.

Energy and the Environment (1): New methods of information sharing may be required to achieve reduced fuel consumption and more efficient ascent and descent profiles, but overall this Challenge would have only a small impact on energy and the environment.

Synergies with National and Homeland Security (9): This Challenge will facilitate the information sharing in situations involving national and homeland security, and will help manage disparate concerns (e.g., closing airspace for security versus maintaining capacity).

Support to Space (3): This Challenge would facilitate space launch applications through civil airspace and could be of value in the future as the number of U.S. space launches—and U.S. space launch facilities—increases.

Why NASA?

Supporting Infrastructure (3): NASA's Ames and Langley Research Centers have expertise in several human factors areas, although additional insight may be required for the organizational, security, and business aspects of information sharing.

Mission Alignment (9): This Challenge will benefit all segments of the civil aviation community and is consistent with ongoing research at several NASA research centers.

Lack of Alternative Sponsors (3): Industry is not motivated to produce methods and protocols for information sharing beyond some limited applications benefiting themselves (e.g., the Collaborative Decision-Making Program). Some work exists at federally funded research and develop centers (FFRDCs) and the FAA Air Traffic Control System Command Center. The large-scale and interdisciplinary aspects of the research are beyond the scope of most individual university research programs.

Appropriate Level of Risk (9): There is moderate risk due to the scale of the problem and the Challenge of responding to different users with different resources and needs.

E6 Vulnerability analysis as an integral element in the architecture design and simulations of the air transportation system

More than three-fourths of air transportation system delays are weather related (Meyer, 2005). Snow or thunderstorms at major hub airports often significantly reduce overall system capacity and efficiency. Abnormal en route winds cause unexpected peaking and depeaking at arrival gateways. En route convective weather causes disruptive and unpredictable rerouting, precipitating en route delays and reducing capacity and efficiency. Disruptions can also be caused by natural disasters (such as volcanoes, hurricanes, tornadoes, and wildfires), electronic attacks (such as power outages, hurricanes, GPS spoofing, spurious communication messages, and hacking into navigation aids), and physical attacks (such as destruction of control facilities and radars). The effects of these disruptions may be local, regional, or national. In all cases, system capacity and efficiency are directly affected, and, more important, the safety of the air transportation system may be compromised by an inadequate response.

Airlines use a variety of techniques to respond to such disruptions. Some reduce schedule to preposition aircraft for the recovery, when the weather abates; others try to fly their full schedule, hoping that the recovery will take care of itself.

System safety impacts of unplanned service disruptions should be evaluated early in the development cycle of new ATM system architectures, operating concepts, and system components. An agile ATM design should include provisions to counter or recover from system disruptions, and the design of the overall air transportation system should be evaluated by research and simulation to develop both system design concepts and/or operational procedures. In addition, quantitative analyses should be used to assess the safety impact of system architecture options. This Challenge would introduce vulnerability analyses as an integral element in the architecture design and simulations of the air transportation system to reduce the likelihood that the system will experience major system disruptions, to mitigate the severity of specific system disruptions, and to facilitate recovery from system disruptions. The result would be an air transportation system that is self-diagnosing and self-healing. Key milestones include

- Complete end-to-end vulnerability analysis of system architecture and signal flow.
- Demonstrate the ability of a more capable model to simulate critical element disruptions as defined by vulnerability analyses.
- Document safety and capacity impacts using modified system simulations.
- Develop changes in system architecture and operational procedures and demonstrate that they can mitigate the effects of specific system disruptions.

Relevance to Strategic Objectives

Capacity (3): This Challenge would minimize the magnitude, geographic extent, and duration of reductions in system capacity during air transportation system disruptions.

Safety and Reliability (9): Both safety and reliability would be significantly and directly impacted if vulnerability analysis is not included as a basic consideration in system design.

Efficiency and Performance (9): Because of the complex nature of the air transportation system, if sufficient attention is not paid to vulnerability analysis, the overall performance of the system could be excessively degraded in response to weather, major system malfunctions, terrorist actions, etc.

Energy and the Environment (1): Delays and diversions caused by disruptions in the air transportation system would likely have only a short-term effect on energy utilization or the environment.

Synergies with National and Homeland Security (9): Vulnerability analysis would improve the ability of the DoD and DHS to prevent disruptions to the air transportation system and to prepare themselves to respond as effectively and quickly as possible when they do occur.

Support to Space (1): This Challenge would have little relevance to the space program.

Why NASA?

Supporting Infrastructure (3): The NASA Ames air traffic simulation facilities would be an excellent evaluation tool in support of this Challenge. The FAA, DoD, and industry also have relevant facilities and capabilities.

Mission Alignment (9): This Challenge is directly tied to NASA's aeronautics role.

Lack of Alternative Sponsors (3): The FAA, DoD, and DHS also have an interest in vulnerability analysis.

Appropriate Level of Risk (9): Conducting comprehensive vulnerability analyses of the air transportation system would be challenging, but it can be accomplished if adequately supported.

E7 Adaptive ATM techniques to minimize the impact of weather by taking better advantage of improved probabilistic forecasts

Adaptive traffic flow management methods are needed to take advantage of recent improvements in automated aviation weather forecasts. About 70 percent of aviation delay is due to operationally significant weather, including thunderstorms, low ceilings and visibilities, high winds, and turbulence. Exploitation of weather data collected from ground sensors and satellites using advanced image processing and machine intelligence has enabled significant improvements in aviation weather forecasts. One- to two-hour storm motion products are now being routinely displayed in key airport and en route air traffic facilities and in airline dispatch centers. Included are automatically updated estimates of the forecast accuracy, expressed as a probability (Robinson et al., 2004). This information is beginning to be used by air traffic managers and dispatchers, but only manually (Wolfson et al., 2004).

Algorithms are needed that automatically translate the weather forecasts into actionable traffic flow recommendations, with the goal of fully incorporating the weather data into air traffic automation designs. A few examples of automation that translate probabilistic weather forecasts into traffic flow recommendations have been developed, and FAA air traffic managers have shown they can reduce delays. For example, the LaGuardia Airport traffic flow managers are using storm motion forecast tools, such as the Route Advisory Planning Tool, to automatically identify safe departure routes (Evans, 2006). However, many automation systems are not incorporating the new weather information into their designs. This Challenge would demonstrate the use of automated weather forecasts in making traffic flow decisions and determine where this capability is cost beneficial. Key milestones include

- Identify potential reductions in weather-induced delays.
- Demonstrate use of automated weather forecasts in making traffic flow decisions.
- Quantify the benefit of using automated weather forecasts in making traffic flow decisions.
- Determine where this capability is cost beneficial.

Relevance to Strategic Objectives

Capacity (9): Significant benefits from the manual use of new aviation weather forecasts have been realized in isolated cases. ATM operations could make use of advanced weather data to significantly reduce delays.

Safety and Reliability (3): Better integration of improved weather forecasts with air traffic control will reduce flight risks and improve traffic reliability (i.e., on-time performance).

Efficiency and Performance (9): Taking better advantage of improved weather forecasts will reduce unnecessary flight delays due to unnecessary holds on the ground or in the air.

Energy and the Environment (3): With improved weather data, fuel consumption could be reduced through better routing and more efficient climb and descent profiles.

Synergies with National and Homeland Security (1): Improved weather forecasts contribute to enhanced situational awareness and the navigation of intercept aircraft, but overall this Challenge would have a small impact on national and homeland security.

Support to Space (1): This Challenge would have little relevance to the space program; advanced weather forecasts are already providing launch-critical information.

Why NASA?

Supporting Infrastructure (3): NASA has several ATM research programs, but NASA's capabilities are not unique.

Mission Alignment (9): This Challenge would benefit all segments of the civil aviation community and is consistent with ongoing research at several NASA research centers.

Lack of Alternative Sponsors (3): Industry is not developing methods to integrate weather with traffic flow management. Some work exists at FAA-sponsored FFRDCs. Little work is being done at universities, because the necessary test facilities are cost-prohibitive.

Appropriate Level of Risk (9): There is moderate risk due to the scale of the problem and the Challenge of responding to many aviation users with different equipment capabilities and levels of pilot expertise and who respond differently to various kinds of adverse weather.

E8a Transparent and collaborative decision support systems

Air traffic operations are enhanced by effective decision support systems that assist pilots, controllers, traffic flow managers, and airline personnel in tasks such as routing, flight planning, scheduling, and traffic separation. These decision support systems contribute to safe and efficient operations by using technology to enhance human capabilities and collaborate with the operator, as opposed to fully automated systems, which use technology rather than an operator to perform tasks. Collaborative decision support systems are most effective when the operators understand the basis for and limitations in the system's reasoning process and can judge the appropriateness of system-generated recommendations. Similarly, the system's recommendations should take into account operators' knowledge and intentions as well as the context in which they operate. Support for reciprocal information sharing and mutual understanding of intentions and actions—a process called grounding—is critical to avoid breakdowns in human–machine collaboration and overall system performance (Sorkin et al., 1988; Lee and Moray, 1994; Smith et al., 2001; McGuirl and Sarter, 2006). This Challenge will identify the type of information to be shared between human operators and automated decision support systems and develop candidate designs for these systems. Key milestones include

- Identify the type of information to be shared between human operators and automated decision support systems and the most appropriate form of information representation and exchange.
- Develop, demonstrate, evaluate, and iteratively refine candidate designs in collaboration with operators.

Relevance to Strategic Objectives

Capacity (3): Collaborative decision support systems will help the air transportation system safely handle the expected growth in demand.

Safety and Reliability (9): Past experience with decision support systems has shown that a lack of transparency can cause users to rely on information provided by decision aids too much or too little, because they do not understand how the decision aids work and what their limitations are. Also, stand-alone systems that perform tasks for, rather than with, human operators have been shown to make it very difficult for operators to monitor their performance and intervene when necessary. Both of these problems can affect safety and can be addressed through improved design of collaborative decision support systems.

Efficiency and Performance (9): Improved system efficiency and performance is a prerequisite for meeting future demands on the air transportation system. Collaborative decision support systems will contribute to this goal by leading to more effective communications, fewer misunderstandings, and more effective operations.

Energy and the Environment (1): More effective and collaborative decision support systems would have only a small impact on energy and the environment.

Synergies with National and Homeland Security (3): National and homeland security would be enhanced by more effective and collaborative decision support systems.

Support to Space (3): This Challenge would facilitate planning and execution of space missions.

Why NASA?

Supporting Infrastructure (3): NASA, especially the Ames and Langley Research Centers, has the resources and expertise to develop improved decision support system designs. NASA is well connected to other organizations conducting research in this field.

Mission Alignment (9): The design of safe and efficient human–machine systems and interfaces is an important part of NASA's aeronautics mission.

Lack of Alternative Sponsors (3): Research associated with this Challenge will likely find a sponsor if NASA does not perform the work. It is not clear, however, that other sponsors are as well qualified.

Appropriate Level of Risk (9). Developing improved collaborative decision support systems is difficult, but NASA has the capability to make substantial progress over current systems.

E8b Using operational and maintenance data to assess leading indicators of safety

Safety analysis is often a reactive, ad hoc process made difficult, in part, by the very high level of safety required of air transportation in the United States. Few unambiguous data points (accidents) are available for analysis, the number of data points continues to decrease because of the success of ongoing safety efforts, and accidents that do occur are increasingly the result of a complex chain of unlikely circumstances, each of them benign (Leiden et al., 2001). While human error is often cited as a major safety concern, successful human performance is also a major (and underreported) contributor to system safety. Thus, a particular concern for safety analysis is the human contribution to safety, especially when predicting the safety impact of dramatic changes to the role of human operators and increased reliance on automation. Likewise, safety analysis must consider individual aircraft as well as systemwide safety, which involves complex interactions among many agents. Using a common set of safety metrics (see R&T Challenge E16), this Challenge would develop methods both for monitoring the current system through ongoing analysis of operational and maintenance data and for predicting potential safety problems associated with proposed changes to the air transportation system. Key milestones include

- Produce a common taxonomy for all safety information acceptable to all stakeholders.
- Demonstrate methodologies to discover and analyze anomalous system, components, and human behavior in nominal and off-nominal conditions.
- Demonstrate methods to integrate system models into analytical processes.
- Demonstrate advanced, affordable methods to analyze anecdotal written reports of safety problems and cross-reference them to operational data from aircraft, ATM, and weather systems.
- Demonstrate methods to cross-reference operational data to certification and training simulator data to determine if aircraft are performing as designers intended and if pilots and controllers are performing as trained.

Relevance to Strategic Objectives

Capacity (3): Safety concerns can pose limits on operating concepts and criteria such as separation standards; however, more effective safety analysis might show that these limits are overly conservative and could be relaxed.

Safety and Reliability (9): Established metrics and methods of assessing and predicting safety issues are fundamental to developing and maintaining safe operations.

Efficiency and Performance (9): Without these developments, efficiency and performance will likely be constrained by overly conservative standards and operational procedures.

Energy and the Environment (1): Methods of ensuring safety should be careful to not conflict with methods of conserving energy and protecting the environment, and vice versa, but overall this Challenge would have only a small impact on energy and the environment.

Synergies with National and Homeland Security (3): Safety concerns can impose limits on unusual types of operations (e.g., UAV operations) within general-use airspace; likewise, in extreme circumstances (e.g., closing the national airspace and grounding all aircraft in an emergency), safe methods are needed to transition between normal and emergency modes of operations, including transfer of operational authority.

Support to Space (3): This Challenge would help improve the safety of the space program.

Why NASA?

Supporting Infrastructure (3): NASA Ames Research Center has expertise in relevant system-safety analysis and monitoring, although additional insight may be required for the organizational, security, and business considerations in safety analysis.

Mission Alignment (9): This Challenge would benefit all segments of the civil aviation community and is consistent with ongoing research at several NASA research centers.

Lack of Alternative Sponsors (3): Industry is not motivated to perform systemwide safety analysis or to share some sensitive data and analyses, which impedes systemwide analyses and the easy sharing of safety-critical information. FAA-sponsored FFRDCs are conducting some related research, but NASA has a unique and objective third-party role given the regulatory function of the FAA. The large scale and interdisciplinary aspect of the research is beyond the scope of most individual university research programs.

Appropriate Level of Risk (9): There is moderate risk due to the scale of the problem and the difficulty of analyzing many types of operations with different characteristics and technologies.

E8c Interfaces and procedures that support human operators in effective task and attention management

The expected growth in air transportation demand will likely require operators to perform a wider range of tasks and to collaborate more closely with one another and with modern technologies. Pilots may begin to play a more active role in traffic separation or spacing and will need to coordinate their activities and intentions with other pilots and controllers. They will need to interact and exchange information, often interrupting each other and creating new tasks for one another. In general, more information will need to be distributed in a timely manner, task sets will increase, interruptions will become more likely, and the tolerance for delayed action or intervention will probably be reduced. It will be critical to ensure that operators are supported in properly scheduling and prioritizing their tasks, to improve attention management and avoid errors caused by unnecessary task switching, unnecessary interruptions, or inappropriate dismissals of demands (i.e., the failure to switch attention when appropriate and necessary) (Woods, 1995; McFarlane and Latorella, 2002; Ho et al., 2004). Major milestones include

- Complete basic research to document how operators absorb information, process information, and prioritize tasks.

- Demonstrate tools to efficiently evaluate operational data and reports of nominal and off-nominal decision making by operators.
- Demonstrate and evaluate candidate designs and procedures in support of preattentive reference, time-sharing among different tasks, and task switching.[4]

Relevance to Strategic Objectives

Capacity (3): In order to handle the expected growth in demand on the future air transportation system, timely interactions among operators and proper task prioritization will be important

Safety and Reliability (9): Increased task and attention demands on pilots and controllers lead to unwarranted interruptions of ongoing tasks and lines of reasoning. Interruptions are known to increase the risk of errors involving both the interrupted and the interrupting task and should therefore be minimized. At the same time, inappropriate dismissals of demands can similarly affect the safety of operations, especially in congested airspace with reduced separation.

Efficiency and Performance (9): Making sure that operators optimally direct their attention will help avoid errors and breakdowns in coordination that reduce system efficiency and performance.

Energy and the Environment (1): This Challenge has little or no relevance to energy or the environment.

Synergies with National and Homeland Security (3): Attention management challenges faced by DoD and DHS are similar to this Challenge faced by civil aviation. All are engaged in operations that require fast and efficient information exchange and interactions among many stakeholders, and research that will reduce unnecessary disruptions and improve attention management in the civil air transportation system will be of some assistance in addressing these challenges in the DoD and DHS.

Support to Space (3): Some space operations require operators to cope with competing demands for their attention, especially in case of system failures. Thus, this Challenge is relevant to the space program.

Why NASA?

Supporting Infrastructure (3): NASA (especially the Ames and Langley Research Centers) has the resources and expertise to conduct research on improved systems and procedures in support of task and attention management. In fact, limited efforts are already under way to address this growing Challenge.

Mission Alignment (9): Human-centered design and the development of safe and efficient human–machine systems for aviation and space operations are an important part of NASA's mission.

Lack of Alternative Sponsors (3): Research on task and attention management may be conducted by other sponsors if NASA does not address this Challenge. It is not clear, however, that research by others would have a sufficient aeronautical focus.

Appropriate Level of Risk (9): This Challenge faces moderate risk but is likely to lead to significant improvements.

E11 Automated systems and dynamic strategies to facilitate allocation of airspace and airport resources

Many major airports have little or no excess capacity. The competition for airspace and airport resources (e.g., airport departure and arrival slots) at major airports will be exacerbated by growth in commercial and private air travel (including, for example, the introduction of VLJs). Automated systems and dynamic strategies driven by economic reasoning would facilitate quick and effective decision making when the transportation system encounters disruptions. They could also be used during normal operations to quickly negotiate and make decisions regarding, for example, real-time allocation of airspace and landing slots among aircraft with diverse size and performance characteristics, while considering the needs of all stakeholders, air transportation system efficiency, and energy conservation (Cramton et al., 2002). Key milestones include

- Modify available models to create dynamic tools for assessing the current and future state of the air transportation system.
- Document decision-making drivers of all users of the air transportation system.
- Create and demonstrate an architecture to allocate landing slots.
- Create and demonstrate an architecture to allocate airspace dynamically.

Relevance to Strategic Objectives

Capacity (9): Fast-response negotiation, quick decision making, and dynamic strategies that adapt themselves to changing conditions could substantially increase capacity at times and places where air transportation resources are constrained, especially during disruptions.

Safety and Reliability (3): Fast-response decision making will increase safety and reliability during system disruptions.

[4] Preattentive reference is supported by presenting partial information about a potentially interrupting task or event to help the operator decide whether a shift in attention is warranted. The information needs to be presented in such a way that it is quickly noticed and easily processed and understood without requiring an interruption of the ongoing task or line of reasoning (Woods, 1995). Operational systems that provide preattentive reference reduce the risk of task-switching errors and improve operator efficiency and performance.

Efficiency and Performance (9): More effective allocation of access to air transportation resources, including quick response to disruptions in the air transportation resources, will greatly improve system efficiency and performance.

Energy and the Environment (3): Dynamic strategies that lead to more effective allocation of airspace and landing slots would have a positive impact on energy conservation and the environment.

Synergies with National and Homeland Security (3): Quick, highly effective decision-making systems suitable for emergency situations would enhance national and homeland security.

Support to Space (1): This Challenge is not directly relevant to the space program.

Why NASA?

Supporting Infrastructure (3): NASA has the resources to develop decision-making tools, methodologies, and systems based on software agents, and it is well connected to other relevant research communities with relevant skills.

Mission Alignment (9): Improving the performance of the air transportation system is part of NASA's mission.

Lack of Alternative Sponsors (1): The FAA is already supporting research related to traffic flow management and collaborative decision making that would contribute to this Challenge. Current funding is probably not at a high enough level to address all of the complexities of this issue, but in any case, solutions to this Challenge are highly dependent on specific FAA implementation plans and architectures (e.g., the Enhanced Traffic Management System).

Appropriate Level of Risk (9): The technology goals are realistic, and the technical risk is moderate.

E12 Autonomous flight monitoring of manned and unmanned aircraft

Safe, high-capacity operations can only be achieved if pilots (and ground-based operators of UAVs) reliably comply with their communicated intentions. This means that each UAV must have numerous fail-safes in case of system failure or loss of the communications link between a UAV and its operator (Sabatini, 2006).

The effects of unexpected deviations can ripple throughout the system, increasing delays and the risk of collision. An autonomous flight monitoring system would identify unanticipated and unauthorized deviations from flight plans, as well as their likely cause (e.g., degraded equipment performance, adverse weather, or unexpected actions by the pilot or autopilot). The system would also disseminate relevant information to agents in the air transportation system that might be affected, with the goal of rapidly initiating analysis and intervention systems to identify and resolve near- and long-term conflicts or hazards. To be useful, flight monitoring technologies will need to avoid false alarms associated with routine deviations to adjust spacing between aircraft or in response to local weather conditions.

Technologies developed in response to this Challenge could also be used to better predict the near-term consequences of flight plan deviations. Data generated by onboard systems could be combined with data provided by ground systems to extend the time horizon of the predictive capabilities to more than just a few minutes. Key milestones include

- Produce a detailed set of requirements and design specification for flight monitoring systems deployed on manned and unmanned aircraft.
- Demonstrate algorithms and knowledge to enable a flight monitoring system that accurately anticipates, detects, and diagnoses flight plan deviations.
- Demonstrate the ability to more accurately project the near-term results of manipulating aircraft controls and inform pilots of likely consequences in terms of aircraft motion, potential collisions, airspace violations, etc.
- Design protocols for disseminating information from flight monitoring systems locally and throughout the air transportation system.
- Specify corrective actions appropriate for manned and unmanned aircraft in response to unplanned deviations detected by a flight monitoring system.

Relevance to Strategic Objectives

Capacity (3): The ability of an autonomous flight monitoring system to provide early warning of unplanned flight deviations could have a modest impact on capacity by enabling reduced separation standards as a result of increased trust that each aircraft will accurately follow its flight plan.

Safety and Reliability (9): An autonomous flight monitoring system would increase safety by providing information on localized and systemwide deviations and their potential to create hazards or conflicts.

Efficiency and Performance (3): An autonomous flight monitoring system would help assess systemwide performance and delays as well as provide warnings of flight plan deviations, which would allow operators to make changes to improve system performance.

Energy and the Environment (1): An autonomous flight monitoring system would have minimal impact on energy and the environment.

Synergies with National and Homeland Security (9): Early warning of flight plan deviation could help identify potentially hostile action. Autonomous flight monitoring could also provide early indication of tampering with air transportation system equipment (e.g., navigation signals and communication channels).

Support to Space (1): Future space vehicles will have extensive onboard monitoring capabilities, but this Challenge focuses on monitoring of the air transportation system and thus would have minimal direct relevance to the space program.

Why NASA?

Supporting Infrastructure (3): NASA has placed significant emphasis on aircraft and spacecraft diagnostics and monitoring, but it has limited expertise in the systemwide monitoring capabilities associated with this Challenge.

Mission Alignment (9): Autonomous flight monitoring has the potential to significantly enhance aviation safety through improved compliance with flight plans and improved situational awareness.

Lack of Alternative Sponsors (3): Development of a systemwide autonomous flight monitoring capability may also be supported by the FAA, and some capabilities would be supported by commercially developed flight management systems that currently monitor and store statistics onboard each aircraft.

Appropriate Level of Risk (9): Systemwide flight monitoring is feasible, but significant challenges remain to reliably detect, diagnose, and project the impact of flight plan deviations.

E13 Feasibility of deploying an affordable broad-area, precision-navigation capability compatible with international standards

Current Global Positioning System (GPS) satellites do not provide navigation data that are precise enough (particularly with regard to altitude) or reliable enough to support precision approach and landing in low visibility conditions (Shively and Hsaio, 2005). In addition, GPS signals are vulnerable to jamming (Volpe, 2001). Installation of ground-based augmentation systems, such as pseudolites, at each landing area would address these shortcomings. However, ground augmentation, as currently envisioned, is not an affordable solution for many small airports.

In cooperation with the Russian and European sponsors of the Global Navigation Satellite System (GLONASS) and Galileo satellite navigation systems, this Challenge would investigate the feasibility of improving existing systems so that measurement geometry, terrestrial coverage, integration of inertial sources, etc. would provide navigation signals with enough integrity and redundancy to enable a broad-area precision navigation capability without the need for ground-based augmentation or independent backup. Key milestones include

- Demonstrate appropriate concepts of operations.
- Review currently planned enhancements to the GPS, Galileo, and GLONASS systems to assess the degree to which performance improvements and/or design modifications of one or more systems would be required.
- Assess the cost, affordability, and technical feasibility of developing and deploying a fully compliant broad-area precision navigation system, including quality assurance methodologies, based on one or more operational concepts.

Relevance to Strategic Objectives

Capacity (9): A broad-area precision navigation capability (including vertical guidance) would open up any site as a potential landing area and facilitate greater reliance on runway-independent operations. This would increase air transportation system capacity, especially at small airports that currently lack precision approaches.

Safety and Reliability (9): Safety is enhanced with the addition of precision landing guidance in all weather conditions.

Efficiency and Performance (3): Efficiency and performance would be enhanced by the availability of Category III landing at any site.

Energy and the Environment (1): The Challenge would have no significant impact on this Objective.

Synergies with National and Homeland Security (3): A broad-area precision navigation capability would facilitate response to natural disasters or terrorist attacks because it would not depend on preexisting ground infrastructure and it would provide coverage to disaster sites wherever they might be.

Support to Space (1): A broad-area precision navigation capability would be of minimal use to space operations.

Why NASA?

Supporting Infrastructure (3): NASA's space program already has a role in GPS spacecraft design and could help develop requirements and design specifications. The DoD, however, has deployment and operational control of the GPS satellite constellation and the FAA is responsible for certification. Assessing the feasibility of a broad-area precision navigation capability for civil aviation is closely tied to regulatory and economic issues that fall outside NASA's area of expertise.

Mission Alignment (3): The technical feasibility issues are well aligned with NASA's aeronautics mission, but economic feasibility issues are better handled by industry, and regulatory feasibility issues are better handled by the FAA.

Lack of Alternative Sponsors (3): DoD has no requirement to provide a GPS system with these capabilities. Industry is in a good position to assess cost, affordability, and economic feasibility.

Appropriate Level of Risk (9): This Challenge has moderate to high risk.

E14 Advanced spacecraft weather imagery and aircraft data for more accurate forecasts

FAA-sponsored weather research has been successfully exploiting data from the National Weather Service and FAA weather radars, as well as space imagery from Geostationary Operational Environmental Satellites (GOES), to enable accurate 1- to 2-hour forecasts of convective storm motion and other weather hazardous to aviation. These forecasts are beginning to provide significant benefits to civil aviation, especially by reducing delays. More forecast improvements are needed but await better understandings of storm growth and decay phenomena (Wolfson et al., 2004). That, in turn, requires improved space imagery with finer resolution and higher update rates than existing Next Generation Weather Radar (NEXRAD) radars. There is also significant interest in using satellite-based multi- and hyperspectral images to better understand cloud formations and develop precursors to storm growth and decay. New imagery that is expected to provide these data will come with the 2012 launches of the GOES-R satellite and the first satellite for the National Polar-Orbiting Operational Environmental Satellite System (NPOESS). Even so, research is needed to understand how to use these data and combine the knowledge they provide with existing terrestrial sensor data. Key milestones include

- Demonstrate the ability to use existing multi- and hyperspectral space imagery for tactical aviation forecasts using data from the following sources:
  —Hyperion imager on the Earth Observing-1 Satellite.
  —Atmospheric Infrared Sounder.
  —the Infrared Atmospheric Sounding Interferometeron on the Meteorological Operational Satellite (METOP).
  —Geosynchronous Imaging Fourier Transform Spectrometer developed by Langley Research Center, an on-the-shelf GOES-R prototype sensor that tracks the three-dimensional movement of water vapor and winds, to accelerate the exploitation of GOES-R imagery.
- Demonstrate methods of predicting the growth and decay of convective systems using space imagery and terrestrial data sources on timescales consistent with aviation needs.

Relevance to Strategic Objectives

Capacity (3): Extending the forecast horizon to beyond 2 hours would increase the capacity of the air transportation system during adverse weather.

Safety and Reliability (3): Improvements in the accuracy and reliability of weather forecasts directly impact both safety and dispatch reliability of flight.

Efficiency and Performance (9): Improved weather forecasts will reduce unnecessary flight delays due to excessive ground holds, reroutes, and airborne holding.

Energy and the Environment (3): With improved weather data, reduced fuel consumption and more efficient climb and descent profiles may be achieved.

Synergies with National and Homeland Security (1): Improved weather forecasts contribute to enhanced situational awareness and the navigation of intercept aircraft, but this would be only a small contribution to the overall state of national and homeland security.

Support to Space (1): Advanced weather forecasts are already providing launch-critical information, and improved forecasts would have only a small impact on the space program.

Why NASA?

Supporting Infrastructure (3): NASA is already supporting research in hyperspectral imagery as well as launch and support systems.

Mission Alignment (9): This Challenge would directly benefit the air transportation system, and it would complement ATM research already under way.

Lack of Alternative Sponsors (3): NASA support will ensure that space imagery research already under way includes a component that is specifically focused on tactical forecasts for civil aviation.

Appropriate Level of Risk (9): This Challenge has moderate to high risk.

E15 Technologies to enable refuse-to-crash and emergency autoland systems

Flight management systems can safely execute flight plans from takeoff through landing under nominal conditions, and UAV autopilots are approaching this level of capability. Less well understood is the response of flight management systems to abnormal operating conditions that may occur on the aircraft or in the portion of the air transportation system in which the aircraft is operating. Emergencies may be caused by miscommunications; errors involving onboard or remote pilots, air traffic controllers, or automated systems; and other irregularities (e.g., airframe damage) that significantly impact aircraft performance. Refuse-to-crash systems are intended to prevent both controlled and uncontrolled flight into terrain and collisions with other aircraft (Croft, 2003). Emergency autoland systems would be activated to safely land an aircraft when a failed system makes the aircraft difficult or impossible to continue powered flight safely. Such systems are likely to be most successful if implemented both onboard and on the ground. Key milestones include

- Demonstrate fundamental flight planning and control algorithms for refuse-to-crash and emergency autoland systems applicable to manned and unmanned aircraft.
- Demonstrate robust integration of algorithms within a distributed air transportation system that enables both onboard and remotely controlled recovery from system failures or hostile activities.

- Specify data requirements and protocols that will enable unambiguous local or remote detection of situations that require response by a refuse-to-crash or emergency autoland agent.

Relevance to Strategic Objectives

Capacity (1): This Challenge focuses on safety and has no direct impact on capacity.

Safety and Reliability (9): Successful development of refuse-to-crash and emergency autoland capabilities would provide another layer of safety.

Efficiency and Performance (1): The responsive algorithms developed by this Challenge apply to important but improbable events, so this Challenge would not significantly affect the overall efficiency or performance of the air transportation system.

Energy and the Environment (1): This Challenge has little impact on this Objective.

Synergies with National and Homeland Security (3): An emergency autoland system can enable a damaged or otherwise compromised aircraft to safely land despite reduced performance capabilities.

Support to Space (1): This Challenge would have little impact on the space program.

Why NASA?

Supporting Infrastructure (3): NASA has an active research program on emergency autoland systems but has placed less emphasis on refuse-to-crash systems. This Challenge requires systemwide implementation of these capabilities, which has not been a major focus of NASA or others.

Mission Alignment (9): This Challenge could broadly benefit commercial and general aviation.

Lack of Alternative Sponsors (3): Although other government and industrial organizations would perform some research related to this Challenge, NASA support will be important for the fundamental research required by this Challenge.

Appropriate Level of Risk (9). This Challenge is feasible, but it has a moderate to high risk.

E16 Appropriate metrics to facilitate analysis and design of the current and future air transportation system and operating concepts

As advanced technologies and procedures are developed to address air transportation needs, it is important that their performance be understood and compared to existing capabilities. Metrics are important because the metrics that are used to measure the performance of a system have a direct impact on the design of the system; parameters that are not measured—or are measured incorrectly or incompletely—will not be fully considered or accounted for in the final design. Hence the importance of identifying appropriate metrics and incorporating them into system analysis and design tools and processes. However, there is no comprehensive, widely held set of metrics to analyze and design the current and future air transportation system, assess related operating concepts, and define bounds on system performance. Key milestones include

- Identify and document objective measures of current capacity, safety, and efficiency.
- Conduct sensitivity analyses of the above factors to determine causality.

Relevance to Strategic Objectives

Capacity (3): Additional capacity is a key requirement that any future air transportation system must meet. However, there are many ways to measure capacity, and many places where capacity could be measured. Thus, it is important to develop composite metrics that capture the overall ability of the system to handle traffic.

Safety and Reliability (3): More effective safety and reliability metrics would help guide system changes to improve performance in these areas.

Efficiency and Performance (9): High efficiency is the basis for cost effectiveness. Thus, it is important that the efficiency of a system be known before and during development.

Energy and the Environment (3): Environmental considerations are often considered only at the end of the design process. However, they will impact the acceptance of any proposed changes as evidenced by the litigation surrounding the expansion of airports and changes to terminal area trajectories in the United States. Therefore, such considerations must be measured directly or by proxy (with appropriate metrics) throughout the design process to ensure that the resulting solution is at the very least environmentally feasible.

Synergies with National and Homeland Security (3): The DoD and DHS use many metrics related to security. Metrics that are developed to evaluate the civil air transportation system would be of some interest to the DoD and DHS, in that they would facilitate military operations in civil airspace and improve the ability of the air transportation system to respond to emergencies.

Support to Space (1): This Challenge is not relevant to this Objective.

Why NASA?

Supporting Infrastructure (3): NASA has some researchers who could address this Challenge, but outside expertise would also be needed.

Mission Alignment (9): This Challenge involves long-term research that should occur before others begin to develop specific components or synthesize components into system prototypes. Thus, it is well aligned with NASA's mission.

Lack of Alternative Sponsors (3): This Challenge might be addressed by the FAA if NASA is not able to perform it.

Appropriate Level of Risk (3): This Challenge involves low risk.

E17 Change management techniques applicable to the U.S. air transportation system

The air transportation system comprises technology as well as large organizations with long-standing institutional cultures and business concerns, which must be motivated to participate in new operating concepts. Changing such a complex interactive system is difficult both in terms of the immensity of changing its individual elements (new technologies, personnel training, etc.) and in terms of motivating the institutional and business changes. Additionally, the end state of the future air transportation system remains undefined, so R&T should create and maintain the flexibility to change the system in any of several different directions. This requires the interdisciplinary application of large-scale system engineering, organization design, economics, and financial analysis, which in some ways is beyond the current state of knowledge. Even so, improved change management techniques are vital to a cost-effective, noncontentious, and safe transition to the air transportation system of the future. Key milestones include

- Demonstrate methods to identify key obstacles to change in large-scale sociotechnical systems.
- Demonstrate methods to describe and predict the impacts of proposed changes on personnel roles, skills, and training; staffing levels; organizational structures; operating procedures, policies, and regulations; and economic and financial concerns, including funding sources for government agencies.
- Create an architecture to apply change management methods to interagency work that is developing the Next Generation Air Transportation System.

Relevance to Strategic Objectives

Capacity (9): Many elements of and organizations involved in the air transportation system must collectively and simultaneously transition to new operating concepts to satisfy future demand for air transportation.

Safety and Reliability (9): Poorly executed transitions to new operating concepts may degrade safety due to unclear responsibility and authority, dissonant work practices between and within organizations, and poor use of resources, including newly available information.

Efficiency and Performance (9): Coordinated, simultaneous shifts to new operating concepts will facilitate multiple agents working together to improve the overall performance of the air transportation system. It will also allow individual entities to use their individual resources effectively and efficiently.

Energy and the Environment (1): Improved change management techniques will not have a substantial impact on energy consumption or the environment.

Synergies with National and Homeland Security (3): Poorly executed transitions to new operating concepts may degrade security due to unclear responsibility and authority, dissonant work practices between and within organizations, and poor use of resources.

Support to Space (1): Improved change management techniques developed for the civil air transportation system are unlikely to be of substantial value to the space program.

Why NASA?

Supporting Infrastructure (1): This Challenge would require NASA to acquire additional expertise in large-scale system engineering, organization design, economics, and financial analysis.

Mission Alignment (3): This Challenge would benefit all aeronautics domains, but many aspects of this Challenge involve organizational and economic issues that fall outside the normal scope of NASA's aeronautics research.

Lack of Alternative Sponsors (3): Many other organizations are already involved in transformation of the air transportation system.

Appropriate Level of Risk (3): The risk associated with this Challenge is low if it is properly incorporated into the overall effort to develop the NGATS.

E18 Certifiable information-sharing protocols that enable exchange of contextual information and coordination of intent and activity among automated systems

Situational awareness is required for safe and efficient operation of individual aircraft and the air transportation system as a whole. Numeric data from radar or onboard instruments is routinely processed by onboard flight management systems for navigation and local deconfliction. ATM tools predict conflicts given information on the current status of aircraft and their flight plans. However, these systems are somewhat rigid and do not always adapt well to contingency system disruptions and frequent flight plan alterations.

The air transportation system of the future may use dynamic routing capabilities to increase capacity and efficiency. This would require the system to understand at a more fundamental level aircraft objectives and the protocols by which flight plans will be changed, rather than rely on static flight plans that will likely become obsolete.

Objectives of this Challenge are to (1) model the information that must be shared in a dynamic air transportation system and the criteria under which data must be explicitly communicated rather than inferred and (2) interact with certification authorities to ensure that it will be feasible to transfer research results to operational systems. Research associated with this Challenge should also take into account limitations on the ability to reconfigure aircraft and the time required for each manned or unmanned aircraft to dynamically alter its flight plan, in response to, for example, changes in the environment and the flight plans of other aircraft. Key milestones include

- Quantify communication bottlenecks in the air transportation system during nominal operation and off-nominal operating conditions.
- Define information-sharing protocols that could enable dynamic routing of all classes of manned and unmanned traffic, considering the performance of different aircraft and their ability to respond to flight plan changes.
- Demonstrate the application of strategies to infer the intent of aircraft locally, to minimize the need for high-bandwidth communication across the air transportation communications networks.

Relevance to Strategic Objectives

Capacity (3): Efficient information sharing is a key component of distributed, dynamic routing, but it does not directly increase capacity.

Safety and Reliability (1): This Challenge is explicitly focused on improving the efficiency of the air transportation system.

Efficiency and Performance (9): Information sharing will become a key enabler in a distributed, dynamic air transportation system. This Challenge would directly increase efficiency and performance of the system through minimal and expressive information exchange.

Energy and the Environment (1): Information sharing has no direct impact on energy or the environment.

Synergies with National and Homeland Security (9): Aircraft coordination and interpretation of intent through information sharing have significant overlap with homeland security. Intent inference could enable early detection of compromised aircraft, and managing such deviations will require elements of the air transportation system to be efficiently coordinated.

Support to Space (1): This Challenge is not relevant to this Objective.

Why NASA?

Supporting Infrastructure (3): NASA has participated in the development of information-sharing protocols present in current and emerging ATM systems.

Mission Alignment (9): This Challenge would improve the safety and (indirectly) the capacity of the air transportation system. Effective information-sharing protocols applicable to all aircraft types (e.g., transport, UAV, and general aviation) will help ensure that all aircraft types will be able to function efficiently within the congested airspace of the future.

Lack of Alternative Sponsors (3): Other government and industrial organizations will likely sponsor related work as it applies to transport operations.

Appropriate Level of Risk (3): The technology risk for this Challenge is low.

E19 Provably correct protocols for fault-tolerant aviation communications systems

The future air transportation system will increasingly rely on information exchange and automation. Thus, the ability to communicate among systems (ground- and airborne-based) must be ensured through the application of fault-tolerant system design. Key milestones include

- Demonstrate new communication protocols and design standards that are widely accepted within the aviation community.
- Demonstrate methods for certifying systems that incorporate the new protocols and standards.

Relevance to Strategic Objectives

Capacity (3): Air-to-ground information sharing and new decision support technologies, which will improve capacity, would benefit from reliable communications systems.

Safety and Reliability (9): Fault-tolerant aviation communications systems will enhance the safety and reliability of the air transportation system.

Efficiency and Performance (3): Fault-tolerant aviation communications systems will be more reliable and thus enhance efficiency.

Energy and the Environment (1): This Challenge is not relevant to this Objective.

Synergies with National and Homeland Security (3): Fault-tolerant networks would be better able to withstand equipment failures and deliberate attacks.

Support to Space (1): This Challenge is not relevant to this Objective.

Why NASA?

Supporting Infrastructure (3): NASA has experience designing fault-tolerant systems.

Mission Alignment (3): NASA does not have an explicit mission to develop communication protocols.

Lack of Alternative Sponsors (1): The FAA and industry organizations are able to take on this Challenge.

Appropriate Level of Risk (1): The risk associated with this Challenge is so low that industry or the FAA will likely achieve success even without NASA's involvement.

E20 Comprehensive models and standards for designing and certifying aviation networking and communications systems

The communications technologies being incorporated into new aircraft and integrated airborne and ground systems constitute a major shift from federated systems connected with dedicated wiring to a network-centered approach, where most data are shared on a common data network. A set of overarching models would facilitate the design and certification of network-based air-ground communications systems. These models should be able to accommodate the increasing number of ATM, dispatch, and other functions that are participating in networked communication, as well as increasingly complex protocols. Key milestones include

- Demonstrate independent models of the safety of networked systems.
- Define and document criteria that could be incorporated into standards used for certification of systems.

Relevance to Strategic Objectives

Capacity (3): A defined and consistent approach to certification of aviation networking and communications systems will facilitate development and deployment of aircraft-to-ground ATM communication systems, which will provide more communication channels, improve the ability of the air transportation system to handle large numbers of aircraft in congested areas, and contribute to increased capacity.

Safety and Reliability (9): Comprehensive models and standards for designing and certifying aviation networking and communications systems could make a significant contribution to safety, because future systems that rely heavily on networking and deterministic means of assuring the safety of these systems might not be available. Single failures of a network could affect the performance of multiple aircraft systems. Improved methods may be required to assure that the ad hoc methods currently used will not result in crucial errors going undetected. A side benefit would be improving the consistency and efficiency of certifying network-based systems.

Efficiency and Performance (3): The efficiency and performance of the air transportation system would be improved by communications and networking systems that reduce the need for voice communications between pilots and controllers. However, lack of certifiable approaches may prevent these systems from being implemented. For example, the FAA Aircraft Certification Service originally prohibited the operational use of these systems because there was no assurance that pilots would receive critical information.

Energy and the Environment (1): This Challenge is not relevant to this Objective.

Synergies with National and Homeland Security (1): This Challenge is not relevant to this Objective.

Support to Space (1): This Challenge is not relevant to this Objective.

Why NASA?

Supporting Infrastructure (3): NASA has communications technology experts, it has conducted research in aviation safety, and it has a background in formal methods that might apply to this Challenge. However, communications networking has not been a focus of NASA research.

Mission Alignment (3): This Challenge is not explicitly within the existing NASA mission.

Lack of Alternative Sponsors (1): The FAA and industry organizations could address this Challenge.

Appropriate Level of Risk (1): There is low risk associated with this Challenge.

REFERENCES

Aerospace Commission. 2002. Final Report of the Commission on the Future of the U.S. Aerospace Industry. Washington, D.C.: U.S. Department of Commerce, International Trade Administration, Office of Aerospace. Available online at <www.faa.gov/programs/oep/v7/library/AeroCommissionFinalReport.pdf>.

Andrews, J., H. Erzberger, and J. Welch. 2005. Safety Analysis for Advanced Separation Concepts, 6th USA/Europe Seminar on Air Traffic Management Research and Development, Baltimore, Md., June 27-30. Available online at <www.eurocontrol.int/eec/public/standard_page/EEC_News_2005_2_Seminar_Best.html>.

Brennan, S.E. 1998. The grounding problem in conversations with and through computers. In S.R. Fussel and R.J. Kreuz (eds.), Social and Cognitive Psychological Approaches to Interpersonal Communication. Hillsdale, N.J.: Lawrence Erlbaum.

Cramton, P., L.M. Ausubel, and P. Milgrom. 2002. Comments on Alternative Policy Options for Managing Capacity and Mitigating Congestion and Delay at LaGuardia Airport before the U.S. Department of Transportation, Washington, D.C., Dockets FAA-2001-9852 and FAA-2001-9854, June 20. Bethesda, Md.: Market Design, Inc. Available online at <www.cramton.umd.edu/papers2000-2004/mdi-comment-to-faa-on-managing-capacity-at-lga.pdf>.

Croft, J.W. 2003. Refuse-to-crash: NASA tackles loss of control. Aerospace America 41(3): 42-45.

Evans, J.E. 2006. Implications of a Successful Benefits Demonstration for Integrated Weather/Air Traffic Management (WX/ATM) System Development and Testing. 12th American Meteorological Society Conference on Aviation, Range and Aerospace Meteorology, Atlanta, Ga., January 29-February 2.

Ho, C.-Y., M. Nikolic, M. Waters, and N.B. Sarter. 2004. Not now: Supporting interruption management by indicating the modality and urgency of pending tasks. Human Factors 46(3). 399-409.

Lee, J., and N. Moray. 1994. Trust, self-confidence, and operators' adaptation to automation. International Journal of Human-Computer Studies, 40: 153-184.

Leiden, K., J. Keller, and J. French. 2001. Context of Human Error in Commercial Aviation. Boulder, Colo.: Micro Analysis and Design, Inc. Available online at <http://human-factors.arc.nasa.gov/ihi/hcsl/HPM_pubs/AvHumanError.pdf>.

Maier, M. 2006. Architecting Principles for Systems-of-Systems. The Information Architects Cooperative (TiAC) White Paper Repository. Available online at <www.infoed.com/Open/PAPERS/systems.htm>.

McFarlane, D.C., and K.A. Latorella. 2002. The scope and importance of human interruption in human-computer interaction design. Human-Computer Interaction 17(1): 1-61.

McGuirl, J., and N.B. Sarter. 2006. Supporting Trust Calibration and the Effective Use of Decision Aids by Presenting Dynamic System Confidence Information. In press.

McLain, T., and R. Beard. 2005. Coordination variables, coordination functions, and cooperative timing missions. AIAA Journal of Guidance, Control, and Dynamics 28(1): 150-161.

Meyer, D. 2005. Corridor Integrated Weather System. Presentation to the National Business Aviation Association Friends/Partners in Aviation Weather Forum, Orlando, Fla., November 11. Available online at <www.ral.ucar.edu/general/fpaw2005/4%20_Segment/meyer.pdf>.

Mundra, A.D. 2001. An Assessment of the Potential for Wake Vortex-Related Enhancements in NAS [national airspace system]. McLean, Va.: MITRE Center for Advanced Aviation System Development.

Olson, G.M., T.W. Malone, and J.B. Smith. 2001. Coordination Theory and Collaboration Technology. Hillsdale, N.J.: Lawrence Erlbaum.

Reynolds, T.G., and R.J. Hansman. 2001. Analysis of Aircraft Separation Minima Using a Surveillance State Vector Approach. Updated version of paper from 3rd International Air Traffic Management R&D Seminar (ATM2000), Naples, Italy, June 13-16, 2000. Available online at <http://icat-server.mit.edu/library/fullRecord.cgi?idDoc=183>.

Robinson, M., J. Evans, B. Crowe, D. Klingle-Wilson, and S. Allan. 2004. Corridor Integrated Weather System Operational Benefits 2002-2003: Initial Estimates of Convective Weather Delay Reduction. Project Report ATC-313. Lexington, Mass.: MIT Lincoln Laboratory.

Sabatini, N.A. 2006. Testimony of the FAA Associate Administrator for Aviation Safety before the House Committee on Transportation and Infrastructure, Subcommittee on Aviation on Unmanned Aircraft Activities, March 29, 2006. Available online at <www.faa.gov/news/news_story.cfm?contentKey=4029>.

Shively, C.A., and T.T. Hsaio. 2005. Error and availability analysis of category IIIb LAAS augmented by radar altimeter. Navigation 52(3).

Smith, P.J., E. McCoy, and J. Orasanu. 2001. Distributed cooperative problem-solving in the air traffic management system. In G. Klein and E. Salas (eds.), Naturalistic Decision Making. Hillsdale, N.J.: Lawrence Erlbaum.

Sorkin, R.D., B.H. Kantowitz, and S.C. Kantowitz. 1988. Likelihood alarm displays. Human Factors 30(4): 445-459.

Volpe National Transportation Systems Center. 2001. Vulnerability Assessment of the Transportation Infrastructure Relying on the Global Positioning System. Cambridge, Mass.: Volpe National Transportation Systems Center. Available online at <http://gps.gov/archive/2001/Oct/FinalReport-v4.6.pdf>.

Wickens, C.D., A.S. Mavor, R. Parasuraman, and J.P. McGee, eds. 1998. The Future of Air Traffic Control: Human Operators and Automation. Washington, D.C.: National Academy Press. Available online at <http://fermat.nap.edu/catalog/6018.html>.

Wolfson, M.M., B.E. Forman, K.T. Calden, W.J. Dupree, R.J. Johnson, R.A. Boldi, C.A. Wilson, P.E. Bieringer, E.B. Mann, and J.P. Morgan. 2004. Tactical 0-2 Hour Convective Weather Forecasts for the FAA. Proceedings of the 11th Conference on Aviation, Range and Aerospace Meteorology, Hyannis, Mass., October 4-8.

Woods, D.D. 1995. The alarm problem and directed attention in dynamic fault management. Ergonomics 38(11): 2371-2393.

Woods, D.D., and E. Hollnagel. 2006. Joint Cognitive Systems: Patterns in Cognitive Systems Engineering. New York, N.Y.: Taylor and Francis.

Woods, D.D., E.S. Patterson, and E.M. Roth. 2002. Can we ever escape from data overload? A cognitive systems diagnosis. Cognition, Technology, and Work 4(1): 22-36.

Xue, M., and E. Atkins. 2006. Noise-minimum runway-independent aircraft approach design for Baltimore-Washington International Airport. AIAA Journal of Aircraft 43(1): 39-51.

F

Lessons Learned from Other Federal Agencies

The statement of task directs the committee to consider lessons learned from the requirements process of the Department of Defense (DoD), from the innovation-driven process used by the Defense Advanced Research Projects Agency (DARPA), and from nonaeronautical research by federal agencies that positively affected areas with strong private sector economic interests. The steering committee did not have time to undertake extensive examinations in these areas, but it did receive presentations from Air Force Headquarters;[1] DARPA;[2] and the Air Force Research Laboratory (AFRL)[3] The committee also reviewed materials on the private sector related to federal research provided to SEMATECH (to support the competitiveness of the U.S. semiconductor industry) and to the Human Genome Project.

SEMATECH was formed in 1987 in response to concerns that the U.S. semiconductor industry was losing market share to international competitors. Initially, 14 high-tech companies, representing 85 percent of the national capacity for semiconductor manufacturing, contributed $100 million. The federal government also established an annual R&D budget of $100 million. Interestingly, DARPA was selected by Congress to be the executive agency for appropriated funds earmarked for SEMATECH. Throughout the 1980s and 1990s, SEMATECH evolved in its role, structure, and orientation in supporting the competitiveness of the U.S. semiconductor industry. It used horizontal and vertical collaboration with industry and government agencies (Carayannis and Alexander, 2004; Spencer and Seidel, 1995).

The steering committee did not determine whether the civil aeronautics industry and the semiconductor industry have enough similarities to justify use of the SEMATECH model for civil aeronautics R&T. However, SEMATECH successfully supported a U.S. industry that was being threatened by foreign competition, as is the U.S. civil aeronautics industry today.

The Human Genome Project is an example of the federal government sponsoring research for the public good. The National Institutes of Health and the Department of Energy brought together international biological and medical research communities in the public and private sectors (Collins et al., 2003; Frazier et al., 2003). Many of the experiences in organizing and managing such a complicated, publicly funded, international effort will be applicable to future large-scale projects in biology. The steering committee did not determine whether these lessons learned would also apply to the civil aeronautics industry. However, the example of federal research funding for the public good is analogous to a great deal of civil aeronautics research, particularly since the government has primary responsibility for providing air traffic control services and also has an interest in improving the efficiency of air transportation to benefit the public and the economy.

DoD's aeronautics R&D is primarily intended to benefit the military services and the defense agencies. The DoD requirements process is derived from a structured dialogue of technology push and mission-requirements pull within the DoD science and technology community, the individual military services, and the unified command user community.

DoD is enhanced by the overlay of the innovation-driven DARPA process. DARPA is organizationally structured to be nimble, and it enjoys certain flexibility and tolerance for risk in expediting prototype demonstrations of fundamental research, cutting-edge discoveries, and new system concepts that have demonstrated some level of success to the science and technology programs of the military services.

---

[1]Ronald M. Sega, under secretary of the Air Force, "Aerospace science and technology update," Briefing to the steering committee on November 8, 2005.

[2]Anthony J. Tether, director, DARPA, "Bridging the gap," Briefing to the steering committee on November 7, 2005.

[3] Walter F. Jones, director, Plans and Programs, AFRL, "Aeronautics at the Air Force Research Laboratory," Briefing to the steering committee on November 8, 2005.

The DARPA model works, in part because successful research products can be handed off directly to a closely associated user community (the military services). Although the NASA aeronautics program does not enjoy this advantage, DARPA's success implies that NASA should strive to nurture an environment that tolerates risk in aeronautics research, as difficult as that would seem to be given the need for risk aversion when it comes to civil aeronautics, where human life is in the balance. The DARPA experience also presents a model for an aeronautics research strategy that combines technology push and mission-requirements pull, with three caveats:

- The U.S. civil aeronautics industry by nature tends to focus on low-risk (and low-cost) solutions to immediate problems and thus is not an ideal source of requirements for a long-term research program with significant risk tolerance.
- Organizations in the U.S. civil aeronautics community have diverse interests and needs and very rarely speak with one voice on the value of or the requirements for any particular aeronautics R&T project.
- Like NASA, DoD maintains a large institutional base of facilities and a civil service workforce as an integral part of its investment decisions and program formulation. However, DARPA does not and is therefore permitted far more freedom in funding and program decisions.

Thus it is unrealistic to expect that NASA's civil aeronautics program will be able to create the same requirements process and constituency support base that DoD has created, nor will it be able to employ the same innovation-driven model that DARPA has enjoyed. It may, however, be possible to create a virtual requirement process guided by decadal surveys of requirements and priorities.

Likewise, NASA does not have the resources to re-create joint government-industry efforts on the scale of SEMATECH or the Human Genome Project, but those two grand efforts do demonstrate the value of joint undertakings, and NASA may wish to pursue a similar model on whatever scale available resources allow.

REFERENCES

Carayannis, E.M., and J. Alexander. 2004. Strategy, structure, and performance issues of precompetitive R&D consortia: Insights and lessons learned from SEMATECH. IEEE Transactions on Engineering Management 51(2).

Collins, F., M. Morgan, and A. Patrinos. 2003. The Human Genome Project: Lessons from large-scale biology. Science 300(April 11).

Frazier, M., G. Johnson, D. Thomassen, C. Oliver, and A. Patrinos. 2003. Realizing the potential of the genome revolution: The genomes to life program. Science 300(April 11).

Spencer, W.J., and T.E. Seidel. 1995. National technology roadmaps: The U.S. semiconductor experience. Proceedings of the 4th International Conference on Solid-State and Integrated Circuit Technology, Beijing, China, October 24-28.

G

Statement of Task and Work Plan

STATEMENT OF TASK

An ad hoc steering committee and supporting panels will be formed under the auspices of the Aeronautics and Space Engineering Board to conduct a study to develop a decadal strategy for federal aeronautics research. The steering committee will provide an overarching roadmap for investment in aeronautics research and technology at NASA. The committee will be supported by five panels, as outlined below.

During the first portion of the study, the steering committee will develop a set of key technical questions, issues, and challenges that federal aeronautics basic and applied research should address over the next 10 years. This broad charge will be the basis for the work of the supporting panels in developing a decadal strategy for national aeronautics. The charge will also identify challenges that will not only have a significant, long-term impact on national aeronautics but that NASA is uniquely suited to address.[1]

As appropriate, the committee will consider the following when laying out its charge to the panels:

- Current plans for national aeronautics research.
- Lessons learned from (1) the requirements process used by the Department of Defense and (2) the innovation-driven process used by the Defense Advanced Research Projects Agency (DARPA) to select, prioritize, plan, and execute its programs.
- Inputs solicited by the committee from government, academia, industry, and other stakeholders in the air transportation community.
- Lessons learned from nonaeronautical research by federal agencies (e.g., the National Institutes of Health) that positively affected areas with strong private sector economic interests.
- The findings of recent studies supported by the federal government and other organizations.

During the second portion of the study, the panels will assess how federal agencies can more effectively answer the key questions, resolve the key issues, and meet the key challenges identified in the first phase. Each panel will be comprised of subject matter experts from the appropriate disciplines and provide the committee with an internal written summary of its results, which will include the following (some topics will not be relevant to the work of all panels):

- Identification of research necessary to further the state of the art in the specific thrust areas identified by the committee.
- A single, prioritized list of goals and objectives in each appropriate technical area suitable for guiding the allocation of federal resources available for national aeronautics research during the next 10 years, including development and maintenance of necessary infrastructure (i.e., research, modeling, simulation, and test facilities); necessary testing and flight demonstrations that will demonstrate the scalability of novel concepts and capabilities to real-world implementation, necessary workforce; and resolution of relevant issues related to safety, security, and the environment.
- Guidance on how federal resources allocated for aeronautics research should be distributed between in-house and out-of-house (academic and industrial) organizations.
- Guidance on how aeronautics research can take advantage of advances in cross-cutting technology funded by federal agencies and private industry.
- Guidance regarding how far along the development and technology readiness spectrum federal agencies should advance key aeronautics technologies.

[1]The result of this portion of the study was a set of six strategic objectives (see Chapter 1) and a modified QFD process, including instructions for defining R&T Challenges and criteria for establishing quantitative scores to prioritize the Challenges (see Chapter 2).

Similarly, the panels will consider the following during their work:

- Worldwide state of the art and state of practice in relevant fields.
- Interdisciplinary research and cross-cutting technologies.
- Systems integration.
- Simulation methods, laboratory and wind tunnel testing, and flight demonstration.
- Special workforce, education, and training issues related to specific areas of expertise.
- Operational requirements of the U.S. air transportation industry, the FAA, airports, the Department of Defense, general aviation, and other users of the national airspace.
- How aeronautics research priorities and endeavors by industry, universities, the Department of Defense (including DARPA), and other government agencies (such as the FAA) should affect aeronautics research by NASA. Areas of particular interest include but are not limited to computational fluid dynamics and turbulence modeling, materials, and networking and information technology.

Based on written internal inputs from the panels, the steering committee will prepare a final report that discusses the framework for current investments and the key issues related to investment in aeronautics R&D, integrates the results of the panels, and provides a set of overall findings and recommendations to provide a cumulative, integrated view of the panel results. The committee will also specifically focus on identifying cross-cutting technologies and broad areas of investment between the various panels' recommendations and highlight areas of possible revolutionary advancement. Neither the committee nor the panels will make specific budget recommendations.

WORK PLAN

The study will begin with a joint kick-off meeting between the steering committee and its supporting panels in order to hear directly from NASA and other federal entities the primary purpose of the study. The steering committee of approximately 15 members will develop an overarching set of principles for investment in national aeronautics research and technology and a set of key challenges (technical thrust areas) that will guide the panels' work. The supporting panels, as overseen by the steering committee, will individually address the key points in the statement of task, meeting approximately three more times. Each panel will provide an examination of the research priorities, possible research plans, and capabilities necessary to undertake the suggested research. Internal working papers and summaries will be provided to the steering committee from the panels outlining their conclusions, findings, and recommendations. The steering committee will prepare a final report that integrates the results of the panels and provides a cumulative, integrated, and prioritized set of overall findings and recommendations.

It is expected that five supporting panels will be formed (see below) composed of approximately 10 members each:

Panel A: Aerodynamics and aeroacoustics
Panel B: Propulsion and power
Panel C: Materials and structures
Panel D: Dynamics, navigation, and control, and avionics
Panel E: Intelligent and autonomous systems, operations and decision making, human integrated systems, and networking and communications

All five panels should address issues related to subsonic, supersonic, and hypersonic flight regimes; infrastructure; transformation of the air transportation system; workforce; and education. As necessary, one or more subgroups of the steering committee will integrate and evaluate the suggestions of the panels in some or all of these areas.

H

Biographies of Committee and Panel Members

STEERING COMMITTEE

PAUL G. KAMINSKI (NAE), *Chair*, is the chairman and chief executive officer of Technovation, Inc., as well as a senior partner in Global Technology Partners. Dr. Kaminski previously served at DoD as the under secretary of defense for acquisition and technology. His government experience also includes a 20-year career as an officer in the U.S. Air Force, where he directed major development programs, including advanced reconnaissance systems and the stealth program. Dr. Kaminski is a member of the National Academy of Engineering (NAE), a fellow of the Institute for Electrical and Electronics Engineers (IEEE), a fellow of the American Institute of Aeronautics and Astronautics (AIAA), and a senior fellow and past chairman of the Defense Science Board. He received a Ph.D. in aeronautics and astronautics from Stanford University, a B.S. from the Air Force Academy, and M.S. degrees in aeronautics and astronautics and in electrical engineering from the Massachusetts Institutes of Technology (MIT).

WILLIAM W. HOOVER, *Vice Chair*, is currently a consultant for aviation, defense, and energy matters. He is the former executive vice president of the Air Transport Association of America, where he represented the interests of major U.S. airlines, particularly as they relate to technical, safety, and security issues. Before that, he served as the assistant secretary, defense programs, U.S. Department of Energy (DOE), where he was responsible for all aspects of the U.S. nuclear weapons development program. He is also a major general, U.S. Air Force (retired), and had responsible positions in the Air Force Space Program, within NATO, at the Pentagon with the secretary of the Air Force, and in Vietnam, where he commanded a combat air wing and flew 97 missions as a fighter pilot. He has served as chairman of the Aeronautics and Space Engineering Board (ASEB) of the National Research Council and as a member of the NASA Advisory Council. He holds a B.S. in engineering from the U.S. Naval Academy, an M.S. in aeronautical engineering from the Air Force Institute of Technology, and is a distinguished graduate of the National War College and a lifetime national associate of the National Academies.

INDERJIT CHOPRA is the Alfred Gessow Professor in Aerospace Engineering and director of the Alfred Gessow Rotorcraft Center at the University of Maryland (UM), where he acted as department chair from 1988 to 1990 and was the Minta Martin Research Professor from 1996 to 2000. He has been working on various fundamental problems related to the aeromechanics of helicopters, including aeromechanical stability, active vibration control, modeling of composite blades, rotor head health monitoring, aeroelastic optimization, smart structures, micro air vehicles, and comprehensive aeromechanics analyses of bearingless, tilt-rotor, servo-flap, compound, teetering, and circulation control rotors. He has been the principal investigator of four major Army research programs: a university research initiative on smart structures technology; a multidisciplinary university research initiative (MURI) on innovative smart technologies for and actively controlled rotorcraft that rides as smoothly as a jet; the Rotary-Wing Center of Excellence (cosponsored by NASA); and a MURI on micro hovering air vehicles. An author of 150 archival journal papers and 234 conference proceedings papers, Dr. Chopra has been an associate editor of the *Journal of the American Helicopter Society*, the *AIAA Journal of Aircraft*, and the *Journal of Intelligent Materials and Systems* and has served on the editorial advisory boards of *VERTICA*, *Smart Materials and Structures*, *SADHANA*, and *Journal of Aircraft*. He received UM's Distinguished Research Professorship in 1992, UM's Presidential Award for Outstanding Service to the Schools in 1995, the AIAA Structures, Structural Dynamics and Materials Award in 2002, the American Helicopter Society (AHS) Grover E. Bell Award in 2002, the American Society of Mechanical

Engineers (ASME) Adaptive Structures and Material Systems Prize in 2001, the A.J. Clark School of Engineering Faculty Outstanding Research Award in 2002, and the SPIE Smart Structures and Materials Lifetime Achievement Award in 2004. He has been a member of the Army Science Board. He is a fellow of the American Institute of Aeronautics and Astronautics (AIAA), a fellow of the American Helicopter Society (AHS), a fellow of the American Society of Mechanical Engineers (ASME), a fellow of the Aeronautical Society of India (ASI), and a fellow of the National Institute of Aerospace (NIA). He received his Sc.D. from MIT.

EUGENE E. COVERT (NAE) is the T. Wilson Professor of Aeronautics, emeritus, at MIT. His long and distinguished career in aerospace has spanned over 60 years in academia and has included additional stints as chief scientist of the U.S. Air Force, member and chairman of the Air Force Scientific Advisory Board, chairman of the Power and Propulsion panel of NATO's Advisory Group for Aerospace Research and Development, director of the Wright Brothers Facility, and chairman of the ASEB. Dr. Covert's experience provides an important perspective on trends in aeronautical research and development, particularly with regard to propulsion.

ALAN C. ECKBRETH is a private consultant who also serves as the vice president/president-elect of the Connecticut Academy of Science and Engineering (CASE). He has extensive industrial experience in complex reacting flows and previous service to NRC/Air Force Office of Scientific Research (AFOSR) panels and many other government committees. In 2004, he chaired the NRC review panel for NASA's Intelligent Propulsion Systems Foundation Technologies. Dr. Eckbreth earned his doctorate in aerospace and mechanical sciences from Princeton University in 1968. He also holds an M.S. in administrative sciences from Rensselaer. After leaving Princeton, he joined United Technologies Research Center (UTRC) and spent 34 years there in both technical research and senior management positions. Initially, he was engaged in research into the properties of high-power electric-discharge CO_2 convection lasers, but the bulk of his career was spent in developing and applying spatially precise laser techniques for gas dynamic and combustion diagnostic purposes, most notably coherent anti-Stokes Raman spectroscopy. His senior management positions at UTRC included director, Fuel Cells Program; director, Aeromechanical, Chemical, and Fluid Systems Program; and director, Pratt & Whitney Program. Dr. Eckbreth is the author of over 60 technical papers and a book on laser diagnostics and has lectured widely on the subject. He is a fellow of the AIAA and of the Optical Society of America (OSA). He served on the U.S. Air Force Scientific Advisory Board from 1995 to 1999 and on numerous other technical advisory panels to NASA and DoD. In 1985, he received the George Mead Medal from United Technologies Corporation for outstanding engineering achievement. After retiring from UTRC and prior to entering consulting, he held the position of vice president and dean, Rensselaer at Hartford, a branch campus of the Rensselaer Polytechnic Institute (RPI), and was also professor of mechanical engineering in the Department of Engineering and Science.

THOMAS M. HARTMANN is a program manager within the Advanced Development Programs (Skunk Works) organization of Lockheed Martin Aeronautics Company. His primary responsibility is for the quiet supersonic transport (QSST) program and related technology initiatives. He joined Lockheed-California Company as a propulsion engineer in 1980. He has worked on the earliest stages of what is now the F/A-22 Raptor, assuming roles of increasing technical responsibility. In 1992, he was instrumental in the capture of a classified program and transitioned into project management. He has managed several technology and development programs within the Skunk Works. Mr. Hartmann received a B.S. from the University of California, Davis in 1980, with a double major in aeronautical and mechanical engineering. He received an M.S. in aerospace engineering from the University of Southern California in 1984. He has attended several management, executive, and program management development programs within Lockheed Martin and at Defense Acquisition University. In 2002 he was recognized by Lockheed Martin with its NOVA award, the corporation's highest employee recognition award. He holds several U.S. patents related to sonic-boom mitigation technologies and their application to supersonic business jets.

ILAN KROO (NAE) is a professor in Stanford University's Department of Aeronautics and Astronautics. He received his B.S. in physics from Stanford in 1978 and then continued studies at Stanford in aeronautics, receiving a Ph.D. in 1983. He worked in the Advanced Aerodynamic Concepts Branch at NASA Ames Research Center for 4 years before returning to Stanford as a member of the aero/astro faculty. Dr. Kroo's research in aerodynamics and multidisciplinary design optimization includes the study of innovative airplane concepts. He has participated in the design of commercial aircraft, UAVs, sailplanes, America's Cup sailboats, and supersonic aircraft. Dr. Kroo, a member of ASEB, has served as a member of the NRC Committee on Review of the Effectiveness of Air Force Science and Technology Program Changes, the NRC Committee on Breakthrough Technology for Commercial Supersonic Aircraft, and the Committee for Materials, Structures, and Aeronautics for Advanced Uninhabited Air Vehicles. In addition to his research and teaching interests, Dr. Kroo is chief scientist of Desktop Aeronautics, Inc., a software and consulting company.

NANCY G. LEVESON (NAE) is professor of aeronautics and astronautics and professor of engineering systems at MIT. Dr. Leveson conducts research on system safety, software engineering, system engineering, and human–computer interaction. In 1999, she received the Association for Computing Machinery's (ACM's) Allen Newell Award for outstanding computer science research and in 1995 the AIAA Information Systems Award for "developing the field of software safety and for promoting responsible software and system engineering practices where life and property are at stake." This year she received the ACM Sigsoft Outstanding Research Award. She has published over 200 research papers and is author of a book, *Safeware: System Safety and Computers,* published by Addison-Wesley. She has served on numerous National Academies committees.

IVETT A. LEYVA is a senior aerodynamicist at Microcosm, Inc., where she is responsible for the development of ablative chambers and also performs CFD simulations of Microcosm's launch vehicles' external aerodynamics. She recently led a testing campaign for a 20k-lbf thrust engine for Microcosm's small launch vehicle for the Defense Advanced Research Projects Agency's (DARPA's) FALCON program. Dr. Leyva worked at the General Electric Global Research Center from 1999 to 2003. There, she worked on pulse detonation engines and led the design, fabrication, and testing of a pulse detonation engine concept at the component and system level. She also initiated and coordinated different research projects with scientists from the former Soviet Union. She has four granted U.S. patents in the area of combustion and four more filed (three of those are also filed in the European Union, Japan, or Canada). Dr. Leyva also worked as a thermal engineer for Exponent, where she investigated the cause, origin, and prevention of aviation accidents as well as fires and explosions on scales ranging from the residential to the industrial. Dr. Leyva graduated from Caltech with a Ph.D. in aeronautics. Her dissertation was a numerical and experimental study on the shock detachment process on cones in hypervelocity flows. Dr. Leyva has been part of several NRC committees for propulsion.

AMY PRITCHETT is the David D. Lewis Associate Professor of Cognitive Engineering in the School of Aerospace Engineering and a joint associate professor in the School of Industrial and Systems Engineering at the Georgia Institute of Technology. Her research encompasses human–automation interaction, including advanced decision aids; procedure design as a mechanism to define and test the operation of complex, multiagent systems (e.g., air traffic control, spacecraft mission control); and simulation of complex systems to assess changes in emergent system behavior in response to implementation of new information technology. She is on the editorial board of the *Journal of Cognitive Engineering and Decision Making*, an area editor of *Transactions of the Society for Computer Simulations*, and associate editor of the *AIAA Journal of Aerospace Computing, Information and Communication*. Her awards include the Jackson Award of the Radio Technical Commission for Aeronautics (RTCA) for contribution to aviation. She previously served on the NRC Committee for Vision 2050 and the NRC committee reviewing the NGATS JPDO plan, where she contributed to the committee's investigation of system modeling and human factors. Dr. Pritchett is currently a member of ASEB.

EDMOND L. SOLIDAY was employed by United Airlines for over 35 years as a pilot, human factors instructor, flight manager, and staff executive, serving the last 11 as vice president for safety, quality assurance, and security. He has served on numerous aviation safety-related advisory boards and commissions, and he chaired the Commercial Aviation Safety Team, the Air Transport Association's (ATA's) Safety Council, the Star Alliance Safety Committee, and the ATA Environmental Committee. Captain Soliday currently serves on the board of governors for the Flight Safety Foundation and on the board for the MIT's Global Airline Industry Program Advisory Group. Among his awards are the Bendix Trophy, the Vanguard Trophy, and the Laura Tabor Barbour International Air Safety Award. Captain Soliday previously served on four NRC study groups.

JOHN VALASEK is director, Flight Simulation Laboratory, and associate professor of aerospace engineering at Texas A&M University. He has been actively conducting flight controls research and configuration design of manned and unmanned air vehicles in both industry and academia for 20 years. His research interests include autonomous intelligent control of unmanned air and ground vehicles, autonomous air refueling, vision-based navigation systems, intelligent cockpit computing and displays, and morphing air vehicles. In industry, he was a flight control engineer for Northrop Corporation's Aircraft Division, where he worked on integrated flight and propulsion control systems in the Flight Controls Research Group and on the AGM-137 tri-services standoff attack missile (TSSAM) in the Flight Controls Analysis Group, where he received the Northrop Corporation Outstanding Contribution to Program Award. He has been an AFOSR summer faculty research fellow for the Flight Dynamics Directorate at Wright Laboratories and a NASA summer faculty researcher in the Guidance and Control Branch at NASA Langley. In addition to university research, he is a consultant on flight control to several companies. He is an associate fellow of the AIAA and a senior member of the IEEE, as well as a chair or member of numerous AIAA and IEEE technical committees. He is a reviewer for the NRC and the AFOSR and a former associate editor for *IEEE Transactions on Education.* He earned a B.S. at California State Polytechnic University and M.S. and Ph.D. degrees from the University of Kansas, all in aerospace engineering.

DAVID VAN WIE is an aerospace engineer in aerospace vehicle design and development, with emphasis on propulsion systems and advanced aerodynamics for supersonic and hypersonic flight vehicles. He has been with the Johns Hopkins University Applied Physics Laboratory since 1983 and is currently a member of the principal professional staff and director of the Precision Engagement Transformation Center. Dr. Van Wie also holds appointments as research professor in the Department of Mechanical Engineering at the Johns Hopkins University and lecturer in the Department of Aerospace Engineering at UM. Dr. Van Wie attended the UM between 1976 and 1986 and received B.S., M.S., and Ph.D. degrees in aerospace engineering. He was also awarded an M.S. in electrical engineering from Johns Hopkins University in 1998. He was awarded the Gene Zara Award for outstanding contributions to the National Aerospace Plane (NASP) program in 1989 and 1992. Dr. Van Wie was a member of the Air Force Scientific Advisory Board's (SAB's) Committee on Hypersonic Airbreathing Vehicles (1991), of the NRC Committee on the Assessment of the Air Force Hypersonic Technology Program (1987), and SAB 2000 summer study on Air Force hypersonics.

ROBERT WHITEHEAD entered government service in 1971 after receiving undergraduate and graduate degrees in engineering mechanics from Virginia Polytechnic Institute and State University and completing a 1-year postdoctoral associateship at the NASA Ames Research Center. Dr. Whitehead began his career with the Department of Navy as a research engineer in the Aviation Department of the David Taylor Naval Ship R&D Center at Carderock. He transferred to the Office of Naval Research (ONR) in 1976 as a scientific officer in applied aerodynamics, managing university and industry research projects. For the next 13 years Dr. Whitehead held a number of positions at ONR, finally as director in the Mechanics Division from 1986 until 1989, when he transferred to NASA Headquarters. Dr. Whitehead began at NASA in the Office of Aeronautics as the assistant director for aeronautics (rotorcraft). He held a variety of other positions, including director of the Subsonic Transportation Division, before becoming the deputy associate administrator for aeronautics in 1994. Dr. Whitehead became associate administrator for aeronautics in 1995 and associate administrator for aeronautics and space transportation technology in 1997. Dr. Whitehead retired from federal service in December 1997. He consulted part-time in the aerospace community and its professional associations, on both a volunteer and a paid basis, until becoming interim president and executive director of the NIA in 2002. With the appointment of a permanent NIA president in October 2003, Dr. Whitehead became NIA vice president for Research and Program Development until June 2004. He currently is a consultant on special projects to NIA.

DIANNE S. WILEY is a Boeing Technical Fellow for Airframe Technology Integration. She serves as the enterprise liaison to the Boeing Technical Fellowship to facilitate technology maturation and technology transition to the space exploration systems business area. In her prior assignment with the Boeing Phantom Works, she was the program manager for airframe technology on the NASA Space Launch Initiative Program, overseeing the development and demonstration of advanced structure and materials technology for next-generation, reusable launch vehicles. Previously, she was with Northrop Grumman for 20 years, where she had been manager of airframe technology. In that position, Dr. Wiley was responsible for research and development and technology transition in structural design and analysis, materials and processes, and manufacturing technology. Dr. Wiley was responsible for developing and implementing innovative structural solutions to ensure the structural integrity of the B-2 aircraft. Dr. Wiley's 25 years of technical experience have involved durability and damage tolerance, advanced composites (organic and ceramic), high-temperature structures, smart structures, low-observable structures, systems engineering, and rapid prototyping. Dr. Wiley holds a Ph.D. in applied mechanics from the University of California-Los Angeles School of Engineering and Applied Science. She attended Defense Systems Management College (1996). She is a graduate of the Center for Creative Leadership (1995), Leadership California Class of 1998, and the Boeing Leadership Center (2002.)

PANEL A: AERODYNAMICS AND ACOUSTICS

DAVID VAN WIE, *Panel Chair* (see biography above).

PAUL BEVILAQUA (NAE) is manager of advanced programs at the Lockheed Martin Aeronautics Company. He joined Lockheed Martin as chief aeronautical scientist of the Lockheed Advanced Aeronautics Company and became chief engineer of advanced development projects in the Lockheed Martin Skunk Works. During this time he played a leading role in creating the Joint Strike Fighter Program and invented the lift fan propulsion system that makes it possible to build variants of a single stealthy, supersonic V/STOL aircraft for the Air Force, Marines, and Navy. Prior to joining Lockheed Martin, he was manager of advanced programs at Rockwell International's Navy aircraft plant. He began his career as a captain in the U.S. Air Force and deputy director of the Energy Conversion Laboratory at Wright Patterson Air Force Base. Dr. Bevilaqua is a member of the NAE and a fellow of the AIAA. He is the recipient of an Air Force Scientific Achievement Award for his contributions to turbulence theory, the AIAA Newbold award for his contributions to V/STOL aircraft technology, the AIAA and SAE aircraft design awards for his contributions to aircraft design, and the Collier Trophy for his lift fan propulsion system. His publications include articles in the *Journal of the AIAA,* the *Journal of the Royal Aeronautical Society,* and the proceedings of many meetings and symposia.

CHARLES BOCCADORO is business area director for Future Strike Systems in the Integrated Systems Western Region, Northrop Grumman, in El Segundo, California. This business area focuses on defining next-generation global attack capability for the Air Force, which currently comprises all Air Force next-generation bomber and related technology studies. His team conducted the DARPA Quiet Supersonic Platform study and shaped sonic boom flight demonstration efforts, which received a 2004 NASA Turning Goals Into Reality Award. His team leads Northrop Grumman's Air Force Next Generation, Long-Range Strike Phase II and Phase III study activities. Previously, he was manager of the aerosciences technology and propulsion advanced design organizations. He has worked for Northrop Grumman since 1980 supporting several air vehicle development programs, including F/A-18E/F, YF-23, and B-2. He is a graduate of MIT and the von Karman Institute for Fluid Dynamics and is an associate fellow of AIAA. He is a patent holder, the recipient of a 2003 *Aviation Week* Laurel, the 2004 AIAA Aircraft Design Award, and the 2005 NASA Vehicle Systems Award.

THOMAS CORKE is the Clark Chair Professor in the Aerospace and Mechanical Engineering Department at the University of Notre Dame. He is also the founding director of the Center for Flow Physics and Control (FlowPAC) and the director of the Hessert Laboratory for Aerospace Research. Dr. Corke is internationally recognized for his research in the areas of fluid instabilities and transition to turbulence, control of turbulent boundary layers, flow visualization techniques, and flow control. He has extensive experimental experience over the full range of Mach numbers, from incompressible to hypersonic flows, in a large number of flow fields, including boundary layers, wakes, and jets. His research also involves computational fluid dynamics especially with regard to acoustic receptivity and plasma actuators. His Ph.D. work on the control of large-scale turbulence in boundary layers for drag reduction led to his receiving a NASA Langley Achievement Award in 1982. He was the first to introduce controlled three-dimensional disturbances to verify subharmonic resonance mechanisms in boundary layers, which was recognized with a NASA Langley Achievement Award in 1995. He was named an associate fellow in AIAA the same year. Dr. Corke wrote the textbook *Design of Aircraft* (Prentice-Hall, 2002), which has to date been adopted by approximately 12 aerospace engineering departments for their capstone design course. He was named a fellow of ASME in 2005.

ILAN KROO (NAE) (see biography above).

ROBERT LIEBECK (NAE) is currently manager of the Blended-Wing-Body Program at Boeing. In his 44 years at Boeing, he has served as program manager on several classified advanced-concept airplane programs, some of which culminated in successful flight vehicles. He has an extensive list of technical publications, and his airfoil work is discussed in several textbooks on aerodynamics. He is also professor of the practice of aeronautics at the Massachusetts Institute of Technology and adjunct professor of mechanical and aerospace engineering at the University of California, Irvine, where he teaches aerodynamics, flight mechanics, and airplane design. He received B.S., M.S., and Ph.D. degrees in aeronautical engineering from the University of Illinois in Urbana Champaign and received the university's College of Engineering Distinguished Alumnus Award in 1996. As a consultant, he designed the wings for racing cars that have won the Indianapolis 500 and Formula 1 races, the keel section for the *America*[3] yacht that won the America's Cup in 1992, and the wing for a World Aerobatic Championship airplane. Dr. Liebeck is a Boeing Senior Technical Fellow, an AIAA fellow, a recipient of the AIAA Aerodynamics Award, a recipient of the AIAA Aircraft Design Award, a recipient of the AIAA Wright Brothers Lectureship in Aeronautics, a fellow of the Royal Aeronautical Society, a recipient of the ASME Spirit of St. Louis Medal, and a member of the NAE.

DAN MARREN is the chief of the Hypersonics Systems Division at the Arnold Engineering Development Center (AEDC). He also serves as the Air Force site director for the AEDC White Oak site in Silver Spring, Maryland, home to the Hypervelocity Wind Tunnel 9. His experience includes technical leadership on over 30 projects relating to advanced hypersonic research and development in the Hypervelocity Wind Tunnel 9 as well as several supersonic projects in Navy supersonic facilities. He had the primary role in the conception, design, and development of a major new facility at White Oak and participated in every other important upgrade to the facility. In his previous experience as ground test coordinator for the Navy reentry special projects office, he managed the ground test program for several Navy programs from the advanced project office. He has chaired panels, studies, and focus groups, providing input as a subject matter expert in hypersonics, and maintains technical positions in the AIAA and the Supersonic Tunnel Association International (STAI). He has designed and taught several short courses on physics, testing, and hypersonics for various audiences, including universities, professional societies, and education and public outreach for younger students. Mr. Marren earned his B.S. in aerospace engineering from the University of Cincinnati and his M.S. in engineering management, specializing in high-temperature gas dynamics, at the University of Maryland.

STEPHEN RUFFIN is an associate professor in the School of Aerospace Engineering at the Georgia Institute of Technology and head of its Aerothermodynamics Research and Technology Laboratory. He leads the computational fluid dynamics (CFD) research thrust and collaborates in its integration with the systems analysis tool and with the aeroelastic analysis. Dr. Ruffin is a specialist in high-temperature gas dynamics, compressible flow aerodynamics, CFD, and airframe-propulsion integration. He is leading the development of a three-dimensional Cartesian-grid-based Navier-Stokes solver for design applications and the development of Cartesian-grid approaches for chemically reacting flows. He has conducted computational and experimental studies of a novel channel concept that provides increased lift/drag ratios for reentry vehicles relative to conventional blunted geometries. Dr. Ruffin is also conducting research on high-speed, high-temperature flows in which vibrational energy modes are substantially excited and in which chemical nonequilibrium exists. He has gained this experience through work in the Thermoscience Division at NASA Ames, NASA Glenn, and during years of high-speed CFD research at the Georgia Institute of Technology, Stanford University (Ph.D., 1993), MIT (M.S., 1987), and Princeton University (B.S.E., 1985).

FREDRIC H. SCHMITZ recently stepped down as the Martin Professor of Rotorcraft Acoustics Research in the Department of Aerospace Engineering at UM and is now working half-time as a senior research professor at UM and is a visiting professor at Stanford University. Dr. Schmitz has 36 years' experience in aeronautics, specializing in rotorcraft aeromechanics with an emphasis on rotorcraft acoustics and low-speed aerodynamics. He has received numerous awards and honors for his management accomplishments and his pioneering research. Since 1998, he has been building a new rotorcraft acoustic program at UM that utilizes fundamental experiments to validate key aspects of acoustic theory. His background includes large-scale and model-scale acoustic testing, rotorcraft impulsive noise theory, development of national and international acoustic test facilities, and other contributions to research in acoustics and to research in aeromechanics of low-speed aircraft. He has led both in the development of novel research programs in wind tunnels throughout the world (DNW, The Netherlands, and CEPR-19, France) and in the management of research and operation at the world's largest wind tunnel, the 40 × 80 × 120 foot wind tunnel (NASA Ames Research Center). Before taking early retirement from NASA Ames in 1998, he served NASA as the director of aeronautics, deputy director of information technology, chief of the Applied Aerodynamics Division, and chief of the Full-Scale Aerodynamics Research Division and the U.S. Army as chief of the Fluid Mechanics Division for the Aeroflightdynamics Directorate at the NASA Ames Research Center. Dr. Schmitz has a bachelor's degree in aerospace engineering from RPI and M.S. and Ph.D. degrees from Princeton University. He taught the rotorcraft aeromechanics classes at Stanford University for 18 years while holding positions at NASA and in the Army, achieving the rank of consulting professor. He is a fellow of the AHS and the AIAA, and has a commercial rotary-wing pilot's license.

JOHN SULLIVAN is a professor in the School of Aeronautics and Astronautics at Purdue University. He received a B.S. in 1967 from the University of Rochester and an M.S. (1969) and a Ph.D. (1973) in aeronautical engineering from MIT. After graduation, he cofounded a small high-technology company in California. In 1975, he joined the faculty of Purdue University. His administrative experiences there include director of the Center for Advanced Manufacturing, Codirector of the Product Lifecycle Management Center of Excellence, head of the School of Aeronautics and Astronautics (1993-1998), and associate head (1991-1993) and director of the Aerospace Sciences Laboratory (1983-1995). He directs graduate student research in the general area of experimental aerodynamics/fluid mechanics. He is the author or coauthor of approximately 115 technical publications and has served as the major professor for 40 M.S. and 12 Ph.D. thesis graduate students. He spent a sabbatical year at ONR in 1989-1990 and a year at the Boeing Company in 2002.

KAREN WILLCOX is associate professor of aeronautics and astronautics in the Aerospace Computational Design Laboratory at MIT. She received a bachelor of engineering degree from the University of Auckland in 1994 and M.S. and Ph.D. degrees in aeronautics and astronautics from MIT in 1996 and 2000, respectively. She spent 1 year as a visiting researcher at Boeing Phantom Works in Long Beach, California, working with the blended-wing-body design team. She joined the MIT faculty in the Department of Aeronautics and Astronautics in the fall of 2001. Dr. Willcox's research interests lie in computational simulation and optimization of engineering systems. Her research focuses are firstly in model reduction for large-scale systems, with applications in active flow control, aeroelasticity, and variable-fidelity design methods, and secondly in multidisciplinary system design and optimization, with particular emphasis on economic and environmental factors in aircraft conceptual design.

PANEL B: PROPULSION AND POWER

ALAN C. ECKBRETH, *Panel Chair* (see biography above).

ROBERT J. BAKOS is the vice president and general manager of ATK GASL. He is responsible for the overall management of the ATK GASL business unit in advanced aeropropulsion and hypersonic technology development. Previously, he has been the vice president of engineering for Allied Aerospace, the vice president of research and development for GASL, and the director of the HYPULSE Laboratory at GASL. Dr. Bakos is the author of over 30 conference and journal papers on the development of hypersonic technologies and test techniques. He is a senior member of the AIAA and serves on the AIAA HyTASP Program Committee. He received his B.S. in civil engineering from Polytechnic University and his M.S. and Ph.D. in mechanical engineering from Cornell University and the University of Queensland, respectively.

MEYER J. "MIKE" BENZAKEIN (NAE) has experience in both industry and academia. Currently the chair of the Department of Aerospace Engineering at the Ohio State University, he has over 37 years of experience with GE Aircraft Engines (GEAE) as the general manager of advanced engineering, where he was responsible for technology maturation and for strengthening the linkage between the preliminary design of engine systems and production hardware design. In a previous assignment, Dr. Benzakein was general manager for engine systems design and integration, where he was responsible for engineering leadership and technical oversight of commercial and military aircraft engine programs. He is a fellow of the AIAA, a fellow of the Royal Aeronautical Society, and was a member of two NRC committees. He was nominated to the NAE for achievements in international technical cooperation and propulsion engine technology. Dr. Benzakein has experience in the design and production process, as well as expertise in engineering materials, noise and emissions, and systems engineering.

JAMES L. BETTNER retired from Rolls-Royce Aero Engines in 2002, where he was the Program Manager for the AE 3007H engine, which is the propulsion system for the Air Force's high-altitude, long-endurance Global Hawk unmanned aerial vehicle. Previously, he was the supervisor of the Preliminary Design Department, where he conducted studies on material properties in advanced engines, convertible engines, gearboxes for high-speed rotorcraft, wave rotors, and fuel cooled engines. Dr. Bettner also directed ERAST studies of optimum propulsion systems for very-high-altitude research aircraft. He directed the preliminary design of a propulsion system for a large fan-in-wing Special Operations Force aircraft, where the engines powered a conventional fan in forward flight but were clutched to the fan-in-wing for vertical takeoff and landing. He directed the preliminary design analysis of developing a 2000-pound-thrust turbofan from the T800 turboshaft engine for a medium-altitude application. Prior to that, Dr. Bettner was a member of the propfan development team, which included the NASA-funded single-rotation propfan test assessment (PTA) and the company-funded counter-rotation PW-Allison 578 projects. He received his Ph.D. from Purdue University. Dr. Bettner has expertise in engine materials, propfans, and other elements of propulsion.

DAVID "ED" CROW (NAE) graduated from the University of Missouri-Rolla with a Ph.D. in mechanical engineering. He joined the faculty of the University of Connecticut as a distinguished professor in residence in the mechanical engineering department after a distinguished career in industry. Dr. Crow joined Pratt & Whitney in 1966, rising to the position of senior vice president of Pratt & Whitney's engineering organization in May 1997, where he was responsible for the design, development, validation, and certification of all Pratt & Whitney large commercial engines, military engines, and rocket products. He also led the research and development of advanced technologies systems to meet future aircraft requirements. Dr. Crow previously held the position of senior vice president for Pratt & Whitney's large commercial engines organization, which included the PW4000 and JT9D high-thrust family of products. He is a past secretary of the Society of Automotive Engineers (SAE) and a member of both ASME and AIAA. In addition to having served as past president of Pi Tau Sigma, he has served on the Engineering Advisory Board at Clarkson University and is an elected member of the University of Missouri-Rolla Academy of Mechanical Engineers. He was named a member of the NAE in 1998. Dr. Crow has expertise in propulsion engineering, thermodynamics, aerodynamics, systems engineering, and rocket propulsion engineering.

MEHRDAD (MARK) EHSANI is Robert M. Kennedy Professor in the Department of Electrical Engineering at Texas A&M University. He is the director of the Power Electronics and Motor Drives Laboratory, the Advanced Vehicle Systems Research Program, and the Texas Applied Power Electronics Consortium (TAPC). Previously, he held positions at the University of Texas Fusion Research Center and at Argonne National Laboratory. Dr. Ehsani is a professional engineer, an IEEE fellow, an SAE fellow, and chairman of the IEEE Vehicular Technology Society's Vehicle Power and Propulsion Committee. He is the author of over 300 publications, 12 books, and over 23 patents. He has won numerous awards for engineering and teaching, including the IEEE Vehicular Technology Society 2001 Avant Garde Award, the BP Amoco Faculty Award for Teaching Excellence in the College of Engineering, and the IEEE Undergraduate Teaching Field Award. He served on the NRC Committee on Assessment of Combat Hybrid Power Systems and the Review of Proposals for NASA's Low Emissions Alternative Power (LEAP) project. He received a B.S. and an M.S. in electrical engineering from the University of Texas at Austin and a Ph.D. from the University of Wisconsin-Madison. Dr. Ehsani has expertise in power electronics, motor drives, vehicle power and propulsion systems, and hybrid vehicles and their control systems.

JEFFREY W. HAMSTRA graduated from the University of Michigan with an M.S. in aerospace engineering. Mr. Hamstra is a Lockheed Martin fellow in propulsion integration at Lockheed Martin Aeronautics Company in Fort Worth, Texas, and is currently assigned to the F-35 Joint Strike Fighter Vehicle Systems team. He has 21 years of experience in jet propulsion systems integration, including program experience from the F/A-22, F-16, and Skunk Works advanced development programs. He has performed as a research and development principal investigator, aircraft project lead, and Propulsion Department manager. He is an associate fellow of the AIAA and deputy director of the AIAA Propulsion and Energy Group. He was inducted as a Lockheed Martin fellow in 2003. He is familiar with the U.S. aircraft engine industry, government propulsion organizations, and propulsion technology development programs and has expertise in propulsion engineering and propulsion systems integration.

IVETT A. LEYVA (see biography above).

TIMOTHY LIEUWEN is an associate professor at the Georgia Institute of Technology. He received his B.S. (1995) from Calvin College and his M.S. (1997) and Ph.D. (1999) from the Georgia Institute of Technology. Dr. Lieuwen is the author of two book chapters, 30 refereed journal articles, and over 100 conference publications. He holds four patents and is an associate editor of the *Journal of Propulsion and Power*. He has consulted with many of the major gas turbine manufacturers. Dr. Lieuwen has held various leadership roles on the Air Breathing Propulsion technical committee of AIAA and the Combustion and Fuels Committee of ASME. Dr. Lieuwen has served on the organizing committees of several major international conferences sponsored by both AIAA and ASME. Dr. Lieuwen's awards include the National Science Foundation's CAREER Award, the AIAA Lawrence Sperry Award, and the ASME/International Gas Turbine Institute (GTI) Turbo Expo Best Paper Award. Dr. Lieuwen has expertise in acoustics, combustion, and stability.

LOURDES QUINTANA MAURICE is the chief scientific and technical advisor for environment in the Federal Aviation Administration's Office of Environment and Energy. She serves as the agency technical expert for basic and exploratory research and advanced technology development focused on aircraft environmental impacts and its application to noise and emissions certification. She previously served as the Air Force Deputy, Basic Research Sciences and Propulsion Science and Technology, in the office of the Deputy Associate Secretary of the Air Force for Science and Technology. She also worked at the Air Force Research Laboratory's Propulsion and Power Directorate from 1983 to 1999, planning and executing basic, exploratory, and advanced development propulsion science and technology programs, focusing on state-of-the-art aviation fuels and propulsion systems. Her areas of expertise include pollutant formation chemistry, combustion kinetics, hypersonic propulsion, and aviation fuels. She received a B.Sc. in chemical engineering and an M.Sc. in aerospace engineering from the University of Dayton in Dayton, Ohio, and a Ph.D. in mechanical engineering from the University of London's Imperial College at London, United Kingdom. She is also a distinguished graduate of National Defense University's Industrial College of the Armed Forces, where she earned an M.Sc. in national resource strategy. Dr. Maurice has served as an advisor to the United Nations' Intergovernmental Panel on Climate Change. She has served on numerous NRC committees, and on the AIAA's Propellants and Combustion Technical Committee and as the U.S. chair for the AIAA/International Council of Aeronautical Sciences' (ICAS's) International Conference in Celebration of the Centennial of Flight. Dr. Maurice is an associate editor for AIAA's *Journal of Propulsion and Power* and serves on the editorial board of the *International Journal of Aeroacoustics*. She has authored over 90 publications and is a 2003 fellow of AIAA.

JAMES C. McDANIEL, JR., is a professor of mechanical and aerospace engineering at the University of Virginia. He is a member of a number of professional societies, including the AIAA, the American Physical Society (Division of Fluid Dynamics), the OSA, and the Combustion Institute of America. He has served on the AIAA Aerodynamic Measurements Technical Committee and currently serves on the AIAA National Aircraft Design Technical Committee and the National SCRAMJET Testing Standards Committee. He has consulted for numerous companies, including Grumman Aircraft, General Electric, Rocketdyne, and Pratt & Whitney. He received a NASP Distinguished Service Award, several aerospace teaching awards, and has advised numerous competition-winning student aircraft design teams. Dr. McDaniel's research interests include fluid mechanics, combustion, laser-based flowfield measurements, and aircraft design. He is the director of the Aerospace Research Laboratory, where basic research in high-speed fuel/air mixing and combustion is conducted using laser-induced fluorescence and other nonintrusive optical measurement techniques. He received a B.S. in aerospace engineering from the University of Virginia as well as an M.S. in electrical engineering and an M.S. and a Ph.D. in aeronautics and astronautics from Stanford. He was also an active-duty pilot in the Air Force and holds commercial, multiengine pilot ratings. Dr. McDaniel has expertise in experimental high-speed propulsion.

TRESA M. POLLOCK (NAE) is the L.H. and F.E. Van Vlack Professor of Materials Science and Engineering at the University of Michigan, Ann Arbor. She received a B.S. from Purdue University in 1984 and a Ph.D. from MIT in 1989. Dr. Pollock was employed at General Electric Aircraft Engines from 1989 to 1991, where she conducted research and development on high-temperature alloys for aircraft turbine engines. She was a professor in the Department of Materials Science and Engineering at Carnegie Mellon University from 1991 to 1999. Her research interests are in the processing and properties of high-temperature structural materials, including nickel-base alloys, intermetallics, coatings, and composites. Professor Pollock is president of The Minerals, Metals and Materials Society (TMS) and associate editor of *Metallurgical and Materials Transactions*. She is a fellow of ASM International and has received the ASM International Research Silver Medal Award. Dr. Pollock was elected to the NAE in 2005. She also served on the NRC's Committee on Material Science and Engineering: Forging Stronger Links to Users. She has expertise in materials for propulsion applications.

WILLIAM TUMAS is a program manager for the Office of Energy and Environment Initiatives at Los Alamos National Laboratory (LANL) as well as director of the Los Alamos Institute for Hydrogen and Fuel Cell Research. He is the lead principal investigator and coordinator of the DOE Center of Excellence for Chemical Hydrogen Storage, which comprises seven universities, four companies, and two national laboratories, including LANL. Prior to joining Los Alamos in 1993, Dr. Tumas was a research chemist, then a project leader in environmental and oxidation catalysis at DuPont Central Research. He was also a member of the DuPont Corporate Catalysis Center and the Corporate Environmental Technology Panel. Dr. Tumas received his B.A. in chemistry summa cum laude from Ithaca College in 1980. He received his Ph.D. in organic chemistry from Stanford University in 1985 as a National Science Foundation (NSF) graduate fellow and a Hertz Foundation fellow, where he studied the dynamics and reaction mechanisms of gas-phase negative ions. He carried out postdoctoral research in organometallic chemistry at the California Institute of Technology from 1985 to 1987 under a National Institutes of Health (NIH)/Chaim Weizmann postdoctoral fellowship. Dr. Tumas's research activities have included chemical hydrogen storage; homogeneous and phase-separable catalysis; catalytic transformations and chemical processing in supercritical fluids and alternate reaction media; green chemistry; and waste treatment technology development and assessment. He is the coeditor of the book *Green Chemistry Using Liquid and Supercritical Carbon Dioxide*. He has over 45 peer-reviewed publications and has given over 40 invited presentations, including invited talks at seven different Gordon Research Conferences. He chaired the 1998 Gordon Conference on Green Chemistry. Dr. Tumas is also the president of Big Rock Consulting, LLC, through which he has carried out technology assessment for the U.S. Army and Science Applications International Corporation (SAIC) for over 4 years. He was also a founding board member of the Green Chemistry Institute (1997-2002), which is now part of the American Chemical Society. He participated on three NRC committees, including 5 years on the Committee on Review and Evaluation of the Army Chemical Stockpile Disposal Program, where he contributed to 10 NRC reports. He has expertise in fuel cells and hydrogen power.

PANEL C: STRUCTURES AND MATERIALS

DIANNE S. WILEY, *Panel Chair* (see biography above).

SATYA N. ATLURI (NAE) is the Samueli/von Karman Chair in Aerospace Engineering at the University of California, Irvine. Previously, he was the Hightower Chair in Engineering at Georgia Tech and the Jerome Clarke Hunsaker Professor of Aeronautics at MIT. He is a member of the NAE, a distinguished alumnus of the Indian Institute of Science, a fellow of the Third World Academy of Sciences, a member of the European Academy of Sciences, a foreign fellow of the Indian National Academy of Engineering, an honorary fellow of the International Congress on Fracture, and a fellow of several learned societies, including the American Academy of Mechanics, AIAA, the Aeronautical Society of India, ASME, the U.S. ACM, and the International Association for Computational Mechanics (IACM). He is also the recipient of numerous awards, including the Hilbert Medal, the National Medal of Technology Citation for Distinguished Service, the FAA Excellence in Aviation Award, the Pendray Aerospace Literature Award, and the A.C. Eringen Medal. He has chaired committees for the National Academy of Engineering, the USA-India Science and Technology Forum, and the U.S. Army, as well as numerous technical conferences. He is the editor of a number of journals, including *Computer Modeling in Engineering and Sciences*, which he also founded. His research includes computational modeling in multidisciplinary engineering and the sciences; structural integrity and damage tolerance of rotorcraft; meshless methods of computational sciences, especially the meshless local Petrov-Galerkin (MLPG) and the meshless local boundary integral equation (LBIE) methods that he and his students recently pioneered; computational nanoengineering and science; wireless virtual airport for enhanced aviation security; device modeling in microelectromechanical systems; and aging and life-enhancement of aircraft, spacecraft, and power-generating systems. He has performed research for the U.S. Rotorcraft Industry Technology Association, the NSF, ONR, AFOSR, the Army Research Office, the Nuclear Regulatory Commission, NASA, and many others, including directing major research efforts such as the FAA-sponsored National Center of Excellence for Aging Aircraft and the SAFPAS remote airport project at UCLA.

GREGORY CARMAN is professor of mechanical and aerospace engineering at UCLA, having been at the university since 1992. He is also head of the Active Materials Laboratory, where research is performed in shape memory alloys, piezoelectric materials, and fiber-optic sensors, focusing on developing and understanding the combined electromagnetothermomechanical response to these active materials. Dr. Carman is a fellow of the ASME, serving in leadership roles within the Adaptive Structures and Materials Systems Committee of the Aerospace Division. He serves on the editorial advisory board of the *Journal of Composite Materials* and has served in various editorial roles on many other materials journals. Dr. Carman is associate editor on the *Journal of Intelligent Material Systems and Structures*. He was awarded ASME's Adaptive Structures and Material Systems Prize in 2004 and was an invited lecturer at the NAE's Annual Frontiers Symposium in 2004. Dr. Carman has spent several summers performing research at government laboratories, including the Jet Propulsion Laboratory and the AFOSR at Wright Patterson Air Force Base. He received a Ph.D. in engineering mechanics from Virginia Polytechnic Instititute, an M.S. in metallurgical and materials engineering from the University of Alabama, and a B.S. in engineering science and mechanics from Virginia Polytechnic Institute. Dr. Carman has prior experience in proposal review for the AFOSR, NSF, and Hong Kong Science Foundation.

INDERJIT CHOPRA (see biography above).

JANET DAVIS is manager of the Composite Structures Department at the Rockwell Scientific Company and has more than 15 years of research experience in materials science. Prior to joining Rockwell Scientific in 1996, she held positions at Lawrence Livermore National Laboratory (LLNL) and Cambridge University. Her work has focused on strengthening, toughening, and improving the reliability of advanced materials, especially fiber-reinforced composites and structural ceramics. Her current responsibility is to guide a team of research scientists to develop advanced ceramics and composites, with an emphasis on microstructure and property relationships and robust processing methods. Dr. Davis has extensive experience in ceramic powder processing, composite fabrication, mechanical properties evaluation, and microstructural analysis. She obtained a B.S. in ceramic engineering from the Ohio State University and a Ph.D. (1993) in materials engineering from the University of California at Santa Barbara, and has authored or coauthored more than 30 research publications and two patents.

RAVI B. DEO is responsible for space research and technology programs at Northrop Grumman Integrated Systems. During his 27 years at Northrop Grumman, he has been a program and functional manager for government- and company-sponsored projects on cryotanks, integrated system health management, aerospace structures, materials, subsystems, avionics, thermal protection systems, and software development. He has extensive experience in roadmapping technologies, program planning, technical program execution, scheduling, budgeting, proposal preparation, and business management of significant technology development contracts. Among his significant accomplishments are the NASA-funded Space Launch Initiative (SLI), Next-Generation Launch Technology (NGLT), the Orbital Space Plane (OSP), and high-speed research (HSR) programs, where he was responsible for the development of multidisciplinary technologies. Dr. Deo has over 50 technical publications and is the editor of one book. He holds a B.S. in aeronautical engineering from the Indian Institute of Technology in Bombay and M.S. and Ph.D. degrees in aerospace engineering from the Georgia Institute of Technology.

PRABHAT HAJELA is a professor of mechanical, aerospace, and nuclear engineering at RPI. Current research interests include analysis and design optimization of multidisciplinary systems; system reliability; emergent computing paradigms for design; artificial intelligence; and machine learning in multidisciplinary analysis and design. Dr. Hajela recently completed a year as an ASME congressional fellow in the office of Senator Conrad Burns, advising on technology policy. Before joining RPI, he was on the faculty at the University of Florida for 7 years. Dr. Hajela is a fellow of the AIAA, the Aeronautical Society of India, and the ASME, and he is vice president of the International Society of Structural and Multidisciplinary Optimization. He has served on the Multidisciplinary Design Optimization Technical Committee of the AIAA and the executive committee for the ASME Aerospace Division (chair, 2001-2002) and was chair of the Division's Technical Committee on Structures and Materials (1999-2002). He is the editor of *Evolutionary Optimization*, has served as an associate editor of the *AIAA Journal*, and is on the editorial board of six other international journals. He has published over 255 papers and articles in the areas of structural and multidisciplinary optimization and is an author or coauthor of four books in these areas. Dr. Hajela has an M.S. and a Ph.D. in aeronautics and astronautics from Stanford University and a B.Tech in aeronautical engineering from IIT at Kanpur, India. He has not previously served on an NRC committee.

MARK K. HINDERS holds B.S., M.S., and Ph.D. degrees in aerospace and mechanical engineering from Boston University and is currently a professor of applied science at the College of William and Mary in Virginia. Before coming to Williamsburg in 1993, Dr. Hinders was senior scientist at Massachusetts Technological Laboratory, Inc., and a research assistant professor at Boston University. Previously, Dr. Hinders was an electromagnetics research engineer at the Air Force Rome Laboratory located at Hanscom Air Force Base. He conducts research in wave propagation and scattering phenomena applied to medical imaging, intelligent robotics, security screening, remote sensing, and nondestructive evaluation. Dr. Hinders has not previously served on an NRC committee.

ROBERT SCHAFRIK is currently the general manager of the Materials and Process Engineering Department at GE Aircraft Engines. He is responsible for developing advanced materials and processes used in GE's aeronautical turbine engines and their marine and industrial derivatives. He oversees materials application engineering activities supporting GEAE's global design engineering, manufacturing, and field support activities. He also operates a state-of-the-art in-house laboratory for advanced materials development, characterization, and failure analysis. Prior to joining GE in November 1997, he served in two concurrent positions at the NRC (the operating arm of the National Academy of Sciences and the NAE), which he joined in 1991: director, National Materials Advisory Board, and director, Board on Manufacturing and Engineering Design. Under his direction, 33 final reports for studies were issued that addressed significant national issues in materials and manufacturing. Dr. Schafrik also served in the U.S. Air Force in a variety of capacities and retired as a lieutenant colonel. He has a Ph.D. in metallurgical engineering from Ohio State University, an M.S. in information systems from George Mason University, an M.S. in aerospace engineering from the Air Force Institute of Technology, and a B.S. in metallurgy from Case Western Reserve University.

NANCY R. SOTTOS is a professor in the Department of Theoretical and Applied Mechanics and the Beckman Institute for Advanced Science and Technology at the University of Illinois at Urbana-Champaign. Her research interests include the mechanics of complex heterogeneous materials (advanced composites, thin-film devices, smart materials); mesoscale characterization; and autonomic materials systems. Her work at the Beckman Institute addresses issues in the development of autonomic materials systems that have the ability to adapt and respond in an independent and automatic fashion. Dr. Sottos's research group is investigating new experimental methods to quantify autonomic response (e.g., the healing efficiency of a self-healing polymer) and understand this response in terms of the materials chemistry, processing, and microstructure. Dr. Sottos began her career at the University of Illinois in 1991, serving as an assistant professor. In 1997 she became an associate professor, in 1998 she served a 1-year rotating term as assistant dean of engineering, and in 2002 she was promoted to full professor. In 2005, she was named the Donald Biggar Willett Professor of Engineering. She serves as an editorial board member for the *Composites Science and Technology* journal, as senior technical editor of *Experimental Mechanics*, and as a technical reviewer for multiple technical journals. Dr. Sottos received a B.S. and a Ph.D. in mechanical engineering from the University of Delaware. She also serves as the faculty advisor for the Student Chapter of the Society of Women Engineers and as national student chapter coordinator for the Society of Engineering Science.

GREGORY WASHINGTON holds the rank of professor of mechanical engineering at the Ohio State University (OSU). He is also the associate dean of research for the College of Engineering at OSU. Dr. Washington has been involved in multidomain research for the last 12 years. His core area of interest is dynamic systems, with an emphasis on modeling and control of smart material systems and devices. During this time he has been involved in the following applications: the design and control of mechanically actuated antennas, the design and control of advanced automotive systems incorporating smart materials, the design and control of hybrid electric vehicles, and structural position and vibration control with smart materials. He is presently working on ultralightweight, structurally active antennae and sensory systems that involve the use of smart materials. His specific area of research lies in the modeling and control of novel systems and devices that incorporate smart materials. He is the author of more than 100 technical publications in journals, edited volumes, and conference proceedings. He is a technical reviewer for ASME, AIAA, and IEEE journals, as well as the NSF. He participated in the 2004 NAE *Frontiers in Engineering* and the Defense Sciences Study Group. He has received multiple research and teaching awards. He received B.S., M.S., and Ph.D. degrees from North Carolina State University.

TERRENCE A. WEISSHAAR is professor of aeronautics and astronautics at Purdue University and is currently a program manager at DARPA. Dr. Weisshaar's research areas center on aircraft design, structural optimization processes, and integration of aerospace technologies into vehicle conceptual and preliminary design. His past research contributions include development of aeroelastic design tailoring with advanced composite materials, and studies that assisted in the development of the DARPA X-29 research aircraft. He led fundamental aeroelastic research efforts for aircraft configurations such as the oblique-wing supersonic aircraft, the X-wing stopped rotor, and the joined-wing Sensorcraft. In addition, over the past decade, he developed smart material aeroelastic control concepts for aircraft structures. Dr. Weisshaar is a fellow of the AIAA, a past member of the U.S. Air Force Scientific Advisory Board, and the 2005 recipient of the AIAA Structures, Structural Dynamics, and Materials Award. He received the Air Force Exceptional Civilian Service Decoration in 1998. Dr. Weisshaar has both research and development experience in integrated aircraft structural design coupled to active control devices systematically interfaced toward optimum aircraft performance. His research skills will assist the committee in identifying opportunities in research and development technologies in the systems control of aircraft performance for advanced UAVs.

PANEL D: DYNAMICS, NAVIGATION, AND CONTROL, AND AVIONICS

NANCY G. LEVESON (NAE), *Panel Chair* (see biography above).

RICHARD ABBOTT is a technical fellow emeritus at Lockheed Martin Aeronautics Company in Palmdale, California. He received a Ph.D. in chemical physics from Northern Illinois University, where his research concentrated on cooperative phenomena in molecular systems and the renormalization group. He continued studies as a research associate in statistical mechanics at the University of Chicago's James Franck Institute, where he contributed to theories of energy relaxation in condensed media using Monte Carlo and molecular dynamics techniques. His career includes over 25 years of experience in the areas of guidance, navigation, and control systems design and analysis, sensor data fusion design, and sensor system simulation and modeling for both manned and unmanned aircraft. He has supervised the development and execution of large-scale simulations of complex air vehicles, led the development of the avionics functional architecture for the demonstration/validation phase of the YF-22 program, and developed fault detection and redundancy management algorithms for navigation systems aboard the X-33 single-stage-to-orbit vehicle. He also has served as principal investigator for the DARPA software-enabled control technologies for reliable autonomous control project and has been the co-chair for the Technologies for Autonomous Control session of the IEEE Aerospace Conferences.

CLARK R. BADIE is the business manager for the Displays and Crew Interface Division of Honeywell Aerospace Marketing and Project Management. In addition, he is the U.S. chairperson for the Avionics Harmonization Working Group, which leads the development of new and modified joint federal-European regulations and advisory material for advanced flight-deck displays. Previously, he was the product portfolio manager for strategic marketing and technology, where he aligned avionics product strategies with customer needs and was involved in strategic planning. Other positions he has held at Honeywell include manager of Honeywell Head-Up Displays Development; manager of product marketing and advanced technology for air transport displays; department manager for legacy and head-up displays systems and software engineering; and principal engineer for commercial electronic displays engineering. He has worked extensively with all avionics product disciplines: displays, flight controls, management systems and sensors, as well as certification issues such as software, flight test, safety, and environmental evaluation. Mr. Badie received a B.E. from the Stevens Institute of Technology and an MBA from Arizona State University.

JEFFERY ERICKSON is a senior technical fellow for human factors and crew system design at the Boeing Company, where he is responsible for technical leadership of advanced human–machine interface technology initiatives and their application to aircraft; spacecraft; command, control, communications, and computers; and intelligence, surveillance, and reconnaissance. He is currently assigned to the Boeing Phantom Works, which develops advanced products and provides enabling technologies, prototypes, engineering processes, and advanced methods. Previously he served as the manager of human–system integration for the Boeing Phantom Works and as the manager of crew systems for McDonnell Douglas Aerospace. He is a fellow of the Royal Aeronautical Society and has served on a number of advisory panels, such as the Air Force Scientific Advisory Board, the Naval Research Advisory Committee, and the Department of Defense Technical Advisory Group for Human Factors, among others. Mr. Erickson has received the Exceptional Civilian Service Medal from the Secretary of the Air Force, the Outstanding Achievement Award from McDonnell Aircraft and Missile Systems, and the Engineering Achievement Award from Douglas Aircraft. He received a B.A. in psychology and an M.S. in industrial psychology from California State University, Long Beach.

EPHRAHIM GARCIA is currently associate professor of mechanical and aerospace engineering at Cornell University, where his interests lie in the development of new types of actuation systems utilizing smart material transducers, system-level demonstrations of smart structures applied to defense platforms, and morphing aircraft systems bioinspired intelligent machines. Dr. Garcia served as a program manager in the Defense Sciences Office at DARPA from 1998 to 2002. From 1991 to 1998, he was an assistant and associate professor of mechanical engineering at Vanderbilt University, where he was director of the Center for Intelligent Mechatronics and the Smart Structures Laboratory. In this capacity he directed research in the areas of smart structures, control structure interaction, and bioinspired robotics. From 1991 to 1997, he owned and operated Garman Systems, Inc. (now Dynamic Structures and Materials, LLC), a small engineering corporation that designed and fabricated devices in adaptive structural systems utilizing piezoelectric, electrostrictive, and shape memory alloy materials. Dr. Garcia has been named an ONR Young Investigator, appointed a 1993 Presidential Faculty Fellow by President Clinton, and twice (1990 and 1991) received Summer Faculty Fellowship awards from the Air Force Office of Scientific Research. In 1995, he was named Most Promising Scientist by *Hispanic Engineer* magazine (now *Technica*) and received this award at the Hispanic Engineer National Achievement Awards Conference. Dr. Garcia is author of more than 140 articles, books, chapters, and edited volumes. He serves on the ASME Aerospace Division's Executive Committee and is on the editorial advisory board of *Smart Materials and Structures*. In 2002, Dr. Garcia received ASME's Adaptive Structures Prize for "significant contributions to the sciences and technologies associated with adaptive structures and/or materials systems."

CHARLES L. GUTHRIE is the director of advanced capabilities development for Northrop Grumman's Western Region within the Integrated Systems Sector. He is responsible for programs in space systems, future strike systems, missile defense systems, and naval system integration. Some of his previous positions include director of unmanned systems rapid prototyping and advanced concepts at the Boeing Phantom Works, director of the Joint Strike Fighter air vehicle IPT for Boeing Military Aircraft and Missiles, and director of air vehicle advance design for the Phantom Works. He is a Boeing technical fellow and was named Manager of the Year in 1993 and 1994 by North American Aircraft and the Southern California Area Council, respectively, and Engineer of the Year in 1987 and 1988 by North American Aircraft/Rockwell. Besides earning a B.S. in aerospace engineering from the University of Kansas, Mr. Guthrie has completed a number of technical short courses in topics such as radar, aircraft design, and engine–airframe integration and employee development courses. He works to support the California Polytechnic San Luis Obispo School of Engineering, the University of Kansas Aerospace Department, the Naval Postgraduate School, and Cal State Long Beach by providing industry feedback, serving on advisory boards, and conducting guest lectures. He is a senior member of the AIAA and has served on its Aircraft Design Technical Committee. He is also a senior member of the Association for Unmanned Vehicle Systems International (AUVSI) and a member of the National Management Association (NMA).

ELLIS F. HITT is president of Strategic Systems Solutions. He is responsible for analysis of alternative systems configurations and determining total life cycle cost for the U.S. Coast Guard's HC-130 fleet. He retired in 2005 as a senior manager for Battelle Corporation and was a chairman of the AIAA Digital Avionics Technical Committee. He has a B.S. in electrical engineering from the University of Kansas and an M.S. in electrical engineering from the Air Force Institute of Technology, along with postgraduate studies at OSU and the University of New Mexico. Mr. Hitt is a nationally recognized authority on avionics and flight control systems. He has extensive experience in conceptual, preliminary, and final design of avionics, including navigation, guidance, control, communications, controls and displays, sensors, stores management, weapons delivery, and electrical power subsystems; integration, testing, and analysis of avionics; development of mathematical models and computer programs for performing error analysis, systems simulation and evaluation, and life-cycle cost analyses; and mission software design, development, validation, and verification. Mr. Hitt's responsibilities before retiring from Battelle included senior marketing manager for the Air Force market sector and technical leader on total ownership cost. Prior to promotion to these positions, he was chief engineer, Design Engineering Program, and manager, Avionics Systems Engineering Business Development.

JAMES C. NEIDHOEFER is the CEO of Aerotonomy, Inc., which specializes in the development of advanced UAVs; UAV guidance, navigation, and control systems; and UAV flight-test-related products and services. He is an associate fellow of the AIAA, deputy director of the AIAA Information Systems Group, a past chairman of the AIAA Intelligent Systems Technical Committee, and associate editor for the AIAA *Journal of Aerospace Computing, Information and Communication* and has chaired a number of AIAA conference sessions. He is also a member of Penn State University's Industrial and Professional Advisory Committee. Dr. Neidhoefer was a recipient of the Best Paper Award at the international conference Artificial Neural Networks in Engineering. In addition, he is the author of numerous conference papers, journal articles, and book chapters. He received his B.S., M.S., and Ph.D. from the University of Alabama.

DARRYLL J. PINES is a professor and associate chair in the Department of Aerospace Engineering at UM, College Park, on loan as a program manager in the DARPA Defense Sciences Office. At DARPA, he is the program manager of the sensor dart, long gun, XNAV, and NAV programs. As a former DOE technical staff member working at the LLNL, Dr. Pines developed advanced guidance algorithms for interceptors and the final approach algorithm for the 1994 Clementine flyby mission, which was the first probe to discover water near the south pole of the Moon. His research interests include smart materials/structures technology, structural health monitoring, structural dynamics, micro and nano air vehicle systems, and vehicle guidance, control, and navigation. He has published over five book chapters and 200 journal/conference articles on topics in structural dynamics, damage detection, and vehicle flight dynamics, control, and navigation. He is an associate fellow of the AIAA, a fellow of the Institute of Physics, and chairs the Adaptive Structures Technical Committee of the AIAA. Dr. Pines graduated from MIT with Ph.D. and M.S. degrees in mechanical engineering and earned his B.S. degree from UC Berkeley in the same discipline.

JAMES RANKIN is the director of the Avionics Engineering Center at Ohio University. He holds a Ph.D. in electrical engineering from Iowa State University. His M.S.E.E. is also from Iowa State University and his B.S.E.E. from the South Dakota School of Mines and Technology. He has more than 25 years of experience in avionics research and design from both academic and industrial perspectives. Dr. Rankin has been involved with the NASA small aircraft transportation system. Previously, he was the PI on terminal area controller-pilot data link communications research, which was integral to NASA's low-visibility landing and surface operations flight test at Atlanta Hartsfield airport (1997) and the NASA runway incursion prevention system test at Dallas-Fort Worth airport (2000). Dr. Rankin was with Rockwell Collins, where his projects included airborne collision avoidance systems, four-dimensional flight management systems, and air transport display systems. As a senior member (2003) of the IEEE, he was twice elected (1999, 2002) to 3-year terms on the IEEE Aerospace and Electronic Systems board of governors. He is a senior member (2003) of the AIAA and was elected chair of the AIAA Digital Avionics Technical Committee in 2003. Dr. Rankin also has memberships in the Institute of Navigation (ION), Air Traffic Control Association, International Loran Association, and the Aircraft Owners and Pilots Association. He is an active member of the aviation community as a certified flight instructor with single-engine, multiengine, and instrument ratings.

JASON L. SPEYER (NAE) is currently a professor and past chairman in the Mechanical, Aerospace and Nuclear Engineering Department (now the Mechanical and Aerospace Engineering Department) at UCLA. He spent a research leave as Lady Davis Visiting Professor at the Technion (Israel Institute of Technology) in 1983 and was the 1990 Jerome C. Hunsaker Visiting Professor of Aeronautics and Astronautics at MIT. His industrial experience includes research at Boeing, Raytheon, Analytical Mechanics Associated, and the Charles Stark Draper Laboratory. He was the Harry H. Power Professor in Engineering Mechanics, University of Texas, Austin. Dr. Speyer was twice elected member of the board of governors of the IEEE Control Systems Society and chairman of the Technical Committee on Aerospace Controls. He served as an associate editor for *Technical Notes and Correspondence* (1975-1976) and *Stochastic Control* (1978-1979), *IEEE Transactions on Automatic Control*, the *AIAA Journal of Spacecraft and Rockets* (1976-1977), the *AIAA Journal of Guidance and Control* (1977-1978), and the *Journal of Optimization Theory and Applications* (1981-present). From October 1987 to October 1991 and from October 1997 to October 2001, he served as a member of the Air Force Scientific Advisory Board. He is a fellow of both the AIAA and the IEEE (life fellow). He was the recipient of the AIAA Mechanics and Control of Flight Award (1985), the AIAA Dryden Lectureship in Research (1995), the IEEE Third Millennium Medal (2000), and the Air Force Exceptional Civilian Decoration (1991 and 2001). Dr. Speyer received his B.S. in aeronautics and astronautics from MIT in 1960 and his Ph.D. in applied mathematics from Harvard University in 1968.

JOHN VALASEK (see biography above).

PANEL E: INTELLIGENT AND AUTONOMOUS SYSTEMS, OPERATIONS AND DECISION MAKING, HUMAN INTEGRATED SYSTEMS, AND NETWORKING AND COMMUNICATIONS

EDMOND L. SOLIDAY, *Panel Chair* (see biography above).

ELLA ATKINS is an assistant professor in the Department of Aerospace Engineering at UM, where she cofounded the Autonomous Vehicles Laboratory and is an active researcher in the Space Systems Laboratory and Alfred Gessow Rotorcraft Center. She holds B.S. and M.S. degrees in aeronautics and astronautics from MIT and M.S. and Ph.D. degrees in computer science and engineering from the University of Michigan. Her research integrates task-level planning and scheduling with trajectory optimization algorithms for safety-critical robotic systems and flight vehicles. Dr. Atkins is an active member of the AIAA Intelligent Systems technical committee, associate editor of the *AIAA Journal of Aerospace Computing, Information, and Communication*, and is a private pilot. She has written more than 50 papers that develop and apply real-time, artificial intelligence and optimization strategies to aerospace applications.

TAMER BASAR (NAE) is the Fredric G. and Elizabeth H. Nearing Professor of Electrical and Computer Engineering at the University of Illinois at Urbana-Champaign (UIUC). He is also a professor in the Center for Advanced Study of UIUC. His research interests include robust nonlinear and adaptive control; routing, pricing, and congestion control in communication networks; control over wired and wireless networks; mobile computing; and risk-sensitive estimation and control. He is a fellow of the IEEE as well as of the International Federation of Automatic Control and a past president of the Control Systems Society and the International Society of Dynamic Games. He has edited a number of books, book series, and journals, including *Automatica* and *IEEE Transactions on Automatic Control*. He has authored or coauthored over 150 journal articles and book chapters, over 200 conference publications and several books. He has received the Giorgio Quazza Medal of the International Federation of Automatic Control, the Hendrik W. Bode Lecture Prize of the IEEE Control Systems Society, the IEEE Millennium Medal, and the Medal of Science of Turkey, among many other awards. Dr. Basar received his B.S.E.E. from Robert College, Istanbul, and M.S., M.Phil., and Ph.D. degrees in engineering and applied science at Yale University.

THOMAS Q. CARNEY is professor of aviation technology and head of the Department of Aviation Technology at Purdue University, where he has taught since 1972. Dr. Carney has over 37 years of experience as a pilot, with more than 10,130 flight hours, and holds the ATP certificate with multiengine, Beechjet, and Mitsubishi Diamond type ratings in addition to the Certified Flight Instructor certificate with airplane single-engine, multiengine, and instrument ratings. Dr. Carney's primary teaching areas in aviation include advanced aviation meteorology, high-performance turbine operations, high-altitude flight, and corporate flight department management. Dr. Carney holds an M.S. and a Ph.D. in atmospheric science; his primary areas of interest in atmospheric science include aviation meteorology and the impact of weather on aviation operations, synoptic-scale dynamics and energetics, and the interactions between synoptic- and mesoscale motion fields. Dr. Carney is the senior editor of the *Collegiate Aviation Review* and a member of the editorial boards of the *Journal of Aviation/Aerospace Education and Research* and the *Journal of Air Transportation*. He serves on the board of directors of the Council on Aviation Accreditation (CAA), the Certified Aviation Manager governing board (and is currently serving as that board's first chairperson), and is chairperson of the CAA Standards Committee. Dr. Carney is an active consultant in corporate flight operations and an expert witness in litigation involving flight operations and aviation meteorology. In 2002, he was awarded the William A. Wheatley award by the University Aviation Association, given annually to a professional educator of more than 10 years' experience, who has made outstanding contributions to aerospace education. In 2004, he was designated a Certified Aviation Manager by the Certified Aviation Manager governing board.

JOHN-PAUL CLARKE is an associate professor in the School of Aerospace Engineering and director of the Air Transportation Laboratory at the Georgia Institute of Technology (Georgia Tech). His research and teaching address issues of optimization and robustness in aircraft and airline operations, air traffic management, and the environmental impact of aviation. He received his S.B. (1991), S.M. (1992), and Sc.D. (1997) from MIT and was a faculty member at MIT prior to moving to Georgia Tech. He has also been a researcher at the NASA Jet Propulsion Laboratory and a visiting scholar at the Boeing Company. Dr. Clarke is a member of the Airline Group of the International Federation of Operations Research Societies, AIAA, the Institute for Operations Research and the Management Sciences (INFORMS), ION, and Sigma Xi, The Scientific Research Society. He serves on several national and international committees, including the FAA Research Engineering and Development Committee (REDAC), the AIAA Air Transportation Systems Technical Committee, and the SAE Aircraft Noise Committee. Dr. Clarke was the first director of the Partnership for AiR Transportation Noise and Emissions Research (PARTNER) at the Center of Excellence for Aviation Noise and Aircraft Emissions Mitigation and is an active researcher in both PARTNER and the National Center of Excellence for Aviation Operations Research (NEXTOR). In 1999, he was awarded the AIAA/AAAE/ACC Jay Hollingsworth Speas Airport Award, and in 2003, he was awarded the FAA Excellence in Aviation Award. Dr. Clarke is currently a member of ASEB.

MICHAEL DeWALT is a former national resource specialist for software for the FAA. Currently, he is the chief scientist of aviation systems for Certification Services, Inc., a consulting firm for electronic equipment in aviation. In the FAA, Mr. DeWalt was responsible for providing technical guidance on policy, training, research, and development in airborne software and its associated ground-based systems; he was the technical focal point for industry and the FAA for evaluation of new technology and interpretation of existing policy as applied to aircraft systems. Prior to becoming the FAA's national resource specialist, Mr. DeWalt worked as a software life-cycle consultant for Telos Consulting Services, as a software control system engineer for Pacific Technologies, Inc., as an avionics certification engineer for the FAA, as software focus for the Boeing 757/767 autopilot, and as a digital and analog avionics engineer for Honeywell Flight Systems. He has given many presentations at national and international conferences on design assurance of safety-critical, software-based systems. He is a member of IEEE and has participated as a member of working groups drafting new standards for safety-critical software and revising IEEE document 1012, "Standards for Verification and Validation," and is a member of the Association of Computing Machinery. He received a B.S. in electrical engineering from the University of Washington and an M.S. in software engineering from Seattle University. Mr. DeWalt previously served on two NRC study groups.

FRANK L. FRISBIE is vice president for strategic planning in the transportation sector of Apptis, Inc. He was a longtime senior executive with the FAA and DoD and was vice president and senior client executive for civil aviation with Northrop Grumman Information Technology before joining Apptis in January 2005. He joined the FAA in 1958, where he held a variety of positions. In his last two posts, he was directly responsible for research, development, system engineering, acquisition, deployment, and maintenance of all 20,000 air traffic control facilities in the United States. Mr. Frisbie was awarded the Glen A. Gilbert Memorial Award in 2002 by the Air Traffic Control Association (ATCA) for his long-standing contributions to the air traffic control and civil aviation communities. He has been involved with the development, deployment, maintenance, and operation of virtually every system employed in the U.S. civil aviation infrastructure. He earned his B.E.E. degree from Manhattan College and his M.B.A. degree from American University. He is a member of the NASA Aeronautics Research Advisory Board, a member of the Russian Academy of Navigation and Motion Control, and he holds a professional engineer's license.

ANDREW LACHER is a research strategist working on system transformation and security for MITRE Corporation's Center for Advanced Aviation Systems Development (CAASD), where he helps coordinate internally directed R&D efforts as well as cross-corporate research issues associated with unmanned aircraft systems. He also manages CAASD's collaboration and interaction with NASA and works closely with the FAA Joint Planning and Development Office (JPDO) on the definition of NGATS. Mr. Lacher is a member of the JPDO's Agile Air Traffic Services integrated product team and its executive committee. He serves on the FAA's RE&D Advisory Committee's Air Traffic Services subcommittee and on the NEXTOR steering committee. He was a leader in formulation and eventual implementation of the collaborative decision-making (CDM) approach for air traffic management and led a number of studies that helped illustrate the benefit and feasibility of CDM and helped define many of the early concepts. Previously, he was a product manager for Orbcomm and a strategic information technology consultant working with small airlines. He is an associate fellow of the American Institute of Aeronautics and Astronautics and a member of the Airline Dispatcher Foundation, INFORMS, and the Airline Group of the International Federation of Operational Research Societies (AGIFORS). Mr. Lacher received both an M.S. in operations research and a B.S. in electrical engineering from the George Washington University.

RAYMOND R. LaFREY retired as manager of the air traffic control mission area at MIT's Lincoln Laboratory in 2003. His responsibilities encompassed surveillance, navigation, communications, and weather sensing and involved 150 staff and support personnel. Key elements include the development of airport surface technology, modern open architecture surveillance systems, and integrated airport and regional weather systems that provide time-critical weather knowledge directly to operational staff at FAA and airline facilities. After receiving a B.S.E.E. and an M.S.E.E. at Michigan State University, Mr. LaFrey served 6 years in the U.S. Army as a Signal Corps officer, installing satellite communications ground stations in Europe, Africa, and Vietnam. He joined MIT Lincoln Laboratory in 1969 and began developing air traffic control technology in 1974. From 1977 to 1982, he led the team that developed the first TCAS II flight hardware and conducted surveillance flight-test activities. During the 1980s he led the development and flight-testing of a GPS navigation set for small aircraft. He also led the Precision Runway Monitor Program, which enabled simultaneous instrument approaches to parallel runways spaced as close as 3,000 feet. He has served on a variety of advisory boards, including the American Astronautical Society's (AAS's) Recovery Team and a Defense Science Board task force on aviation safety. Mr. LaFrey is currently a member of the FAA's REDAC and the REDAC Air Traffic Services Subcommittee, and he chaired a REDAC study on transitioning research to operational capabilities. Mr. LaFrey has received FAA awards for his work on the traffic collision avoidance system, the precision runway monitor, and the ASR-9. He is also an inactive instrument-rated pilot.

CARL McCULLOUGH retired from the federal service in November 2005. His last position, as a member of the Senior Executive Service, was associate director for airspace, ranges, and airfield operations, Office of the Deputy Chief of Staff for Air and Space Operations, Headquarters of the U.S. Air Force, Washington, D.C. In addition, he is the executive director for the DoD Policy Board on Federal Aviation. Mr. McCullough is responsible for providing worldwide access to airspace and ranges, as well as deployable combat-capable air traffic control, airfield management, and base operations personnel and equipment. He also represents DoD positions in support of the U.S. National Airspace System as a seamless partner with the FAA. In addition, he provides strategic vision for Air Force and DoD participation and partnering in modernization of U.S. and global air transportation systems, as well as civil aviation policy formulation, airspace and aircraft access, air traffic control infrastructure, and international cooperation, to include all regional airspace initiatives. Mr. McCullough is the primary point of contact between the DoD and the Department of Transportation on domestic and international civil aviation issues with potential impact on military flying operations and air defense. Previously, Mr. McCullough served 24 years as a naval aviator. In his final tour he commanded the Naval Plant Representative Office at McDonnell Douglas Corporation in St. Louis. Following his retirement from the Navy in 1990, Mr. McCullough served with McDonnell Douglas Helicopter Company as general manager of its MD-500 program and then as vice president of the RAIL Company's Eastern Region. From 1993 to 2002, Mr. McCullough held numerous managerial and executive assignments with the FAA, including program manager for the wind shear and weather radar programs, program manager for satellite navigation systems, and director of the Office of Communication, Navigation, and Surveillance Systems. In May 2002, he was assigned to the White House Office of Science and Technology Policy as a Department of Transportation representative to the National Science and Technology Council. He is a graduate of the U.S. Naval Academy and the Naval Postgraduate School.

AMY PRITCHETT (see biography above).

DONALD W. RICHARDSON is a fellow of the AIAA and has been a member of AIAA continuously for 57 years. He is currently the immediate past president of the AIAA and served as the president of AIAA in 2004 and 2005. He has been named as a fellow of the Royal Aeronautical Society and was recently co-opted by the Royal Aeronautical Society to serve on its Engineering Council for 2003-2004. He was awarded the NASA Public Service Medal in 2002 for his work in reinvigorating U.S. federal funding for R&D in aeronautics. He holds bachelor's, master's, and Ph.D. degrees in aeronautical and mechanical engineering. A commercial instrument pilot with multiengine land and seaplane ratings, he has been an active pilot for 58 years. His engineering career included assignments as an aerodynamics and flight test engineer, research pilot, and engineering manager. He is presently employed as a vice president of SAIC, where he is responsible for all FAA and civil aviation corporate activities.

NADINE SARTER is currently an associate professor in the Department of Industrial and Operations Engineering and the Center for Ergonomics at the University of Michigan. She received her M.S. degree in experimental/applied psychology from the University of Hamburg (Germany) in 1983 and her Ph.D. in industrial and systems engineering from OSU in 1994. Dr. Sarter's primary research interests include (1) the design of multimodal interfaces in support of effective human–machine communication and coordination and computer-supported collaborative work, (2) the development of robust and transparent decision support systems, and (3) the use of design and training to support error management in a variety of complex event-driven domains, such as aviation and military operations. From 1994 to 1996, she served as technical advisor to the FAA Human Factors Team to provide recommendations for the design and operation of and training for advanced glass cockpit aircraft. For her research in the aviation domain, she received *Aviation Week and Space Technology*'s Aerospace Laurels Award for Outstanding Achievement in the Field of Commercial Air Transport in 1996 and the TGIR (Turning Goals Into Reality) Award as member of the UIUC Aircraft Icing Project Team from NASA Glenn Research Center in 2001. Her aviation-related research was supported by NASA, the FAA, and an NSF CAREER award.

I

Speakers

The National Research Council appreciates the efforts of and the information provided by the individuals who volunteered their time to speak at the meetings of the steering committee and panels:

Damodar Ambur, NASA Glenn Research Center
Dominick Andrisani, Purdue University
P. Douglas Arbuckle, Next Generation Air Transportation System, Joint Planning and Development Office
Steven Arnold, NASA Glenn Research Center
Aaron Auslender, NASA Langley Research Center
Milind Bakhle, NASA Glenn Research Center
Michael Ball, University of Maryland
Joan Bauerlein, Federal Aviation Administration
Chris Benich, Honeywell Aerospace
Isaiah Blankson, NASA Glenn Research Center
Iain Boyd, University of Michigan
Jack Boyd, Cytec Engineered Materials
Irene Brahmakulam, Office of Management and Budget
Charles Browning, University of Dayton
Sam Bruner, Raytheon Aircraft
Wayne Bryant, NASA Langley Research Center
Robert Bucci, Alcoa Technical Center
Carl Burleson, Federal Aviation Administration
Dennis Bushnell, NASA Langley Research Center
Jay Carter, Jr., CarterCopters, LLC
Fu-Kuo Chang, Stanford University
Ravi Chona, Air Force Research Laboratory
Stanley Cole, NASA Langley Research Center
Bradford Cowles, Pratt & Whitney
Scott Cruzen, Williams International
Leo Dadone, The Boeing Company
James DeBonis, NASA Glenn Research Center
Russ Duren, Baylor University
Gregg Dvorak, Office of Science and Technology Policy
Andrew Eckel, NASA Glenn Research Center
Barry Farmer, Air Force Research Laboratory
Tom Farris, Purdue University
Ed Feddeman, U.S. Congress, House Science Committee, Subcommittee on Space and Aeronautics
Michael Francis, Defense Advanced Research Projects Agency
Mike Freeman, Northrop Grumman Corporation
Peretz Friedmann, University of Michigan
Bob Friend, The Boeing Company
Sanjay Garg, NASA Glenn Research Center
Thomas Gates, NASA Langley Research Center
Billy Glover, The Boeing Company
Marvin Goldstein, NASA Glenn Research Center
Kenneth Goodrich, NASA Langley Research Center
Joe Grady, NASA Glenn Research Center
H. Thomas Hahn, University of California, Los Angeles
Richard Hallion, U.S. Air Force
Dave Halstead, GE Aircraft Engines
Awatef Hamed, University of Cincinnati
Hans Hornung, California Institute of Technology
Dennis Huff, NASA Glenn Research Center
Michael Hyer, Virginia Polytechnic Institute and State University
Thomas Jackson, Air Force Research Laboratory
Dale Johnson, U.S. Army
Dick Johnson, Gulfstream Aerospace Corporation
Sylvia Johnson, NASA Ames Research Center
Walter Jones, Air Force Research Laboratory
Suresh Joshi, NASA Langley Research Center
Joel Kaplan, U.S. Congress, House Appropriations Committee, Subcommittee on Science, State, Justice, and Commerce
Fred Krause, GE Aircraft Engines
Robert Krieger, Boeing Phantom Works
John LaGraff, Syracuse University
Sanjiva Lele, Stanford University
Anita Liang, NASA Glenn Research Center
Richard Lind, University of Florida

Kathryn Logan, National Institute of Aerospace, Virginia Polytechnic Institute and State University
Christopher Lynch, Georgia Institute of Technology
John Malone, NASA Langley Research Center
Bob Mercier, Aerospace Corporation
Alan Miller, The Boeing Company
Larry Moody, Boeing Phantom Works
Joseph Mook, University of Buffalo
Mark Moore, NASA Langley Research Center
Michael Nathal, NASA Glenn Research Center
Dick Obermann, U.S. Congress, House Science Committee, Subcommittee on Space and Aeronautics
Walter O'Brien, Virginia Polytechnic Institute and State University
Ozden Ochoa, Air Force Office of Scientific Research
Lisa Porter, NASA Headquarters
Mark Potapczuk, NASA Glenn Research Center
Louis Povinelli, NASA Glenn Research Center
David Radzanowski, Office of Management and Budget
Jonathan Ransom, NASA Langley Research Center
Vince Rausch, NASA Langley Research Center
Eli Reshotko, Case Western Reserve University
David Rhodes, Next Generation Air Transportation System, Joint Planning and Development Office
Kevin Rivers, NASA Langley Research Center
Gary Roberge, Pratt & Whitney
Kenneth Rosen, Sikorsky Aircraft Corporation (retired)
Carl Rousseau, Lockheed Martin Corporation
Jayant Sabnis, Pratt & Whitney
John Schmisseur, Air Force Research Laboratory
Jeffery Schroeder, NASA Ames Research Center
Corey Schumacher, Air Force Research Laboratory
Ron Sega, U.S. Air Force
Arun Sehra, NASA Glenn Research Center
Gary Seng, NASA Glenn Research Center
Ashok Singhal, CFD Research Corporation
Agam Sinha, MITRE Corporation
Charles Smith, NASA Ames Research Center
Robert Smith, Honeywell Defense and Space Company
Rick Stanley, GE Aircraft Engines
David Stark, NASA Glenn Research Center
Anthony Strazisar, NASA Glenn Research Center
Kenneth Suder, NASA Glenn Research Center
Paul Tan, FAA, William J. Hughes Technical Center
Anthony Tether, Defense Advanced Research Projects Agency
Charles Trefny, NASA Glenn Research Center
Juris Vagners, University of Washington
William Wallace, Textron, Inc.
William Welsh, Sikorsky Aircraft Corporation
William Winfree, NASA Langley Research Center
Richard Wlezien, NASA Headquarters
Rob Wolz, Gulfstream Aerospace Corporation
Erik Zahn, The Boeing Company
John Zuk, NASA Ames Research Center

J

Acronyms and Abbreviations

3-D	three-dimensional
ACARE	Advisory Council for Aeronautics Research in Europe
ADS-B	automatic dependent surveillance broadcast
AFRL	Air Force Research Laboratory
Al-Li	aluminum-lithium alloys
ATM	air traffic management
CBM	condition-based maintenance
CFD	computational fluid dynamics
CMC	ceramic matrix composite
CNS	communications, navigation, and surveillance
CO	carbon monoxide
CO_2	carbon dioxide
COTS	commercial off the shelf
DARPA	Defense Advanced Research Projects Agency
DoD	Department of Defense
DOE	Department of Energy
DHS	Department of Homeland Security
DNL	day-night average sound level
ESTOL	extremely short takeoff and landing
FAA	Federal Aviation Administration
FFRDC	federally funded research and development center
FOD	foreign object damage
GLONASS	Global Navigation Satellite System
GNC	guidance, navigation, and control
GOES	Geostationary Operational Environmental Satellites
GPS	Global Positioning System
H_2O	water

ICAO	International Civil Aviation Organization
ILS	instrument landing system
ITAR	International Traffic in Arms Regulations
IVHM	integrated vehicle health management
JPDO	Joint Planning and Development Office
JPL	Jet Propulsion Laboratory
JSF	Joint Strike Fighter
L/D	lift:drag ratio
LDI	lean direct injection
LES	large eddy simulation
LPP	lean, premixed, prevaporized
MDO	multidisciplinary design optimization
MEA	more-electric aircraft
MEANS	Materials Engineering for Affordable New Systems (Program)
METAR	meteorological aviation report
MURI	Multidisciplinary University Research Initiative
NASA	National Aeronautics and Space Administration
NCAR	National Center for Atmospheric Research
NDE	nondestructive evaluation
NGATS	Next Generation Air Transportation System
NOAA	National Oceanic and Atmospheric Agency
NO_x	oxides of nitrogen
NRC	National Research Council
PIREP	pilot report
PM	particulate matter
PMAD	power management and distribution
QFD	quality function deployment
R&D	research and development
R&T	research and technology
RANS	Reynolds-averaged Navier-Stokes
RNP	required navigational performance
RVSM	Reduced Vertical Separation Minima
SIGMET	significant meteorological information
SMA	shape memory alloys
SO_x	oxides of sulfur
SSTO	single stage to orbit
STOL	short takeoff and landing
TAF	terminal area forecast
TBCC	turbine-based combined cycle
TCP	Transmission Control Protocol
TPS	thermal protection system
TRL	technology readiness level
TSTO	two stage to orbit
T/W	thrust to weight ratio

UAV	unmanned air vehicle
UHC	unburned hydrocarbons
USAF	United States Air Force
VGC	variable geometry chevron
VLJ	very light jet
VMS	vehicle management systems
VOR	very high frequency (VHF) omnidirectional range
V/STOL	vertical and short takeoff and landing
VTOL	vertical takeoff and landing